Soil Management for Sustainable Agriculture

Soil Management for Sustainable Agriculture

Guest Editors

Anetta Siwik-Ziomek
Anna Figas

MDPI

Basel • Beijing • Wuhan • Barcelona • Belgrade • Novi Sad • Cluj • Manchester

Guest Editors
Anetta Siwik-Ziomek
Department of Biogeochemistry and Soil Science
Faculty of Agriculture and Biotechnology
Bydgoszcz University of Science and Technology
Bydgoszcz
Poland

Anna Figas
Department of Agricultural Biotechnology
Faculty of Agriculture and Biotechnology
Bydgoszcz University of Science and Technology
Bydgoszcz
Poland

Editorial Office
MDPI AG
Grosspeteranlage 5
4052 Basel, Switzerland

This is a reprint of the Special Issue, published open access by the journal *Agriculture* (ISSN 2077-0472), freely accessible at: https://www.mdpi.com/journal/agriculture/special_issues/SWWK95Z7H3.

For citation purposes, cite each article independently as indicated on the article page online and as indicated below:

Lastname, A.A.; Lastname, B.B. Article Title. *Journal Name* **Year**, *Volume Number*, Page Range.

ISBN 978-3-7258-3321-4 (Hbk)
ISBN 978-3-7258-3322-1 (PDF)
https://doi.org/10.3390/books978-3-7258-3322-1

Contents

About the Editors

Anetta Siwik-Ziomek

Anetta Siwik-Ziomek has been employed at the Faculty of Agriculture and Biotechnology of the Bydgoszcz University of Technology since 1997, and is currently a university professor. She obtained her Master of Science in Agricultural Sciences in 1996, becoming a Doctor of Agricultural Sciences in 2005, and a Doctor of Habilitation in Agricultural and Horticultural Sciences in 2019 at her home faculty. She conducts research on the influence of natural and anthropogenic factors on the activity of various soil enzymes involved in the transformation of C, N, and S elements in soil.

Anna Figas

Anna Figas has been employed at the Faculty of Agriculture and Biotechnology of the Bydgoszcz University of Technology since 1999, currently serving as a university professor. She obtained her Masters of Science in Agricultural Sciences in 1996, and became Doctor of Agricultural Sciences (2007) and Habilitation in Agricultural and Horticultural Sciences (2022) at her home Faculty. She conducts research on the physiological reactions of plants to abiotics stress factors and on the in vitro micropropagation of plants with medicinal properties.

Preface

This Special Issue features research and review articles focused on the characterization of soil, which provides the essential nutrients, water, and root support that food-producing plants need to grow and thrive. Through this collection, we aim to provide soil scientists with up-to-date information on the influence of various soil fertilizers (including mineral and natural fertilizers, e.g., straw and manure) using novel approaches. The featured studies will focus on the analysis dynamics of organic carbon and stable humic compounds in long-term experiments, as well as different land uses. This Special Issue encompasses the impact of irrigation and fertilization on soil enzyme activities. Recent studies on the effects of tillage and sowing methods on the physical and chemical properties of soils are also within the scope of this Special Issue. In this reprint, soil resistance to root penetration and water infiltration are expanded upon. Soil conditions determine the effectiveness of any crop. This Special Issue evaluates the effect of fumigation on soil properties, soil-borne pathogens, and microbiological and chemical properties of soils as a result of anthropogenic denudation. We hope that the publication of this Special Issue will support sustainable agriculture efforts to meet society's food needs without compromising the ability of future generations to meet their own needs.

Anetta Siwik-Ziomek and Anna Figas
Guest Editors

agriculture

Editorial

Soil Management for Sustainable Agriculture

Anetta Siwik-Ziomek [1,*] and Anna Figas [2]

1 Department of Biogeochemistry, Soil Science, Irrigation and Drainage, Faculty of Agriculture and Biotechnology, Bydgoszcz University of Science and Technology, Bernardyńska 6, 85-029 Bydgoszcz, Poland
2 Department of Biotechnology, Faculty of Agriculture and Biotechnology, Bydgoszcz University of Science and Technology, Bernardyńska 6, 85-029 Bydgoszcz, Poland; figasanna@pbs.edu.pl
* Correspondence: ziomek@pbs.edu.pl

Soil provides the essential nutrients, water, and root support that food-producing plants need to grow and thrive. The quality of soil can be changed by factors such as climatic conditions, time, and the impact of living organisms, and especially human management of soil [1–3]. The condition of the soil determines the effectiveness of any crop. In the last century, the use of mineral fertilizers and plant protection products has steadily increased, negatively affecting the condition of soils and the environment. The ineffective use of nutrients applied with fertilizers has resulted in the migration of harmful compounds. The use of mineral fertilizers and intensive cultivation has had a negative impact on biodiversity, significantly worsening the physical and chemical properties of soils [4,5]. Sustainable soil management not only reverses this trend but also improves soil fertility. With this Special Issue, "Soil Management for Sustainable Agriculture", we aimed to curate a collection of articles focusing on landscape ecology, applications in conservation ecology and the preservation of biodiversity in agriculture, soil as a key component in carbon sequestration, the use of irrigation and fertigation to relieve water stress, the transformation of nutrients and their interaction with soil enzymes, methods for the effective management of ingredients, the use of intelligent soil monitoring tools, the sustainable use of useful microorganisms in the rhizosphere, and the reclamation of soils contaminated with metals and pesticides. This Special Issue includes twelve research articles and one review. They have already attracted great interest, with approximately 26,472 reads and more than 50 citations recorded. The paper exploring nitrogen (N) fertilizer application as one of the causes of soil acidification at tea plantations [6] has attracted particular interest.

Most of the articles focus on the Earth's system; soil organic carbon (SOC) constitutes up to 75% of the terrestrial carbon pool, representing the largest carbon pool, which is approximately three times greater than the amount of carbon stored in the atmosphere or vegetation [7]. Straw and manure play a critical role in soil organic carbon (SOC) sequestration and crop yield in China. Zhao et al. [8], in their meta-analysis, evaluated the impact of straw and manure amendments, both individually and combined, on crop yield, SOC, and soil nutrients in China by collecting 173 studies. Their findings reveal that straw return and manure application increased crop yields by 14.4% and 70.4%, respectively, overall. Combined straw and manure application achieved a better improvement effect than straw alone but was less effective than manure alone. In productive practice, the research presented in this study can guide farmers in applying straw or manure precisely and reduce waste. Šimon et al. [9] analyze the trends in soil organic carbon (SOC) observed in soils from the oldest long-term Czech field experiment, the Prague–Ruzyně Long-Term Fertilizer Experiment, which has been conducted in Haplic Luvisol since 1955. Through the evaluation of SOC dynamics in the topsoil in this field experiment, the authors identify

Received: 16 January 2025
Revised: 22 January 2025
Accepted: 26 January 2025
Published: 6 February 2025

Citation: Siwik-Ziomek, A.; Figas, A. Soil Management for Sustainable Agriculture. *Agriculture* **2025**, *15*, 345. https://doi.org/10.3390/agriculture15030345

multi-crop rotation with fodder as the key management strategy for increasing or at least maintaining SOC stocks in the long term. On the contrary, a simple crop rotation with only two crops appears more sensitive to rising temperatures and shows higher SOC losses. These losses can be partially mitigated via FYM fertilization. The results of this year-long research study show that a combination of appropriate management strategies, i.e., diversified crop rotation and FYM fertilization, can attenuate the negative impacts of environmental conditions on SOC stock or amplify their positive effects. At present, the theory of green development is being incorporated around the world; the studies in this Special Issue not only offer benefits for sustainable agriculture with this trend in global development. Amaleviciute-Volunge et al. [10] share the results of a long-term (1995–2022) study designed to determine differences in the formation of humic compounds in natural and agricultural arenosols ecosystems. After 27 years, afforestation had significantly increased the soil organic carbon (SOC) and influenced the qualitative composition of humic substances into the SOC. Grassland cultivation showed faster SOC sequestration, a higher humic acid/fulvic acid (HA/FA) ratio, and an increased HD. To maintain a stable humus balance, arenosols should be used in crop rotation with approximately 40% leguminous plants.

On the other hand, Li et al. [11] observed that the effect of fumigation combined with soil amendments was better than that of fumigation alone, and silicon fertilizer had the best effect. Their results suggest that dazomet fumigation combined with soil amendments can improve soil nutrient supply, activate soil enzyme activities, enhance the control effect of soil-borne pathogens, and thus promote strawberry growth. The work of Siwik-Ziomek and Kuśmierek-Tomaszewska [12] discusses the enzymes activity in Alfisol soil. The work shows the effect of sprinkler irrigation on the activity of selected soil enzymes in terms of nitrogen metabolism and oxidation–reduction processes in soil with different doses of inorganic nitrogen fertilizers. The enzymatic activities changed throughout the research years. During the maturity stage, the lower ammonium nitrogen content in the soil was the result of a higher spring barley uptake due to drought stress. Irrigation probably contributed to the increased leaching of nitrate in the soil.

Ma et al. [13] assessed the impact of soil quality in mesofauna communities, the abundance and functional traits of which could be used to assess soil quality. Greater richness was observed in areas with relatively weaker soil quality, suggesting that the consequences of soil quality decline in soil-dwelling mesofauna were not exclusively negative. Various taxa of soil-dwelling mesofauna exhibited varying degrees of response to the decline in soil quality. Oribatida were overwhelmingly dominant in the sampling fields with medium soil quality, and most Entomobryidae were found in agricultural lands with very weak soil quality.

During soil quality decline, soil nutrients were observed to correlate positively with the density of soil-dwelling mesofauna. Soil acidification is a serious aspect of soil degradation worldwide and has been reported in various ecosystems and regions [13]. Liu et al. [14] studied the characteristics of soil acidification after nitrogen fertilizer treatment in a tea plantation. The loss of exchangeable base cations owing to N application is the main mechanism behind soil acidification at tea plantations. In agricultural production, the amount of nitrogen fertilizer should be strictly controlled for tea plantations that are not seriously acidified, and measures such as applying nitrogen fertilizer synergists and organic fertilizers should be taken to prevent further acidification of the soil. For severely acidified tea plantations, alkaline biomass materials should be appropriately applied to improve soil acidification.

Kanarek et al. [15] studied and evaluated the microbiological and chemical properties of soils as a result of anthropogenic denudation. The study found that the over-intensification of agriculture negatively affected soil in moraine areas, leading to erosion. Cultivation operations move soil components from the tops of hills to the foothills. These findings underscore the need to adapt agricultural practices to the terrain, especially in hilly areas, and to use sustainable agriculture for long-term soil and environmental protection. This research indicates that there is a clear tendency toward the accumulation of higher levels of organic carbon, total nitrogen, and plant-available forms of P and K at the foot of the slope, likely resulting from the transfer of soil material due to tillage and water erosion. Moreover, for most of the microbial groups studied, more stimulating conditions for their development were observed at the foot than at the summit, associated with the accumulation of larger quantities of, and easier access to, nutritional substrates in these areas. Kanarek et al. [15] highlighted that monitoring microbiological and chemical changes in soil quality can be utilized as a significant prognostic tool in assessing further erosive changes.

In their article, Cui et al. [16] review research on the quantity and quality of humic substances following different land uses in karst peak cluster depressions in Guangxi, China. The authors studied soil under five major land uses (grassland, afforestation, sugarcane field, corn field, and pitaya field). The results show that the soil organic carbon contents in both afforestation and naturally restored grassland are significantly higher than in the sugarcane field, corn field, and pitaya field. Different land uses in the same area also have a certain impact on the amounts of soil humic fractions. The structural characteristics of the humic fractions of five different land uses show that different land uses directly affect the structure of humic fractions. The article by Dias et al. [17] reviewed soil attributes and their inter-relationships with resistance to root penetration and water infiltration in areas with different land uses in the Apodi Plateau, a semiarid region of Brazil. Different soils were analyzed for their resistance to root penetration, water infiltration, inorganic fractions, soil density, total porosity, the potential of hydrogen, electrical conductivity, total organic carbon, potential acidity, and the sum of bases. Textural classification was an important factor in the analysis of soil resistance to root penetration (Q) and the infiltration rate, as evidenced in the cluster analysis, allowing the formation of two groups, one for the surface layers of the areas and another for the subsurface layers, with the inorganic sand and clay fractions standing out with the greatest dissimilarity.

Dias et al. [17] suggest implementing conservation practices in soil management to correct pore space problems and the degradation of agroecosystems in areas with semiarid soils. Wang et al. [18] studied the effects of tillage and sowing methods on soil physical properties and corn plant characteristics. This study consisted of four tillage and sowing methods: plow tillage and precision seeder sowing (PTS), rotary tillage and precision seeder sowing (RTS), no tillage and no tillage seeder sowing (NTS), and no tillage and precise sowing in a stubble field (STS) (all four treatments involved total straw return). The results of this study indicate that the short-term application of different tillage and sowing methods has a significant impact on soil physical properties and plant growth.

Li et al. [11] studied the effects of soil amendments on soil properties, soil-borne pathogens, and strawberry growth after dazomet fumigation. In this study, silicon fertilizer, potassium humate, Bacillus biofertilizer, and a combination of the latter two were added to soil after DZ fumigation. The results showed that DZ fumigation combined with soil amendments significantly increased the activities of catalase, sucrase, and urease in the soil by varying degrees. In particular, silicon fertilizer significantly increased soil nutrients, as well as the contents of nitrogen, phosphorus, potassium, and organic matter. At the same time, the soil amendments further improved the control effect of soil-borne pathogens and significantly promoted the growth of strawberries, and the effect was better than that of DZ fumigation alone. The improvement in enzyme activities in the soil indicated an increase in soil biological activity, which directly or indirectly promoted strawberry growth. The results show that DZ fumigation combined with silicon fertilizer had the best effect, not only improving the soil environment but also reducing the number of soil-borne pathogens and promoting the growth of strawberry plants. Furthermore, soil amendments could replace some fertilizers and promote green, efficient, and sustainable development. Li et al. [11] note that the experiments were conducted in a controlled environment; when these amendments are applied to the field, a small-scale test should be carried out first to avoid economic losses due to the complexity of the field environment.

Dobrzynski at all [19] share a review paper exploring bacteria of the genus *Bacillus* and related genera (e.g., *Paenibacillus, Alicyclobacillus*, or *Brevibacillus*) belonging to the phylum Firmicutes. The group consists of aerobic and relatively anaerobic bacteria capable of spore-forming. *Bacillus* spp. and related genera are widely distributed in the environment, with a particular role in soil. Their abundance in agricultural environments mainly depends on fertilization but can also depend on soil cultivation methods, i.e., whether plants are grown in monoculture or rotation systems. The highest abundance of the phylum Firmicutes is usually recorded in soil fertilized with manure. Due to the great abundance of cellulose in the environment, one of the most important physiological groups among these spore-forming bacteria is that of cellulolytic bacteria. Three key cellulases produced by *Bacillus* spp. and related genera are required for complete cellulose degradation: endoglucanases, exoglucanases, and β-glucosidases.

We hope that the reader will find this Special Issue a helpful reference in further understanding soil management techniques for sustainable agriculture.

Author Contributions: Conceptualization, A.S.-Z. and A.F.; writing—original draft preparation, A.S.-Z. and A.F.; writing—review and editing, A.S.-Z. All authors have read and agreed to the published version of the manuscript.

Funding: This research received no external funding.

Conflicts of Interest: The authors declare no conflicts of interest.

References

1. Havlin, J.L. Soil: Fertility and nutrient management. In *Landscape and Land Capacity*; CRC Press: Boca Raton, FL, USA, 2020; pp. 251–265.
2. Elkhlifi, Z.; Iftikhar, J.; Sarraf, M.; Ali, B.; Saleem, M.H.; Ibranshahib, I.; Chen, Z. Potential role of biochar on capturing soil nutrients, carbon sequestration and managing environmental challenges: A review. *Sustainability* **2023**, *15*, 2527. [CrossRef]
3. Marschner, P.; Rengel, Z. Nutrient availability in soils. In *Marschner's Mineral Nutrition of Plants*; Academic Press: Cambridge, MA, USA, 2023; pp. 499–522.
4. Khan, M.N.; Mobin, M.; Abbas, Z.K.; Alamri, S.A. Fertilizers and their contaminants in soils, surface and groundwater. *Encycl. Anthr.* **2018**, *5*, 225–240.
5. Pahalvi, H.N.; Rafiya, L.; Rashid, S.; Nisar, B.; Kamili, A.N. Chemical fertilizers and their impact on soil health. In *Microbiota and Biofertilizers, Ecofriendly Tools for Reclamation of Degraded Soil Environs*; Springer: Cham, Switzerland, 2021; Volume 2, pp. 1–20.
6. Liu, Y.; Zhang, M.; Li, Y.; Zhang, Y.; Huang, X.; Yang, Y.; Zhu, H.; Xiong, H.; Jiang, T. Influence of Nitrogen Fertilizer Application on Soil Acidification Characteristics of Tea Plantations in Karst Areas of Southwest China. *Agriculture* **2023**, *13*, 849. [CrossRef]
7. Wiesmeier, M.; Poeplau, C.; Sierra, C.A.; Maier, H.; Frühauf, C.; Hübner, R.; Kühnel, A.; Spörlein, P.; Geuß, U.; Hangen, E.; et al. Projected loss of soil organic carbon in temperate agricultural soils in the 21st century: Effects of climate change and carbon input trends. *Sci. Rep.* **2016**, *6*, 32525. [CrossRef] [PubMed]
8. Zhao, Z.; Yang, Y.; Xie, H.; Zhang, Y.; He, H.; Zhang, X.; Sun, S. Enhancing Sustainable Agriculture in China: A Meta-Analysis of the Impact of Straw and Manure on Crop Yield and Soil Fertility. *Agriculture* **2024**, *14*, 480. [CrossRef]
9. Šimon, T.; Madaras, M.; Mayerová, M.; Kunzová, E. Soil Organic Carbon Dynamics in the Long-Term Field Experiments with Contrasting Crop Rotations. *Agriculture* **2024**, *14*, 818. [CrossRef]
10. Amaleviciute-Volunge, K.; Tripolskaja, L.; Kazlauskaite-Jadzevice, A.; Slepetiene, A.; Baksiene, E. Effect of Long-Term Different Land Uses on Improving Stable Humic Compounds in Arenosol. *Agriculture* **2024**, *14*, 250. [CrossRef]

11. Li, Q.; Andom, O.; Fang, W.; Yan, D.; Li, Y.; Wang, Q.; Jin, X.; Cao, A. Effects of Soil Amendments on Soil Properties, Soil-Borne Pathogens, and Strawberry Growth after Dazomet Fumigation. *Agriculture* **2024**, *14*, 9. [CrossRef]
12. Siwik-Ziomek, A.; Kuśmierek-Tomaszewska, R. Responses of Soil Enzymes Activities to Sprinkler Irrigation and Differentiated Nitrogen Fertilization in Barley Cultivation. *Agriculture* **2024**, *14*, 1255. [CrossRef]
13. Ma, C.; Yao, X.; Du, G. Effects of Soil Quality Decline on Soil-Dwelling Mesofaunal Communities in Agricultural Lands of the Mollisols Region, China. *Agriculture* **2024**, *14*, 766. [CrossRef]
14. Wu, Z.; Sun, X.; Sun, Y.; Yan, J.; Zhao, Y.; Chen, J. Soil acidification and factors controlling topsoil pH shift of cropland in central China from 2008 to 2018. *Geoderma* **2022**, *408*, 115586. [CrossRef]
15. Kanarek, P.; Breza-Boruta, B.; Pawłowski, M.; Kobierski, M. Evaluation of Microbiological and Chemical Properties of Soils as a Result of Anthropogenic Denudation. *Agriculture* **2023**, *13*, 2247. [CrossRef]
16. Cui, T.; Zhang, J.; Luo, W. The Quantity and Quality of Humic Substances following Different Land Uses in Karst Peak-Cluster Depression in Guangxi, China. *Agriculture* **2023**, *13*, 2246. [CrossRef]
17. Dias, P.M.S.; Portela, J.C.; Gondim, J.E.F.; Batista, R.O.; Rossi, L.S.; Medeiros, J.L.F.; Farias, P.K.P.; Mota, P.J.; Bandeira, D.J.d.C.; Filho, L.C.d.A.L.; et al. Soil Attributes and Their Interrelationships with Resistance to Root Penetration and Water Infiltration in Areas with Different Land Uses in the Apodi Plateau, Semiarid Region of Brazil. *Agriculture* **2023**, *13*, 1921. [CrossRef]
18. Wang, Y.; Yang, S.; Sun, J.; Liu, Z.; He, X.; Qiao, J. Effects of Tillage and Sowing Methods on Soil Physical Properties and Corn Plant Characters. *Agriculture* **2023**, *13*, 600. [CrossRef]
19. Dobrzyński, J.; Wróbel, B.; Górska, E.B. Taxonomy, ecology, and cellulolytic properties of the genus Bacillus and related genera. *Agriculture* **2023**, *13*, 1979. [CrossRef]

agriculture

Article

Responses of Soil Enzymes Activities to Sprinkler Irrigation and Differentiated Nitrogen Fertilization in Barley Cultivation

Anetta Siwik-Ziomek [1,*] and Renata Kuśmierek-Tomaszewska [2]

[1] Laboratory of Soil Science and Biochemistry, Department of Biogeochemistry and Soil Science, Faculty of Agriculture and Biotechnology, Bydgoszcz University of Sciences and Technology, 6 Bernardyńska Street, 85-029 Bydgoszcz, Poland

[2] Laboratory of Agrometeorology, Plant Irrigation and Drainage, Faculty of Agriculture and Biotechnology, Bydgoszcz University of Sciences and Technology, 7 Profesora Sylwestra Kaliskiego Av., 85-796 Bydgoszcz, Poland; renata.kusmierek-tomaszewska@pbs.edu.pl

* Correspondence: ziomek@pbs.edu.pl

Abstract: Our study aimed to assess the impact of sprinkler irrigation on the activity of selected soil enzymes in terms of nitrogen metabolism and oxidation–reduction processes in soil with different doses of inorganic nitrogen fertilizers. An Alfisol was sampled from an experimental field of spring barley within the University Research Center in the central part of Poland, namely the village of Mochełek with a moderate transitory climate, during the growing seasons of 2015–2017. The soil resistance (RS) was derived to recognize the resistance enzymes during drought. In the maturity phase, nitrate reductase activity was 18% higher in irrigated soil and the activities of other enzymes were higher than in the non-irrigated plots by 25% for dehydrogenase, 22% for peroxidase, 33% for catalase, and 17% for urease. The development phase in the barley influenced nitrate reductase activity. Enzymatic activities changed throughout the research years. During the maturity stage, a lower ammonium nitrogen content in the soil resulted from a higher spring barley uptake due to drought stress. Irrigation probably contributed to increased leaching of nitrate in the soil. The highest index of resilience was found in the soil catalase activity.

Keywords: urease; nitrate reductase; dehydrogenase; peroxidase; catalase; moderate transitory climate; soil; index of resilience; soil activity monitoring

Citation: Siwik-Ziomek, A.; Kuśmierek-Tomaszewska, R. Responses of Soil Enzymes Activities to Sprinkler Irrigation and Differentiated Nitrogen Fertilization in Barley Cultivation. *Agriculture* **2024**, *14*, 1255. https://doi.org/10.3390/agriculture14081255

Academic Editor: Eric C. Brevik

Received: 26 June 2024
Revised: 20 July 2024
Accepted: 24 July 2024
Published: 30 July 2024

1. Introduction

The forms of mineral and organic nitrogen (N) undergo several transformations throughout the N cycle. This element is easily transformed from a reduced form to an oxidized form, which results in a free migration of nitrogen in hydrological and atmospheric processes. The amount of nitrogen available to plants is positively correlated with the mineralization of organic matter in the soil, biological nitrogen fixation, fertilization, and the total and distribution of atmospheric precipitation [1]. However, due to such processes as immobilization, harvesting and removal, denitrification, volatilization, leaching, runoff, and erosion, nitrogen loss from the soil occurs. The intensity of these processes is influenced by environmental factors such as the soil's pH, the soil's texture, its density, aeration, the water content, and thermal conditions, but also the management of crop residue, the method and timing of fertilization, agricultural treatments such as irrigation, and changes in land use. It is assumed that in most cases, less than five percent of nitrogen in the soil is directly available to plants from the total nitrogen content. This nitrogen is mainly in the form of nitrates NO_3^--N and ammonium NH_4^+-N, with organic N being the residue, which gradually becomes available due to mineralization [2,3]. What is characteristic of arable soils is exceptionally high dynamics of mineral nitrogen forms in the growing season, which results from the microbiological nature of nitrogen transformations in the soil. Nitrogen occurs in many forms, covering the range of valence states from –3 (in NH_4^+) to +5 (in NO_3^--) in both agricultural and natural ecosystems. The change of one valence state into another is mainly biologically mediated and depends primarily on environmental conditions [4]. Soil oxidoreductase enzymes take part in oxidoreductive processes. Dehydrogenases (E.C.1.1.) are extracellular enzymes that can be considered a helpful indicator of microbial activity and oxidative metabolism in soil [5]. Another intracellular enzyme from the oxidoreductase class is catalase (EC 1.11.1.6), which manages oxidative stress in the soil by catalyzing the decomposition of hydrogen peroxide into water and oxygen [6]. Peroxidases (EC 1.11.1) use H_2O_2 as an electron acceptor, and their activity in soil results in the depolymerization of lignin [7]. Urease activity (EC 3.5.1.5) can result in an increase in soil pH and loss of nitrogen to the atmosphere due to the release of NH_3 as a result of the hydrolysis of urea to CO_2 and NH_3 [8]. The activity of this enzyme can be viewed as a desirable indicator of soil quality due to its role in regulating plant nitrogen supply. In turn, the enzyme responsible for catalyzing the reduction of NO_3^- to NO_2^- in anaerobic conditions in soil is nitroreductase (EC 1.7.99.4) [9]. It has been proven that changes in soil use and management affect soil enzymes that actively participate in metabolic processes [10,11]. Enzymes indicate the metabolic level of the microbial community in soil and catalyze specific reactions in the carbon and nutrient metabolism cycle [12,13]. Free enzymes excreted by plants and animals and associated mainly with or within cellular structures are called exoenzymes. Later, they are released into soil after cell lysis and death [14]. Therefore, if soil use and management influence the soil's microbial environment, changes in the activity of soil enzymes can also be observed [15]. The biochemical properties of soil, which are indicators of its quality, are highly variable depending on climatic, weather, and geographical conditions; pedogenic factors; fertilization; and irrigation. Microorganisms living in soil are important factors that determine the nutrient metabolism cycle. They also interact intricately with plant organisms. Land-use systems that improve soil microbiological properties can result in higher yields with better raw material quality while reducing production costs. Similarly, by limiting the use of mineral fertilizers and plant protection agents, these systems support the sustainable development of agricultural areas. Therefore, to improve the condition of soil, it is necessary to constantly monitor and evaluate the physicochemical and biological processes in the soil and to examine changes in its physicochemical properties. Diverse soil use in agricultural systems in terms of crop rotation and plant protection treatments results in changes in soil properties, both physical and chemical, but mostly affects biological activity. This, in turn, affects both productivity and environmental quality, and hence affects human and animal health. Multi-year studies on the impact of agriculture on the biological and biochemical properties of soil bring valuable information on the transformation of nutrients in soils [16,17]. The definition of soil quality indicates the ability of soil to operate within an ecosystem, as well as its ability to support biological productivity, to maintain the quality of the environment, and to encourage the sanitary conditions of plants and animals [16].

The stability (resistance and action) of a soil system is a consequence of the influence of microorganisms on the properties and processes in the ecosystem. To define different systems, it is important to select appropriate indicators that will quantify the relative value of how the system will respond to specific soil-use scenarios. In our paper, we compare our indices with previously published stability indices and test their performance against a real dataset. One of the indicators that quantifies the relative value of the microbiological response is the resistance index, according to Orwin and Wardle [18].

In this study, we aimed to evaluate the following: (1) the responses of N-related properties of an Alfisol, such as the forms of N in the soil and the activity of enzymes involved in the metabolism of nitrogen in the soil; (2) the reaction of enzymatic activity related to the transformation of soil nitrogen depending on soil moisture under sprinkler irrigation during a growing season in spring barley in a warm temperate climate zone; (3) the impact of irrigation on the activities of enzymes related to nitrogen metabolism and oxidation–reduction processes in soil during varied growth stages with various doses of inorganic nitrogen fertilizers; and (4) whether the calculated resistance ratio (RS) can be used to find an effective solution to enzymatic stress.

2. Materials and Methods

2.1. Study Area and Soil Sampling

A carefully controlled field experiment was conducted at the Research Center of the Bydgoszcz University of Science and Technology in the village of Mochełek (53°130 N, 17°510 E). The experiment site was located in the Kujawsko–Pomorskie province, in central Poland. The plant that was investigated in this experiment was spring barley cv. 'Signora', cultivated in three consecutive growing seasons, 2015–2017.

The soil, according to the USDA soil taxonomy, was defined as a typical Alfisol made of sandy loam (clay 6%, sand 79%, loam 15%) [19]. It was found that the reaction of the topsoil was slightly acidic: the pH in 1 M KCl was 5.7–6.1. The topsoil showed relatively low contents of total organic carbon (TOC) (7.60–7.70 $g{\cdot}kg^{-1}$) and total nitrogen (TN) (0.70–0.76 $g{\cdot}kg^{-1}$). The contents of other available nutrients were as follows: the phosphorus (P) (64.0 mg kg^{-1}) and sulfur (S) (13 mg S kg^{-1}) contents were average, and the potassium (K) content was high (126.0 mg^{-1}). The subsoil comprised light loamy sand on shallow medium loam. The soil properties were determined before the experiment and they are presented below (in Table 1). The water content in the soil, corresponding to the water content in 1 m of the soil layer for field capacity, was 215 mm.

Table 1. Properties of soil in the experimental plot.

Soil Property	Content
TOC	7.60–7.70 $g{\cdot}kg^{-1}$
TN	0.70–0.76 $g{\cdot}kg^{-1}$
pH KCl	5.8–6.2
Available P	64.0 $mg{\cdot}kg^{-1}$
Available K	125.0 $mg{\cdot}kg^{-1}$
SO_4^{2-}	12 $mg{\cdot}kg^{-1}$

2.2. Experimental Design and Weather Conditions

This study had a two-factor split-plot design with four replications. The first factor (i) was sprinkler irrigation (where W_0 meant no irrigation, and W_1 meant optimal irrigation with 100% coverage of the water requirements of plants in the period of high water needs). The second factor (ii) was a differentiated level of nitrogen fertilizer application in the crystalline form of ammonium nitrate (three doses: N_1, N_2, and N_3; see Table 2). This fertilizer was applied before sowing at different doses for all groups except N_0 (control) (Table 2). For N_4, an additional dose of fertilizer was top-dressed during the barley's shooting stage. The second factor was static and remained constant throughout the whole experiment. However, the first factor, irrigation, was dynamic and was scheduled according to weather conditions. The spring barley was provided with optimal irrigation. Throughout the period of high water requirements in plants in the rhizosphere, there was a constant reserve of readily available water (RAW). The number of single irrigation doses and the total number of seasonal doses (Table 3) were established based on the amount and distribution of atmospheric precipitation, according to Żarski et al. [20].

Table 2. Description of experimental factors.

Irrigation Factor	Fertigation Factor	Nitrogen Fertigation Level
W_0—no irrigation W_1—optimal irrigation	N_0	Control
	N_1	30 $kg{\cdot}ha^{-1}$ pre sowing
	N_2	60 $kg{\cdot}ha^{-1}$ pre sowing
	N_3	90 $kg{\cdot}ha^{-1}$ (60 $kg{\cdot}ha^{-1}$ pre sowing and 30 $kg{\cdot}ha^{-1}$ top-dressed during the shooting stage)

Table 3. Weather conditions and irrigation doses applied in the growing seasons of 2015–2017.

Growing Season	t (°C)	P (mm)	Irrigation Application Dates	Irrigation Doses (mm)
2015	13.8	193.3	26 May 3 June 10 June 1 July 6 July	30 30 25 30 20
			total applications:	135
2016	14.3	386.7	24 May 8 June	35 32
			total applications:	77
2017	13.1	474.8	29 May 9 June 28 June	20 20 15
			total applications:	55
Average for 1991–2020	14.8	324.5	–	–

The climate conditions of this study area represent a temperate transitory zone in Central Europe. The mean annual temperature and rainfall conditions for the growing season from April to September are 14.8 °C and 324.5 mm, respectively. In the growing season of 2015, classified as dry, as much as 135 mm of water was applied in 4 single doses. For the other two seasons, classified as moist, a total of 77 mm was applied in two doses in 2016 and only 55 mm was applied in three doses in 2017. For the whole experimental period of 2015–2017, the temperature conditions in the area were similar to the climate norm for 1991–2020 (Table 3) (Figure 1). However, the atmospheric precipitation totals from April to September were considerably higher in 2016 and 2017 when compared to the many-year average (Table 3) (Figure 1). The dates of barley sowing were as follows: 23 March 2015, 1 April 2016, and 31 March 2017. Barley was grown according to the recommendations of the State Plant Health and Seed Inspection Service regarding the optimization of phosphorus and potassium fertilization and chemical plant protection. The doses of macroelements for the barley were 60 kg $P_2O_5{\cdot}ha^{-1}$ for phosphorus and 75 $K_2O{\cdot}ha^{-1}$ for potassium, and these were applied before sowing. In all of the study years, due to the use of properly selected herbicide mixtures, weed infestation was minimal. The herbicides used from the 3-leaf stage until the end of tillering (BBCH 1–29) were Pike 20 WG at a dose of 30 $g/dm^3{\cdot}ha^{-1}$ (metsulfuron-methyl) and Aurora 40 WG at a dose of 50 $g/dm^3{\cdot}ha^{-1}$ (carfentrazone-ethyl). To combat fungal diseases in the T1 and T2 periods, a factory mixture of fenpropimorph and epoxiconazole in the Duett Star 334 SE preparation was used in the amount of 1 $l{\cdot}ha^{-1}$, and on the ears, we used epoxiconazole with kresoxim-methyl of the Tocata Duo preparation in the amount of 0.8 $l{\cdot}ha^{-1}$. In all of the study years, grain moths (*Oulema melanopus* L.) were observed on the barley, which was controlled with Bi 58 Nowy (dimethoate) at a dose of 0.5 $l{\cdot}ha^{-1}$. The harvesting area covered 10 m^2. Grain was harvested on 3 August 2015, on 23 July 2016, and on 8 July 2017.

2.3. Irrigation System and Schedule

For the irrigation, a portable sprinkler irrigation system equipped with low-pressure Nelson-type sector sprinkler heads was used. The unit efficiency was 200 $dm^3{\cdot}h^{-1}$. The irrigation system was connected to the municipal waterworks network.

We scheduled the dates of irrigation treatments based on weather monitoring from an automatic weather station set in the vicinity of the experimental plot. Daily precipitation and the content of readily available water (RAW) in the soil were established. The soil water storage from the topsoil to a one-meter depth of the soil profile was 215 mm according to the field capacity. To conduct constant rhizosphere moisture monitoring, we applied the method of readily available water balance, commonly used for irrigation scheduling [20]. Moreover, direct measurements of the soil water content at the depth of 20 cm were conducted with the TDR method using the portable Fieldscout TDR 300 Soil Moisture Meter (Spectrum Technologies, Inc. 3600 Thayer Court, Aurora, IL 60504, USA) for each plot every day. Barley water requirements were met by maintaining soil moisture in the range of RAW in the plant rhizosphere. For the barley grown on irrigated plots, soil moisture in the rhizosphere was maintained in the RAW range from 0 to 30 mm according to field capacity.

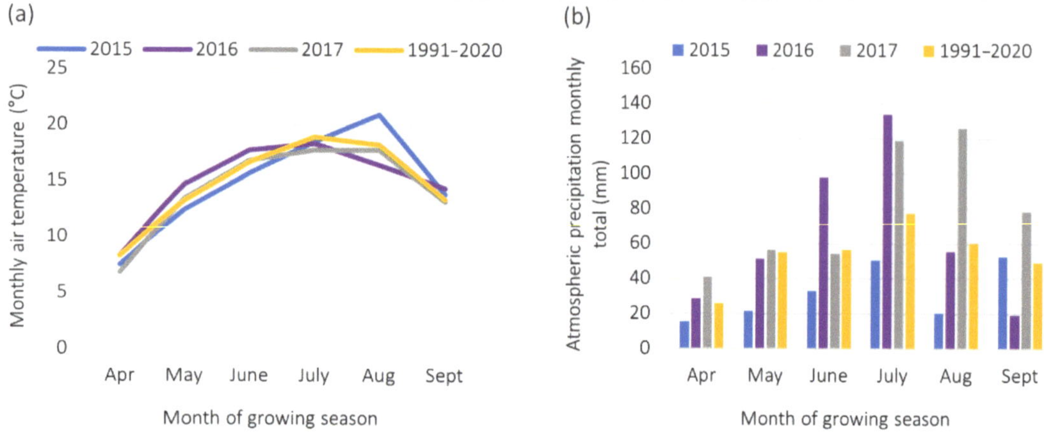

Figure 1. Monthly air temperature (**a**) and the distribution of monthly atmospheric precipitation totals (**b**) in the growing seasons of 2015–2017 compared to the climate norms for 1991–2020.

The method for establishing irrigation deadlines was based on simple weather measurements. It involved balancing the income and expenditure of water from the soil layer with controlled moisture (rhizosphere) during the period of high water requirements in plants that covers May–July. The income side of the balance sheet included the following elements: effective water retention of soil for crops (for barley grown on light soil in compact substrate, this is 30 mm); precipitation; and net water doses from irrigation. On the expenditure side, there was the amount of daily water consumption, which depended on the average daily air temperature; the type of crop; and the type of soil. This balance method is beneficial for controlling irrigation at the point scale. It was tested many times in carefully controlled field experiments, in which it almost perfectly signaled the need to start irrigation, and it was consistent with the methods directly determining soil moisture. A balance was struck during the barley's phase of high water requirements by assuming a value of initial effective water retention of the soil equal to the value of total atmospheric precipitation in the 10 days preceding the beginning of the critical crop period, which was established as 20 May.

2.4. Chemical and Biochemical Analyses

Soil was sampled from 0 to 20 cm of the topsoil three times at the following development phases: I—during spring germination (BBCH 9–19); II—after fertilization/ripening (BBCH 71–78); and III—before harvest/maturity (BBCH 86–87). At each development phase, soil was sampled in four replications for all of the treatments. Material from the field-sampled soil was sieved (2 mm mesh) and kept in a plastic box at 4 °C. After two days, the microbial activity stabilized in the soil and the enzymatic activity was studied.

N-NO_3^- and N-NH_4^+ contents were extracted from moist field soil samples using KCl and K_2SO_4, respectively. The nitrate nitrogen content was determined using the phenol disulfonic acid method and the ammonium nitrogen content was determined with the indophenol method [21].

Urease activity (UR activity; EC 3.5.1.5) in the soil was assayed according to Kandeler and Gerber [22]: An amount of 1 g of soil was incubated with 4 mL of borate buffer (pH 10.0) and a 0.5 mL solution of urea at 37 °C for 2 h. Later, it was filtered after adding 6 mL of 1 M KCl and the solution and then diluted with water. With the spectrophotometric method, the urease activity was evaluated 30 min after adding NaOH salicylate and acid dichloroisocyanide at 690 nm. The UR activity is presented in mg N-NH_4^+ $kg^{-1}{\cdot}h^{-1}$.

Nitrate reductase activity (NR activity; EC. 1.7.99.4) was assayed as described by Kandeler [23]: Soil samples were incubated with KNO_3 (substrate) and a solution of 2,4-DNP at 25° C for 24 h. These samples were provided with KCl solution and filtered, and 5 mL of solution was provided with 3 mL of ammonium chloride buffer and reagent for staining; then, the samples were mixed and measured at 520 nm. The unit of NR activity was mg N-NO_2^- $kg^{-1}{\cdot}24\ h^{-1}$.

The activity of dehydrogenase (DH; EC 1.1.) is presented in mg TPFg$\cdot^{-1}$ h$\cdot^{-1}$, and was determined according to Thalmann [24]. Soil samples were mixed with buffered tetrazolium salts (TTC) and glucose and incubated at 30oC for 24 h, and the activity of the DH oxidoreductase was assayed with the spectrophotometric method at 546 nm.

The catalase activity (CAT activity; EC 1.11.1.6) was determined using the method by Johnson and Temple [25]. The soil was incubated for 20 min with hydrogen peroxide and then, in an acidic environment, titrated with potassium permanganate. The catalase activity was calculated against the control samples in μmol $H_2O_2{\cdot}g{\cdot}^{-1}{\cdot}min^{-1}$.

Peroxidase activity (PER activity; EC 1.11.1.7) was quantified in accordance with Ladd [26]. The substrates consisted of pyrogallol and hydrogen peroxide, and the peroxidase is presented in mmol of purpurogallin $g^{-1}{\cdot}h^{-1}$.

2.5. Data Analyses

The resistance of the soil (RS) was derived from the formulas suggested by Orwin and Wardle [18]:

$$RS(t_{0)} = 1 - \frac{2D_0}{(C_0+|D_0|)} \tag{1}$$

where $|D_0|$ is the difference between the control soil (C_0) and performing soil (P_0) at the end of irrigation (t_0).

The enzymatic activity results and chemical analysis results were subjected to analyses of variance via Tukey's test with a 5% level of significance using statistics software analysis of variance for orthogonal experiments of the Bydgoszcz University of Science and Technology: ANALWAR–5.1. FR software package,. Pearson's linear correlation coefficients of the biometric feature were calculated using Statistica 13.1 for Windows 10 software.

3. Results

The NO_3^--N and NH_4^+-N contents in the Alfisol and their dynamics during the growing season significantly depended on the conditions of the experiment from irrigation to nitrogen fertilization (Table 4). The content of NH_4^+ depended on the interaction of irrigation during the development phases (Table 4). At the second date (after fertilization), during ripening, the content of ammonium ions was higher; in no-irrigation plots, it was on average 13% less than in the irrigated plots. Before harvest, a higher content of the ions was observed in irrigated plots, especially with the N_1 and N_3 doses. In the plots fertilized with nitrogen, the lowest content of NO_3^--N⁻ occurred in spring (germination). After applying mineral fertilization, the content of those ions increased considerably, and then slightly decreased at the end of vegetation. The content of mineral nitrogen N_{min} depended also on nitrogen fertilization. Differences in contents were found depending on the irrigation before the harvesting of the spring barley. In the plots without irrigation, on average, 34% more mineral nitrogen was recorded than in the plots with irrigation.

Table 4. Contents of nitrate, ammonium, and mineral nitrogen in soil under barley (mean values for 2015–2017).

		NH_4^+			NO_3^-			N_{min}		
Date	**N dose**	**IRR**	**NIRR**	**Mean**	**IRR**	**NIRR**	**Mean**	**IRR**	**NIRR**	**Mean**
Germination	N_0	6.107b	6.107b	6.107	2.657	2.657	2.657a	39.437	39.437	39.437a
Ripening	N_0	4.310a	3.95a7	4.133	10.023	6.497	8.260a	64.502	47.040	55.771a
	N_1	4.513a	4.383a	4.448	7.703	13.147	10.425a	54.977	78.885	66.931a
	N_2	5.217b	8.040b	6.628	21.930	29.850	25.890b	122.16	125.51	123.83b
	N_3	6.060b	6.357b	6.208	23.777	19.637	21.707ab	134.27	116.97	125.62b
Average		5.025	5.684	5.355	15.858	17.283	16.570	93.976	92.100	93.038
Maturity	N_0	4.413a	3.777a	4.095	6.553	6.937	6.745a	32.683	31.545	32.114a
	N_1	4.667a	2.960a	3.813	3.150	20.260	11.705a	35.175	104.49	69.833a
	N_2	3.030a	4.397a	3.713	10.763	20.953	15.525ab	67.235	106.10	86.670b
	N_3	5.603b	3.573a	4.588	25.330	19.753	22.542b	101.87	118.31	110.09b
Average		4.428	3.677	4.053	11.449	16.976	14.213	59.240	90.111	74.676

IRR—irrigation, NIRR—no irrigation, different letters following the mean values indicate significant differences with the Tukey test at $p \leq 0.05$.

The contents of NH_4^+-N and NO_3^--N$^-$ during the experimental years depending on nitrogen fertilization and irrigation are shown in Figure 1. The content of NH_4^+-N ranged from 1.187 to 6.867 mg·kg^{-1} of soil and it did not depend on irrigation; it increased only slightly with increasing doses of the nitrogen fertilizer. However, the content of NO_3^--N$^-$ fell within a wider range from 1.50 to 33.23 mg·kg^{-1} of soil (Figure 2). In all of the experimental years, a higher content of this nitrogen fraction was found in samples from non-irrigated plots when compared to samples from the irrigated ones, and this difference in each year chronologically was 50%, 30%, and 12%.

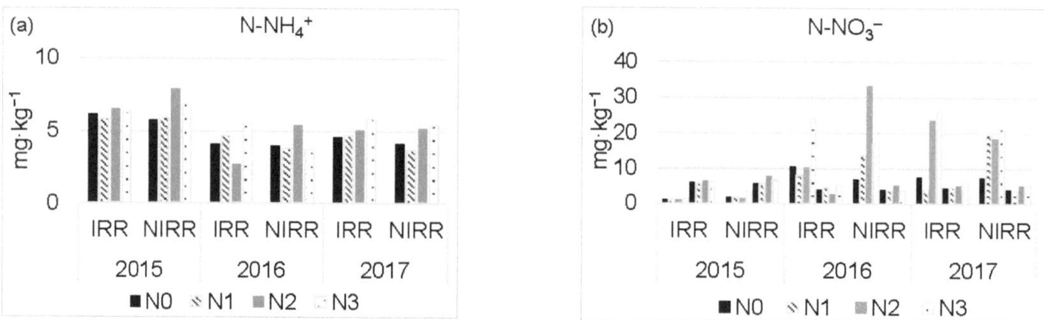

Figure 2. Contents of (**a**) ammonium and (**b**) nitrate nitrogen in soil under barley depending on fertilization in 2015–2017.

However, at maturity, only NR activity was 18% higher in irrigated soil (Table 5). The activities of the other enzymes were higher in no-irrigation treatments by 25% for DH, 22% for PER, 33% for CAT, and 17% for UR compared to the irrigated soil. The statistical analysis showed the effect of irrigation on the PER, CAT, and NR activity. As for NR, the activity was influenced, apart from the influence of irrigation, by the barley's development phase. A significantly higher NR activity was found in the soil sampled in 2016—on average about four times higher compared to the average activity determined for the samples taken in 2016 and three times higher compared to the average soil activity in 2017. However, the activities of the other oxidoreductases developed differently over the experimental years.

The highest catalase activity was found in the samples from 2015, where it was 29% higher compared to the average from 2017.

Table 5. Enzymatic activity at ripening and maturity in barley in 2015, 2016, and 2017.

Treatment		Ripening					Maturity				
		DH ‡	PER	CAT	NR	UR	DH	PER	CAT	NR	UR
Irrigation	N_0	13.55	7.717	3.192	4.800	5.032	18.21	4.819	2.887	6.260	8.144
	N_1	33.38	9.394	3.641	4.884	4.030	18.03	4.606	2.825	5.248	6.370
	N_2	29.51	8.205	3.783	3.174	3.646	24.35	5.643	3.103	4.077	4.551
	N_3	23.60	8.266	4.799	5.471	4.980	57.86	3.843	4.261	6.067	7.758
	Mean	25.01	8.395	3.853	4.582	4.422	29.61	4.728	3.269	5.413	6.706
No irrigation	N_0	27.25	7.198	3.574	7.134	4.364	16.92	6.710	2.284	2.486	6.623
	N_1	27.58	9.242	3.368	2.970	5.959	37.26	4.209	1.897	5.211	7.808
	N_2	27.33	8.601	4.310	7.901	3.195	55.62	8.235	3.069	3.888	8.040
	N_3	53.08	7.473	3.843	7.927	3.242	45.51	5.185	2.595	4.568	9.758
	Mean	33.79	8.128	3.774	6.483	4.190	38.83	6.085	2.462	4.038	8.057
Development phase		ns	ns	ns	ns	ns	ns	ns	ns	1.013	ns
Irrigation		ns	ns	ns	ns	ns	ns	1.813	0.970	0.132	ns
N fertilization		ns	ns	ns	ns	ns	ns	ns	ns	ns	ns
Development phase× irrigation		ns	ns	ns	1.559	ns	ns	ns	ns	1.724	ns

‡ DH—dehydrogenase activity per mg TPFg $^{-1}$ h $^{-1}$. PER—peroxidase activity per mmol of purpurogallin $g^{-1} \cdot h^{-1}$. CAT—catalase activity per µmol $H_2O_2 \cdot g^{-1} \cdot min^{-1}$. NR—nitroreductase activity per mg N-NO_2^- $kg^{-1} \cdot 24\ h^{-1}$. UR—urease activity per mg N-NH_4^+ $kg^{-1} \cdot h^{-1}$, ns—not significant.

Peroxidase, on the other hand, showed 70% higher activity in samples taken in 2017 when compared to that in the 2015 samples. Regarding the influence of fertilization on the enzymatic activity, DH and CAT activity increased with increasing fertilizer doses. As for PER, the highest dose of fertilizer resulted in a 14% reduction in its activity when compared to N_2. The activities of enzymes involved in nitrogen metabolism in the soil were different compared to the oxidoreductases (Figure 3). The activities of both enzymes were highest at the third date of soil sampling. As for UR, the activity in this period was on average 43% higher than at the beginning of the season (germination), and nitrogenase showed 27% higher activity when compared to the lowest activity at the second sampling date (Table 6). The influence of nitrogen fertilization on the activities of these enzymes was also recorded, and on average, the activity of UR was reduced by 13% when fertilized with a dose of N_2 and that of NR was reduced by 7% when fertilized with a dose of N_1, as compared to the control.

Table 6. Enzymatic activity during germination in barley in 2015, 2016, and 2017.

Year	Germination				
	DH ‡	PER	CAT	NR	UR
2015	6.930	4.340	5.120	0.311	4.780
2016	25.30	4.490	2.420	3.452	6.890
2017	18.40	8.970	2.021	7.890	6.590
Mean	16.88	5.930	3.187	3.884	6.087

‡ designations the same as in Table 5.

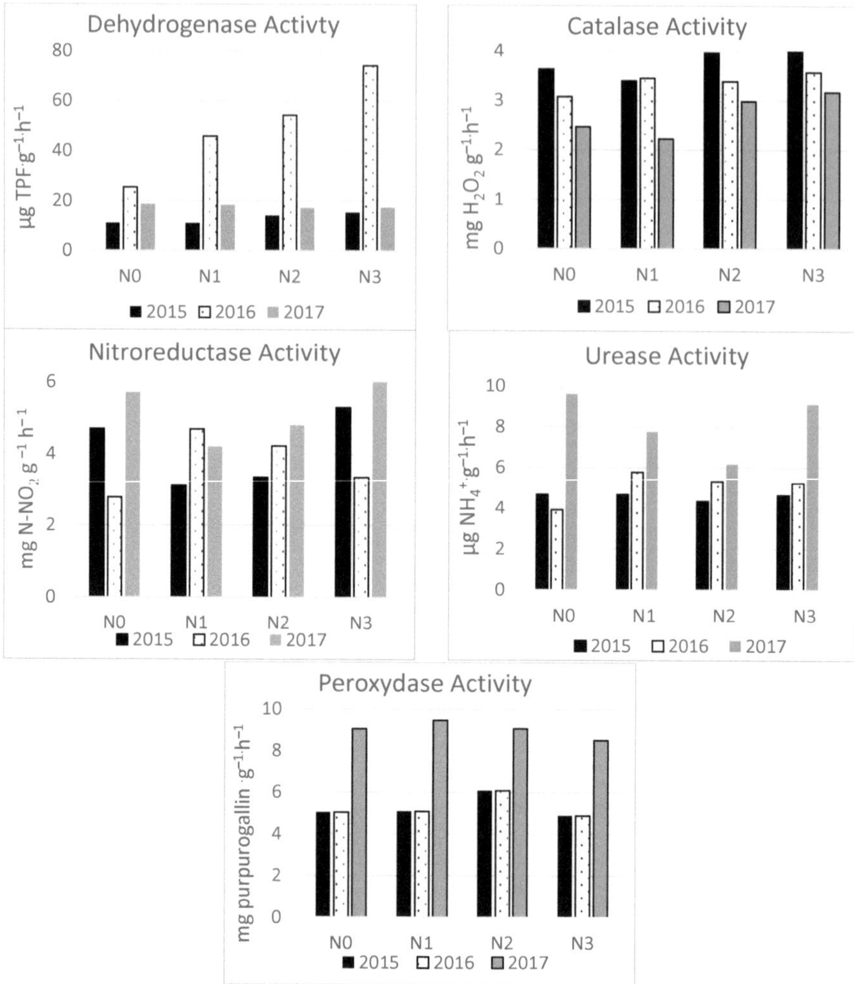

Figure 3. Enzymatic activities in barley depending on nitrogen doses in 2015, 2016, and 2017.

Sprinkler irrigation applied in the barley experiment did not have a significant impact on soil enzymatic activity. This was proven by the non-significant coefficients of correlation between the content of RAW and the enzymatic activity both for the irrigated and non-irrigated plots (Table 7). The most sensitive enzyme to soil water content was peroxidase ($r = -0.1652$), while the other ones demonstrated a similar level of response (r between -0.0712 and 0.0735). As for the second factor, there was no response of urease to the nitrogen fertilizer level, while catalase and dehydrogenase responded with positive, yet non-significant, values ($r = 0.2001$ and $r = 0.2576$, respectively). Importantly, the values of coefficients were bidirectional, depending on the type of enzyme, which confirms that the reactions of these enzymes to irrigation and N fertilizer dose were ambiguous.

Table 7. Coefficients of correlation (r) between soil enzymatic activity and the content of readily available water (RAW) and differentiated N fertilization level.

Type of Soil Enzyme	RAW	N Fertilization
Catalase	−0.0712	0.2001
Dehydrogenase	0.0735	0.2576
Peroxidase	−0.1652	−0.0087
Urease	0.0532	0.0000
Nitroreductase	0.0676	0.0711

RAW—readily available water in soil, N fertilization—nitrogen fertilization.

Regarding the contents and enzymatic activities determined in this study (Table 8), catalase, peroxidase, and urease activities were significantly correlated with the NH_4^+-N content (r = 0.299, r = 0.331, and r = −0.297, respectively). However, dehydrogenase and peroxidase were positively correlated with the NO_3^--N content in the soil. The urease activity was significantly negatively correlated with the soil enzymatic activities of nitroreductase (r = −0.340) and peroxidase (r = −0.245).

Table 8. Relationship between selected soil properties.

Dependent Variable (y)	Independent Variable (x)	Equation	Correlation Coefficient (r)
Catalase activity	NH_4^+	y = 2.8422 + 0.64010x	0.299
Dehydrogenase activity	NO_3^-	y = 3.8906 + 0.28549x	0.523
Peroxidase activity	NH_4^+	y = 3.6836 + 0.54160x	0.331
Urease activity	NH_4	y = 7.6386 – 0.3937x	−0.297
Nitroreductase activity	Urease activity	y = 6.2506 − 0.2875x	−0.340
Peroxidase activity	NO_3^-	y = 3.6337 + 1.2621x	0.336
Urease activity	Peroxidase	y = 6.9388 − 0.1986x	−0.245

The resistance of soil (RS) data are presented in Table 9. Differences in resistance to irrigation across the enzymes were observed for nitrogen doses. Oxidoreductases (PER, CAT, DH) with the highest RS value were observed for N_0 and N_1. The highest RS values (0.991 and 0.934) were calculated for CAT activity and for N_0 and N_1. For these N doses, high values of RS for DH activity (0.859 for N_0) and PX activity (0.828 for N_1) were recorded. For UR, the highest RS values were found for N_1 (0.986) and N_3 (0.907). The RS values for the activities of UR and NR were negative: N_1 (−0.597) and N_0 (−0.206).

Table 9. Resistance of soil (RS) for enzymatic activities depending on nitrogen doses during vegetation in spring barley.

N Doses	Resistance of Soil (RS)				
	NR	UR	CT	PX	DH
N_0	−0.206	0.627	0.934	0.560	0.859
N_1	0.986	−0.597	0.991	0.828	0.319
N_2	0.907	0.395	0.580	0.521	0.425
N_3	0.506	0.660	0.444	0.589	0.573

4. Discussion

Water and nitrogen are in charge of limiting rural production in most parts of the world [27]. Transforming nitrogen in soil is essential for nitrogen metabolism and crop tolerance to drought stress, and it is engaged in nearly all of the physiological transformations in plants and microorganisms [28]. According to Wang et al. [29], NH_4^+-N uptake is universally enhanced in the majority of plants during drought stress, and superior nitrogen uptake may increase plant drought hardiness. The outcome of the present experiment identified the impact of irrigation on the development phases in spring barley. During barley vegetation, it was found that with the development of plants, the NH_4^+-N content in lessive soil showed a decreasing trend, especially in non-irrigated soil; the ammonium content decreased significantly, which could have been due to the fact that at maturity, spring barley has a higher NH_4^+-NNH_4^+-N uptake due to drought stress. This result is consistent with the work of Lawlor et al. [30], who recorded an increasing effective NH_4^+ nitrogen uptake and increasing activity of NR in plants during drought stress. As compared to the non-irrigated soil, the content of NO_3^- and N_{min} in soil exposed to irrigation decreased during spring barley vegetation. The lowest contents of NO_3^--N and N_{min} at the third sampling date suggest NO_3^-—N leaching. Similar results were obtained by Wu et al. [31], who reported that the mineral nutrient content in soil changed depending on irrigation and nitrogen fertilization and that a high irrigation water content can increase nutrient leaching and reduce the soil nutrient content. Muhammad et al. [32] show that the mechanisms of NO_3^--N leaching depend on the physical properties of soil, especially its water capacity. A higher amount of N (300 kg N ha^{-1}) resulted in a higher soil SOC and higher total and mineral N under low (60%) irrigation. Nitrogen in the form of nitrate is highly mobile in soil and its content depends on soil water conditions [33]. Irrigation probably contributes to increased leaching of nitrate in soil. The results of the present experiment show that doses of nitrogen fertilizers have an impact on the contents of NO_3^--N and N_{min}. These findings are consistent with those of Jia [34], who report that NO_3^--N leaching increases even with the same N fertilizer rate due to a vast amount of total irrigation. The effects of temperature and moisture on enzyme diffusion and substrate availability are all critical factors influencing soil enzymes' activities [35]. Drought greatly influences almost all of the physiological and biochemical transformations of plants: growth, development, and productivity. The nitrogen content and transformation in soil are decisive during drought stress for plant and microorganism metabolism. The present study demonstrates that enzymes are soil components that are strictly connected with the physicochemical and biological properties of soil. The reactions of enzymes depend on their origin and features [36]. In this study, we demonstrated a lack of responses of all five types of soil enzymes to different levels of nitrogen fertilization in barley. Cui et al. [37] suggest that monoculture and fertilization processes can increase enzyme activity by improving soil nutrients and microbial richness. In 2020, Zhang et al. [38] determined that nitrogen fertilizer lowered both β-1,4-glucosidase and acid phosphatase, while water from irrigation inhibited acid phosphatase only. Additional nitrogen and irrigated water affected enzyme activity mainly by affecting the soil microbial biomass carbon and NH_4^+-N. In general, the addition of nitrogen or water over a long time did not affect β-1,4-glucosidase, which implies that this enzyme is resilient and stays unaffected by modifications in the environment. In contrast, acid phosphatase showed sensitivity to nitrogen and irrigation and responded to seasonal fluctuations in precipitation. This suggests that this enzyme could be an indicator of transformations in soil nutrient cycling. Many field studies have examined the effects of added nitrogen on the activity of enzymes in soil. The results of those studies were inconclusive. Some results suggested that the addition of a nitrogen fertilizer caused soil acidification and inhibited soil enzymatic activity [39]. Other studies indicated a stimulating effect of nitrogen on enzyme activity, or no such effect at all [40–44].

Urease hydrolysis of small organic substrates containing nitrogen into inorganic compounds (ammonia) is used to supply nitrogen for the normal growth and development of plants [45]. In our study, the development phase, irrigation, and N mineral fertilization showed no statistical impact on urease activity. Similar results were obtained by Zhao et al. [46] in their research: single-nitrogen nor mixed-nitrogen applications did not affect urease activity significantly. However, the present study reveals that the activity of hydrolase in the soil increased and later decreased the urease activity, and it hit its maximum at the maturity phase and increased with the increasing doses of N fertilizer, especially in non-irrigated plots. Weng et al. [47] and Gong et al. [44] note that mineral nitrogen fertilizer often increases urease activity. Fortification of urease activity due to natural or organic nitrogen addition was observed by Nayak et al. [15] and Iovieno et al. [48].

Higher hydrolase activities may be due to an increase in carbon and nitrogen in the soil and improvements in the soil physicochemical properties, as well as a more appropriate soil environment for microbial growth and proliferation, which stimulates microbial and enzymatic activity. Negative effects resulting from lower pH have also been observed with the long-term use of nitrogen fertilizers [49].

The activity of enzymes depends on several factors, especially on the presence of a substrate; as for NR, the substrate is nitrate in soil. Nitrate reductase is an enzyme that controls and reduces nitrate assimilation in plants, which is not only responsive to external nitrogen but also indirectly creates a difference in the uptake and utilization of nitrogen by plants [50]. Waraich et al. [51] and Sardans and Peñuelas [52] report on drought stress reducing plant N uptake and assimilation by reducing both nutrient diffusion and N supply via mineralization [53]. In our study, the lower NR activity at maturity in the plots with no irrigation may have resulted from a reaction of the plants and microorganisms to long drought stress. The NR activity increased at ripening and then decreased at maturity when not exposed to irrigation. At maturity in the spring barley, the CAT activity increased in irrigated soil. However, PER activity presented a different reaction and reached its highest in non-irrigated soil and depended significantly on the rates of water. Peroxidase is an enzyme that is expressed for a variety of reasons, including carbon and nitrogen and protection. This enzyme moves into soil via excretion or lysis, where it mediates the ecosystem functions of lignin degradation, humification, carbon mineralization, and dissolved organic carbon release [7]. A higher PER activity in non-irrigated soil indicates high oxygen availability, optimal pH conditions, and optimal mineral activity, and hence a high oxidative activity and limited accumulation of organic matter in the soil [7].

We present interesting observations regarding dehydrogenases, which are one of the most important oxidoreductases, and which are used as an indicator of the total soil microbial activity since they are closely linked with microbial oxidoreduction processes, as they occur in all living microbial cells and as such can indicate the microorganism activity in soil [54]. Our research has shown that soil moisture influences dehydrogenase activity. The high DHA activity observed in the soil during spring barley vegetation in 2016, which was the highest-rainfall year throughout our experiment, coincides with the results reported by Gu et al. [55], who observed an increase in DH in high-moisture soil. A high dehydrogenase activity can be due to two factors: flooding, with the release and spread of soluble organic compounds in the soil, which contributes to the development of a larger number of bacteria that secrete dehydrogenases; and/or the change of oxygen conditions to anaerobic conditions and the proliferation of anaerobic microorganisms [56]. Also, Dora [57] indicates that dehydrogenase and catalase activities are higher in irrigated soil. Tan et al. [58] found that long-term mulched drip irrigation (8, 12, 16, and 22 years) tends to accumulate soil nutrients and rebuild enzyme conditions. Soil enzymes such as catalase and urease were more active in the subsoil than in the topsoil. Also, Liang et al. [59] confirmed that long-term irrigation strongly increased the activity of dehydrogenase as well as urease in soil. Núñez et al. [60] indicated that a reduction in enzyme activity after irrigation termination in corn may point to changes in biogeochemical cycling and even a potential reduction in the decomposition of leftovers [11,61]. However, enzyme activity can also be affected by changes in soil environmental circumstances [61,62], such as reduced water availability, which can increase enzyme immobilization and decrease the diffusion rates, decreasing enzyme efficiency and affecting residue decomposition independent of changes in potential enzyme activity [63]. A negligible effect of irrigation on the activity of soil enzymes was also reported in grassland ecosystems [64]. Also, it has been shown that additional water application can mitigate the effects of nitrogen enrichment on microorganisms by leaching or reducing the accumulation of inorganic nitrogen [65,66], and it can also have a significant effect on soil enzyme activity.

The present study identifies that catalase, peroxidase, and urease were correlated significantly with NH_4^+-N content (appropriately, $r = 0.299$, $r = 0.331$, and $r = 0.297$; $p = 0.05$), and dehydrogenase and peroxidase activity were correlated significantly with

NO_3^--N content (r = 0.523 and r = 0.336; p = 0.05). Nitroreductase was negatively correlated significantly with the activities of urease (r = −0.340; p = 0.05) and peroxidase (r = −0.254; p = 0.05), indicating that some enzyme activities may affect and induce other enzyme activities in the soil considerably.

The resistance of the enzymes to drought was different depending on the doses of nitrogen fertilization. Catalase showed the highest resistance to drought stress, followed by NR and PER. Urease and the dehydrogenases demonstrated a lower resistance to soil drought. The resistance of the enzymes to drought was different depending on the doses of nitrogen fertilization. Catalase showed the highest resistance to drought stress, followed by NR and PER; urease and the dehydrogenases showed a lower resistance to soil drought. The results reported by Lemanowicz [67] point to catalase activity having a strong resistance also to salinity stress.

5. Conclusions

In conclusion, our results suggest that no irrigation influences the NH_4^+-N content in Alfisols with spring barley at maturity due to its low uptake being the consequence of drought stress. Irrigation may contribute to increased nitrate leaching in the soil profile. The results of our experiment show that different doses of nitrogen fertilizer influence the contents of NO_3^--N and N_{min}. The nitrogen fertilization of 60 t ha^{-1} was optimal for achieving the ideal contents of NO_3^-—N and NH_4^+-N available to plants. The present study indicates that enzymes are sensitive to soil properties, which are closely related to the contents of NH_4^+-N and NO_3^--N in the soil. The enzymatic activity changed over the studied years, depending on the weather conditions. The resistance of soil could be used for an enzymatic water stress solution. The highest index of resilience was presented by catalase. These results suggest a need for further research on selected physicochemical and biochemical parameters, as well as on other types of soil and under other crops, especially in areas that have a moderate transitional climate with varied weather conditions.

Author Contributions: Conceptualization, A.S.-Z. and R.K.-T.; methodology, A.S.-Z. and R.K.-T.; software, A.S.-Z. and R.K.-T.; validation, A.S.-Z. and R.K.-T.; formal analysis, A.S.-Z. and R.K.-T.; investigation, A.S.-Z. and R.K.-T.; resources, A.S.-Z. and R.K.-T.; data curation, A.S.-Z. and R.K.-T.; writing, A.S.-Z. and R.K.-T.; supervision, A.S.-Z. and R.K.-T. All authors have read and agreed to the published version of the manuscript.

Funding: This research received no external funding.

Institutional Review Board Statement: Not applicable.

Data Availability Statement: Data are contained within the article.

Acknowledgments: The authors express words of thanks to Stanisław Dudek for his commitment and valuable help in conducting the field experiment.

Conflicts of Interest: The authors declare no conflicts of interest.

References

1. Hofman, G.; Van Cleemput, O. *Soil and Plant Nitrogen*; International Fertilizer Industry Association: Paris, France, 2004; pp. 1–49. Available online: http://www.betuco.be/compost/Soil%20and%20plant%20nitrogen.pdf (accessed on 28 April 2024).
2. Cowden, C.C.; Shefferson, R.P.; Mohan, J.E. Mycorrhizal mediation of plant and ecosystem responses to soil warming. In *Ecosystem Consequences of Soil Warming*; Mohan, J.E., Ed.; Elsevier Academic Press: Amsterdam, The Netherlands, 2019; pp. 157–173. [CrossRef]
3. Girkin, N.T.; Cooper, H.V. Nitrogen and ammonia in soils. In *Encyclopedia of Soils in the Environment*, 2nd ed.; Goss, M.J., Oliver, M., Eds.; Elsevier Academic Press: Amsterdam, The Netherlands, 2023; pp. 142–151. [CrossRef]
4. Shafreen, M.; Vishwakarma, K.; Shrivastava, N.; Kumar, N. Physiology and Distribution of Nitrogen in Soils. In *Soil Nitrogen Ecology, Soil Biology*; Cruz, C., Vishwakarma, K., Choudhary, D.K., Varma, A., Eds.; Springer Nature: Cham, Switzerland, 2021; Volume 2, pp. 8–13. [CrossRef]
5. Moeskops, B.; Buchan, D.; Sleutel, S.; Herawaty, L.; Husen, E.; Saraswati, R.; Setyorini, D.; De Neve, S. Soil microbial communities and activities under intensive organic and conventional vegetable farming in West Java, Indonesia. *Appl. Soil Ecol.* **2010**, *45*, 112–120. [CrossRef]

6. Chabot, M.; Morales, E.; Cummings, J.; Nicholas Rios, N.; Giatpaiboon, S.; Mogul, R. Simple kinetics, assay, and trends for soil microbial catalases. *Anal. Biochem.* **2020**, *610*, 113901. [CrossRef]
7. Sinsabaugh, R.L. Phenol oxidase, peroxidase and organic matter dynamics of soil. *Soil Biol. Biochem.* **2010**, *42*, 391–404. [CrossRef]
8. Das, S.K.; Varma, A. Role of Enzymes in Maintaining Soil Health. In *Soil Enzymology*; Springer: Berlin/Heidelberg, Germany, 2010.
9. Abdelmagid, H.M.; Tabatabai, M.A. Nitrate reductase activity in soils. *Soil Biol. Biochem.* **1987**, *19*, 421–427. [CrossRef]
10. Nannipieri, P.; Kandeler, E.; Ruggiero, P. Enzyme Activities and Microbiological and Biochemical Processes in Soil. In *Activity, Ecology and Applications*; Burns, R.G., Dick, R.P., Eds.; Marcel Dekker: New York, NY, USA, 2002; pp. 1–36.
11. Acosta-Martínez, V.; Cano, A.; Johnson, J. Simultaneous determination of multiple soil enzyme activities for soil health-biogeochemical indices. *Appl. Soil Ecol.* **2018**, *126*, 121–128. [CrossRef]
12. Balota, E.L.; Colozzi-Filho, A.; Andrade, D.S.; Dick, R.P. Long-term tillage and crop rotation effects on microbial biomass and C and N mineralization in a Brazilian Oxisol. *Soil Till. Res.* **2004**, *77*, 137–145. [CrossRef]
13. Sicardi, M.; Garcia-Prechac, F.; Frioni, L. Soil microbial indicators sensitive to land use conversion from pastures. to commercial Eucalyptus grandis (Hill ex Maiden) plantations in Uruguay. *Appl. Soil Ecol.* **2004**, *27*, 125–133. [CrossRef]
14. Gianfreda, L.; Ruggiero, P. Enzyme Activities in Soil. In *Soil Biology*; Nannipieri, P., Smalla, K., Eds.; Springer: Berlin/Heidelberg, Germany, 2006; Volume 8, pp. 257–311.
15. Nayak, D.R.; Babu, Y.J.; Adhyaa, T. Long-term application of compost influences microbial biomass and enzyme activities in a tropical Aeric Endoaquept planted to rice under flooded condition. *Soil Biol. Biochem.* **2007**, *39*, 1897–1906. [CrossRef]
16. Doran, J.W.; Parkin, T.B. Defining and Assessing Soil Quality. In *Defining Soil Quality for a Sustainable Environment*; Doran, J.W., Coleman, D.C., Bezdicek, D.F., Stewart, B.A., Haynes, R.J., Eds.; Soil Science Society of America Special Publication No. 15; American Society of Agronomy: Madison, WI, USA, 1994; pp. 3–21.
17. Samuel, A.D.; Bungau, S.; Tit, D.M.; Melinte, C.E.; Purza, L.; Badea, G.E. Effects of long term application of organic and mineral fertilizers on soil enzymes. *Rev. Chim.* **2018**, *69*, 2608–2612. [CrossRef]
18. Orwin, K.H.; Wardle, D.A. New indices for quantifying the resistance and resilience of soil biota to exogenous disturbances. *Soil Biol. Biochem.* **2004**, *36*, 1907–1912. [CrossRef]
19. USDA. *Soil Survey Staff Keys to Soil Taxonomy*, 11th ed.; USDA Natural Resources Conservation Service: Washington, DC, USA, 2010; pp. 1–346.
20. Żarski, J.; Treder, W.; Dudek, S.; Kuśmierek-Tomaszewska, R. Establish irrigation deadlines on the basis of simple meteorological measurements. (Ustalanie terminów nawadniania na podstawie prostych pomiarów meteorologicznych). *Infrat. Ecol. Rural Area* **2011**, *6*, 101–108, (In Polish with English abstract and captions of tables and figures).
21. Bashour, I.I.; Sayegh, A.H. *Methods of Analysis for Soils of Arid and Semi-Arid Regions*; Food and Agriculture Organization of the United Nations: Rome, Italy, 2007.
22. Kandeler, E.; Gerber, H. Short-term assay of soil urease activity using colorimetric determination of ammonium. *Biol. Fertil. Soil* **1988**, *6*, 68–72. [CrossRef]
23. Kandeler, E. Enzymes Involved in Nitrogen Metabolism. In *Methods in Soil Biology*; Scinner, F., Öhlinger, R., Kandeler, E., Mrgesin, R., Eds.; Springer: Berlin/Heidelberg, Germany, 1995; pp. 163–184.
24. Thalmann, A. Zur Methodik derestimmung der Dehydrodgenaseaktivität in Boden mittels Triphenytetrazoliumchlorid (TTC). *Landwirtsch. Forsch.* **1968**, *21*, 249–258.
25. Johnson, J.L.; Temple, K.L. Some variables affecting the measurement of "catalase activity" in soil. *Soil Sci. Soc. Am. J.* **1964**, *28*, 207–209. [CrossRef]
26. Ladd, J.N. Origin and Range of Enzymes in Soil. In *Soil Enzymes*; Academic R.G. Burns Press: London, UK, 1978.
27. Gonzalez-Dugo, V.; Durand, J.L.; Gastal, F. Water deficit and nitrogen nutrition of crops. *Agron. Sustain. Dev.* **2010**, *30*, 529–544. [CrossRef]
28. Otori, K.; Tanabe, N.; Maruyama, T.; Sato, S.; Yanagisawa, S.; Tamoi, M.; Shigeoka, S. Enhanced photosynthetic capacity increases nitrogen metabolism through the coordinated regulation of carbon and nitrogen assimilation in Arabidopsis thaliana. *J. Plant Res.* **2017**, *130*, 909–927. [CrossRef] [PubMed]
29. Wang, H.; Yang, Z.; Yu, Y.; Chen, S.; He, Z.; Wanga, Y.; Jianga, L.; Wanga, G.; Yang, C.H.; Liu, B.; et al. Drought enhances nitrogen uptake and assimilation in maize roots. *Agron. J.* **2016**, *109*, 39–46. [CrossRef]
30. Lawlor, D.W. Carbon and nitrogen assimilation in relation to yield: Mechanisms are the key to understanding production systems. *J. Exp. Bot.* **2002**, *53*, 773–787. [CrossRef] [PubMed]
31. Wu, H.; Du, S.; Zhang, Y.; An, J.; Zou, H.; Zhang, Y.; Yu, N. Effects of irrigation and nitrogen fertilization on greenhouse soil organic nitrogen fractions and soil-soluble nitrogen pools. *Agric. Water Manag.* **2019**, *216*, 415–424. [CrossRef]
32. Muhammad, I.; Lv, J.Z.; Yang, L.; Ahmad, S.; Farooq, S.; Zeeshan, M.; Zho, X.B. Low irrigation water minimizes the nitrate nitrogen losses without compromising the soil fertility, enzymatic activities and maize growth. *BMC Plant Biol.* **2022**, *22*, 159. [CrossRef]
33. Sanchez-Martín, L.; Meijide, A.; Garcia-Torres, L.; Vallejo, A. Combination of drip irrigation and organic fertilizer for mitigating emissions of nitrogen oxides in semiarid climate. *Agric. Ecosyst. Environ.* **2010**, *137*, 99–107. [CrossRef]

34. Jia, X.; Shao, L.; Liu, P.; Zhao, B.; Gu, L.; Dong, S.; Bing, H.; Zhang, J.; Zhao, B. Effect of different nitrogen and irrigation treatments on yield and nitrate leaching of summer maize (*Zea mays* L.) under lysimeter conditions. *Agric. Water Manag.* **2014**, *137*, 92–103. [CrossRef]
35. Steinweg, J.M.; Dukes, J.S.; Wallenstein, M.D. Modeling the effects of temperature and moisture on soil enzyme activity: Linking laboratory assays to continuous field data. *Soil Biol. Biochem.* **2012**, *55*, 85–92. [CrossRef]
36. Karaca, A.; Cetin, S.C.; Turgay, O.C.; Kizilkaya, R. Soil Enzymes as Indication of Soil Quality. In *Soil Enzymology*; Springer: Berlin/Heidelberg, Germany, 2010.
37. Cui, Y.; Fang, L.; Deng, L.; Guo, X.; Han, F.; Ju, W.; Wang, X.; Chen, H.; Tan, W.; Zhang, X. Patterns of soil microbial nutrient limitations and their roles in the variation of soil organic carbon across a precipitation gradient in an arid and semi-arid region. *Sci. Total Environ.* **2019**, *658*, 1440–1451. [CrossRef] [PubMed]
38. Zhang, J.; Jin, K.; Luo, Y.; Du, L.; Tian, R.; Wang, S.; Shen, Y.; Zhang, J.; Li, N.; Shao, W. Responses of Soil Enzyme Activity to Long-Term Nitrogen Enrichment and Water Addition in a Typical Steppe. *Agronomy* **2023**, *13*, 1920. [CrossRef]
39. Ren, C.J.; Zhao, F.Z.; Shi, Z.; Chen, J.; Han, X.H.; Yang, G.H.; Feng, Y.Z.; Ren, G.X. Differential responses of soil microbial biomass and carbon-degrading enzyme activities to altered precipitation. *Soil Biol. Biochem.* **2017**, *115*, 1–10. [CrossRef]
40. Chen, J.; Luo, Y.; Li, J.; Zhou, X.; Zhou, L. Costimulation of soil glycosidase activity and soil respiration by nitrogen addition. *Glob. Change Biol.* **2016**, *23*, 1328–1337. [CrossRef]
41. Xiao, W.; Chen, X.; Jing, X.; Zhu, B.A. A meta-analysis of soil extracellular enzyme activities in response to global change. *Soil Biol. Biochem.* **2018**, *123*, 21–32. [CrossRef]
42. Ma, W.; Li, J.; Gao, Y.; Xing, F.; Sun, S.; Zhang, T.; Zhu, X.; Chen, C.; Li, Z. Responses of soil extracellular enzyme activities and microbial community properties to interaction between nitrogen addition and increased precipitation in a semi-arid grassland ecosystem. *Sci. Total Environ.* **2020**, *703*, 134691. [CrossRef]
43. Keeler, B.L.; Hobbie, S.E.; Kellogg, L.E. Effects of Long-Term Nitrogen Addition on Microbial Enzyme Activity in Eight Forested and Grassland Sites: Implications for Litter and Soil Organic Matter Decomposition. *Ecosystems* **2009**, *12*, 1–15. [CrossRef]
44. Gong, S.; Zhang, T.; Guo, R.; Cao, H.; Shi, L.; Guo, J.; Sun, W. Response of soil enzyme activity to warming and nitrogen addition in a meadow steppe. *Soil Res.* **2015**, *53*, 242. [CrossRef]
45. Cordero, I.; Snell, H.; Bardgett, R.D. High throughput method for measuring urease activity in soil. *Soil Biol. Biochem.* **2019**, *134*, 72–77. [CrossRef] [PubMed]
46. Zhao, Y.; Wang, Y.; Sun, S.; Liu, W.; Zhu, L.; Yan, X. Different forms and proportions of exogenous nitrogen promote the growth of alfalfa by increasing soil enzyme activity. *Plants* **2022**, *11*, 1057. [CrossRef] [PubMed]
47. Weng, B.; Xie, X.; Yang, J.; Liu, J.; Lu, H.; Yan, C. Research on the nitrogen cycle in rhizosphere of Kandelia obovata under ammonium 425 and nitrate addition. *Mar. Pollut.* **2013**, *76*, 227–240. [CrossRef] [PubMed]
48. Iovieno, P.; Morra, L.; Leone, A.; Pagano, L.; Alfani, A. Effect of organic and mineral fertilizers on soil respiration and enzyme activities 434 of two Mediterranean horticultural soils. *Biol. Fertil. Soils* **2009**, *45*, 555–561. [CrossRef]
49. Ajawa, H.A.; Dell, C.J.; Rice, C.W. Changes in enzyme activities and microbial biomass of tallgrass prairie soil as related to burning and 437 nitrogen fertilization. *Soil Biol. Biochem.* **1999**, *31*, 769–777. [CrossRef]
50. Pu, Y.; Zhu, B.; Dong, Z.; Liu, Y.; Wang, C.; Ye, C. Soil N_2O and NO_x emissions are directly linked with *N*-cycling enzymatic activities. *Appl. Soil Ecol.* **2019**, *139*, 15–24. [CrossRef]
51. Waraich, E.A.; Ahmad, R.; Ashraf, M.Y. Role of mineral nutrition in alleviation of drought stress in plants. *Aust. J. Crop Sci.* **2011**, *5*, 764–777.
52. Sardans, J.; Peñuelas, J. The role of plants in the effects of global change on nutrient availability and stoichiometry in the plant-soil system. *Plant Physiol.* **2012**, *160*, 1741–1761. [CrossRef]
53. Schimel, J.; Balser, T.C.; Wallenstein, M. Microbial stress-response physiology and its implications for cosystem function. *Ecology* **2007**, *88*, 1386–1394. [CrossRef]
54. Wolińska, A.; Stępniewska, Z. Dehydrogenase Activity in the Soil Environment. In *Dehydrogenases*; Canuta, R.A., Ed.; IntechOpen: London, UK, 2012; pp. 1–183.
55. Gu, Y.; Wang, P.; Kong, C. Urease, invertase, dehydrogenase and polyphenoloxidase activities in paddy soils influenced by allelopathic rice variety. *Eur. J. Soil Biol.* **2009**, *45*, 436. [CrossRef]
56. Furtak, K.; Gałązka, A.; Niedźwiecki, J. Changes in soil enzymatic activity caused by hydric stress. *Pol. J. Environ. Stud.* **2020**, *29*, 2653–2660. [CrossRef]
57. Dora, S.A. Effect of irrigation on the biology of soil. *Nat. Resour. Sustain. Dev.* **2013**, *3*, 69–74.
58. Tan, M.; Zong, R.; Lin, H.; Dhital, Y.D.; Ayantobo, O.O.; Chen, P.; Li, H.; Chen, R.; Wang, Z. Responses of soil nutrient and enzyme activities to long-term mulched drip irrigation (MDI) after the conversion of wasteland to cropland. *Appl. Soil Ecol.* **2023**, *190*, 104976. [CrossRef]
59. Liang, Q.; Gao, R.; Xi, B.; Zhang, Y.; Zhang, H. Long-term effects of irrigation using water from the river receiving treated industrial wastewater on soil organic carbon fractions and enzyme activities. *Agric. Water Manag.* **2014**, *135*, 100–108. [CrossRef]
60. Núñez, A.; Ball, R.; Schipansk, M. Plant and soil microbial responses to irrigation retirement in semiarid cropping systems. *Environ. Res. Commun.* **2022**, *4*, 035004. [CrossRef]
61. Alster, C.J.; German, D.P.; Lu, Y.; Allison, S.D. Microbial enzymatic responses to drought and to nitrogen addition in a southern California grassland. *Soil Biol. Biochem.* **2013**, *64*, 68–79. [CrossRef]

62. Schimel, J.P. Life in dry soils: Effects of drought on soil microbial communities and processes. *Annu. Rev. Ecol. Evol. Syst.* **2018**, *49*, 409–432. [CrossRef]
63. Schimel, J.P.; Becerra, C.A.; Blankinship, J. Estimating decay dynamics for enzyme activities in soils from different ecosystems. *Soil Biol. Biochem.* **2017**, *114*, 5–11. [CrossRef]
64. Xiao, L.; Liu, G.; Li, P.; Xue, S. Dynamics of soil specific enzyme activities and temperature sensitivities during grassland succession after farmland abandonment. *Catena* **2021**, *199*, 105081. [CrossRef]
65. Peng, X.Q.; Wang, W. Stoichiometry of soil extracellular enzyme activity along a climatic transect in temperate grasslands of northern China. *Soil Biol. Biochem.* **2016**, *98*, 74–84. [CrossRef]
66. Xu, Z.W.; Wan, S.Q.; Zhu, G.L.; Ren, H.Y.; Han, X.G. The Influence of Historical Land Use and Water Availability on Grassland Restoration. *Restor. Ecol.* **2010**, *18*, 217–225. [CrossRef]
67. Lemanowicz, J. Activity of selected enzymes as markers of ecotoxicity in technogenic salinization soils. *Environ. Sci. Pollut. Res.* **2019**, *26*, 13014–13024. [CrossRef] [PubMed]

agriculture

MDPI

Article

Soil Organic Carbon Dynamics in the Long-Term Field Experiments with Contrasting Crop Rotations

Tomáš Šimon, Mikuláš Madaras *, Markéta Mayerová and Eva Kunzová

Crop Research Institute, Department of Crop Management Systems, 161 06 Prague, Czech Republic; simont@vurv.cz (T.Š.); mayerova@vurv.cz (M.M.); kunzova@vurv.cz (E.K.)
* Correspondence: madaras@vurv.cz

Abstract: Trends in soil organic carbon (SOC) were analyzed in the soils from the oldest Czech long-term field experiment, the Prague-Ruzyně Long-Term Fertilizer Experiment, conducted on Haplic Luvisol since 1955. The aim of the work was to compare the long-term dynamics of SOC in contrasting crop rotations and different fertilization regimes. The trial design includes two crop rotations (CR): simple CR with two-year rotation of sugar beet and spring wheat, and multi-crop rotation (MCR) with nine crops. Four fertilization treatments were chosen for SOC analysis: unfertilized control, only mineral fertilization (NPK), farmyard manure application (FYM), as well as FYM and NPK application. SOC content was significantly affected by both fertilization and crop rotation practices. In the simple CR, both the unfertilized control and the NPK treatment exhibited a consistent decline in SOC content over the study period, with percentages decreasing from an initial 1.33% in 1955 to 1.15% and 1.14%, respectively. Although the FYM and FYM + NPK treatments showed an increase in SOC content in the 1990s, a gradual decline was recorded in the last two decades. This decrease was not observed in MCR: positive C balances were recorded in all treatments within MCR, with the largest increase in SOC stock occurring when NPK was combined with FYM. In contrast, over the last decade, C balances have decreased in simple CR for all treatments except FYM. This trend coincides with changes in the local climate, particularly rising temperatures. The results indicate that diversified crop rotations and FYM fertilization are effective in mitigating the negative impacts of changing environmental conditions on SOC stocks.

Keywords: soil organic carbon; monitoring; temperature; fertilization; crop rotation

Citation: Šimon, T.; Madaras, M.; Mayerová, M.; Kunzová, E. Soil Organic Carbon Dynamics in the Long-Term Field Experiments with Contrasting Crop Rotations. *Agriculture* **2024**, *14*, 818. https://doi.org/10.3390/agriculture14060818

Academic Editors: Ryusuke Hatano, Anetta Siwik-Ziomek and Anna Figas

Received: 20 March 2024
Revised: 29 April 2024
Accepted: 22 May 2024
Published: 24 May 2024

1. Introduction

Soil organic matter (SOM) determines soil fertility, regulates most soil functions, and controls the productivity of agricultural soils. Under stable environmental conditions and agronomy practices, soil organic carbon (SOC) stocks are in a dynamic balance between carbon (C) inputs, mainly in the form of crop residues and organic fertilizers, and C loss due to decomposition of SOM [1]. Long-term field experiments conducted at Rothamsted Experimental Station showed that SOM does not increase above the certain equilibrium level specific to the local conditions of soil type, cropping, and fertilization. These changes often occur slowly in temperate climates [2]. However, with ongoing climate change accompanied by rising temperatures and changes in precipitation and the soil moisture regime, the balance between organic inputs and their decomposition may change.

Some recent studies suggested that the content of SOC in European agricultural soils was declining [3]. The rate of change appears to be proportional to the initial SOC content. SOC changes in the topsoil of Swiss cropland at well-defined monitoring sites every five years from 1990 to 2014 were studied in [4]. SOC remained stable for the set of monitoring sites, although increasing and decreasing trends ranging from −11 to +16% relative change per decade were observed for individual sites. An average -0.29 t C ha^{-1} year^{-1} loss of SOC stocks was found in Swiss long-term field experiments, where permanent grasslands tended to lose less or even gain SOC (0.09 t C ha^{-1} year^{-1} in average) compared to croplands, with an average loss of −0.34 t C ha^{-1} year^{-1} [5].

Bogusz et al. [6] studied SOC content in Polish soils from 2015 to 2021, examining its relationship with agronomic categories and drought intensity. The analysis concentrated on mineral soils, encompassing a wide range of soil textures. Soil samples were collected from 1011 farms, with C content ranging from 0.85% to 2.35%. The analysis revealed that fluctuations in C content were largely influenced by soil management practices and the occurrence of drought during the study period. It was noted that soil moisture conditions significantly impacted the C accumulation. In areas affected by drought, a decrease in SOC content was observed. Szatmári et al. [7] predicted the SOC stock change between 1992 and 2010 in Hungarian soils at various aggregation levels. The total SOC stock in the topsoil was reported at 424.41 Tg in 1992 and 451.59 Tg in 2010. In areas where land use types remained unchanged, it was observed that the SOC stock increased under forests by 16.29 Tg and under pastures by 2.48 Tg, while it decreased under wetlands by 0.49 Tg. No change in SOC stock was noted under agricultural areas.

Management practices, including various tillage systems, significantly influence the quantity of SOC and composition of SOM. Traditionally, tillage has been used to mechanically prepare soils for seeding and to minimize the negative effects of weeds. The impact of tillage on SOC has been extensively reviewed by several authors. These reviews and meta-analyses have demonstrated both beneficial effects on SOC from no-tillage practices compared to conventional tillage [8,9], as well as null effects, indicating no significant difference between the two methods [10,11]. Shen et al. [12] demonstrated that no tillage and deep plowing have positive effects on soil aggregate stability and labile C fractions, enhancing SOC and dissolved and particulate organic C contents. In recent years, less intensive tillage practices and no tillage agricultural management have been promoted to mitigate negative impacts on soil quality and to preserve SOC [13].

To assess the impact of climate change on SOC, it is crucial to have access to long-term series of C measurements, as well as precise data on weather conditions and management practices, such as fertilization at specific sites. On the basis of these data, trends in soil C dynamics are being intensively modeled, both at local and global levels. Using global-scale modeling, Hertzfeld et al. [14] found that global cropland SOC stocks decline until the end of the century by only 1.0% to 1.4% if residue retention management systems are generally applied and by 26.7% to 27.3% in the case of residue harvest. Bruni et al. [15] analyzed data from 11 long-term experiments involving the addition of exogenous organic matter to estimate the amount of C input needed to reach the 0.1 and 0.4% SOC stock increase. They found that to reach this increased target relative to the onset of the experiment, 2.51 and 2.61 t C ha^{-1} year^{-1} of additional C input was necessary, respectively.

Long-term field experiments are used to evaluate the long-term dynamics of SOC in the Czech Republic, as well as in Europe. For example, Balík et al. [16] showed an increase in SOC content of 19% on Luvisol and 15.9% on Chernozem due to the application of farmyard manure in field experiments lasting more than 20 years. Similarly, the authors of [17] demonstrated an increase in SOC due to farmyard manure or farmyard manure with mineral fertilizers compared to the unfertilized control based on 13 long-term experiments. In addition to the effect of organic fertilization, crop rotation (CR) influences SOC stock. Higher crop species diversity in CR than continuous crop monoculture changed the quantity and quality of residue-derived C input into soil systems [18]. The global synthesis revealed that CR overall enhanced SOC content by 6.6% compared to monocultures. SOC content under CR increased more in regions with intermediate mean annual temperature (8–15 °C) and precipitation (600–1000 mm) than in regions with other climate types [19].

The aim of this study is to assess and discuss the dynamics of the SOC content in the soils of the oldest Czech long-term field experiment on Luvisol—the Prague-Ruzyně Long-Term Fertilizer Experiment (RFE). Contrasting fertilization treatments and crop rotations were selected. Consistent with results of numerous studies, we hypothesize that the rate and dynamics of SOC content and resulting C sequestration are predominantly shaped by crop structure and fertilization practices, with notable modulation attributed to evolving climatic factors. Our research focus lies in investigating the extent of this modulation, particularly discerning the direction and rate of SOC changes under combinations of fertilization and crop rotation.

2. Materials and Methods

The study site is located in Prague-Ruzyně, the Czech Republic (latitude 50°05′15″ N, longitude 14°17′27″ E). Altitude of the site is 370 m a.s.l. The long-term temperature norm for the period 1961–1990 is 8 °C, and for 1991–2020 is 9.6 °C. The long-term precipitation norm for the period 1961–1990 is 427 mm, and for 1991–2020 it is 497.5 mm.

The taxonomical soil unit is Haplic Luvisol, clay loam, developed on diluvial sediments mixed with loess (plough layer properties—sand content: 14%, silt: 59%, clay: 27%, pH/KCl = 5.8–7.1, plant available P: 12–100 mg kg^{-1}, and plant available K: 160–250 mg kg^{-1}).

The RFE was established in 1955 with the objective of studying the effects of various fertilization systems on crop yields, nutrient uptake, and soil quality. The experiment comprises several blocks that differ in crop rotation systems and employ a combination of mineral and organic fertilization. A detailed description of the trial design is presented in [20,21]. Two blocks with different crop rotations were chosen for the study.

Field B—simple crop rotation (SCR) with only two crops: sugar beet and spring wheat. It includes 19 treatments in 4 replications (96 individual plots), and the plot size is 12 × 12 m. Four fertilization treatments were selected for the study: control—unfertilized since 1955, NPK—only mineral fertilization, FYM—farmyard manure (25% dry matter and 0.5% N content), 21 t ha^{-1} each second year before sugar beet, and NPK + FYM.

Field IV—multi-crop rotation (MCR) with nine-year crop rotation: alfalfa/alfalfa/winter wheat/sugar beet/spring barley/potato/winter wheat/sugar beet/spring barley under sown with alfalfa. It was designed the same as SCR. Four fertilization treatments were selected: control—unfertilized since 1955, NPK—only mineral fertilization, FYM—farmyard manure, 21 t ha^{-1} before sugar beet and 15 t ha^{-1} before potato, and NPK + FYM (Table 1).

Table 1. Fertilization in the long-term field experiments (mean application rates for crop rotation).

Block/Crop Rotation	N	P	K	Farmyard Manure
		kg ha^{-1} $year^{-1}$		t ha^{-1} $year^{-1}$
Field B—simple crop rotation (SCR)	100	26	75	10.5
Field IV—multi-crop rotation (MCR)	63	24	109	6.5

In both blocks of the trial, most crop residues were removed after harvest. Field trial management involved mold-board ploughing to a depth of 20–25 cm, followed by standard seedbed preparation and sowing procedures.

From each treatment, one of four experimental plots was sampled annually, except some years when sampling was omitted. Soil sampling was carried out before mineral fertilization at the beginning of April in the periods 1970–2022 (SCR) and 1990–2022 (MCR) from topsoil at depths of 0–20 cm at four points of each individual plot. Partial soil samples were combined in 2 kg lots, homogenized, air-dried at room temperature, and run through a 2 mm sieve. SOC content in current and archived soil samples was measured by dry combustion using a Vario MAX CNS/CN analyzer (Elementar Analysensysteme GmbH, Hanau, Germany).

The SOC stock in the soil plough layer (0–20 cm) was calculated using the equivalent soil mass (ESM) method [22]. For calculations, a uniform bulk density of 1.35 g cm^{-1} (typical for studied soils) was considered, as multiple occasional measurements at the time of soil sampling did not confirm statistically significant differences in bulk density between the treatments (unpublished data). C sequestration was calculated as the difference between the actual SOC stock of individual treatments and the (calculated) original SOC stock at the beginning of the experiment.

Although the experiment started in 1955, SOC data collection began much later. Therefore, comprehensive analysis was performed to estimate the initial soil C content at the experimental site based on the observed patterns [23]. This task required trend modeling and the estimation of experimental effects of the block (field), plot of the block, and standardized year, corresponding to the position in the crop rotation of the block. The model assumed that the effects were additive with respect to the soil C content. Baseline soil C contents of 1.33% for SCR (95% probability within the interval 1.29–1.36%) and 1.29% for MCR (1.23–1.37%) were estimated by substituting the standardized year 1955 to the obtained models, with the terms relating to the experimental factors removed [23].

Annual temperature and precipitation data were obtained from a meteorological station located at the Crop Research Institute in Prague-Ruzyně (Figures 1 and 2). A detailed analysis of the course of the weather at the investigated site from 1954 to 2022 is described in [24].

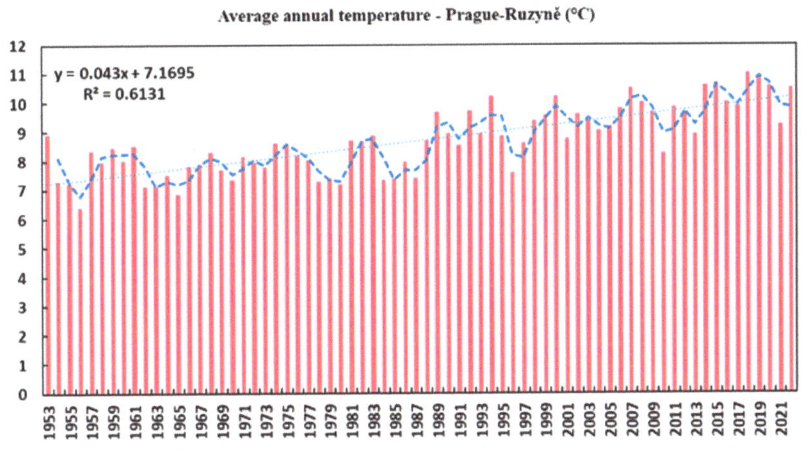

Figure 1. The course of temperatures at the Prague-Ruzyně site for the period 1953–2022. Average values for individual years and moving average.

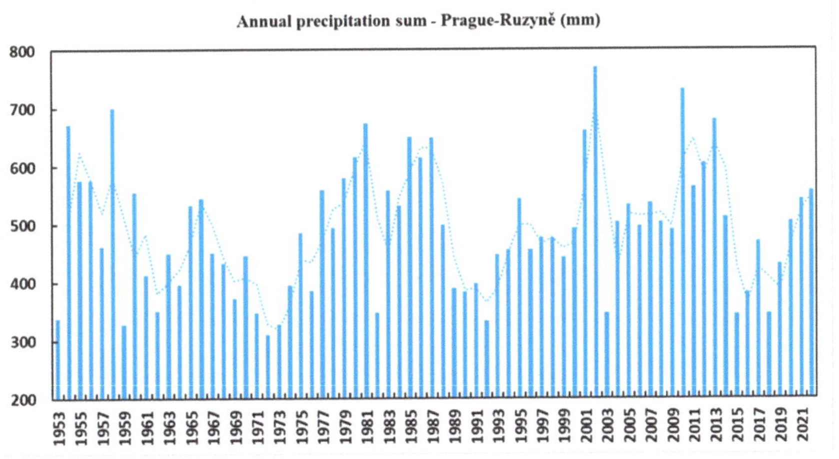

Figure 2. The course of precipitation at the Prague-Ruzyně site for the period 1953–2022. Average values for individual years and moving average.

The basic statistical values were calculated by Microsoft Excel (Microsoft Corporation, Redmund, WA, USA) and STATISTICA 14.0.0.14 software (TIBCO Software Inc., Santa Clara, CA, USA). Samples for SOC measurements from the SCR were available from 1970 and from the MCR from 1990. Therefore, trend evaluation was performed in two ways: for the whole monitoring period in the case of SCR and separately for the periods 1970–1990 and 1990–2022 to compare trends of both fields in the second time period.

3. Results and Discussion

3.1. Simple Crop Rotation

The unfertilized control of the SCR, where a rotation of two crops was operated, showed a consistent decrease in SOC content during the entire monitored period. The SOC contents dropped from 1.33% in 1955 to 1.15% in 2022. The treatment fertilized with mineral NPK had a similar tendency of decrease to SOC = 1.14%. Trends of both treatments were not significantly different ($p > 0.05$). On the contrary, the FYM treatment showed a constant gradual increase of SOC during the entire monitored period, resulting in SOC = 1.43% in 2022. The difference between FYM and unfertilized/NPK treatments was statistically significant (Figures 3 and S1A,B; Table 2).

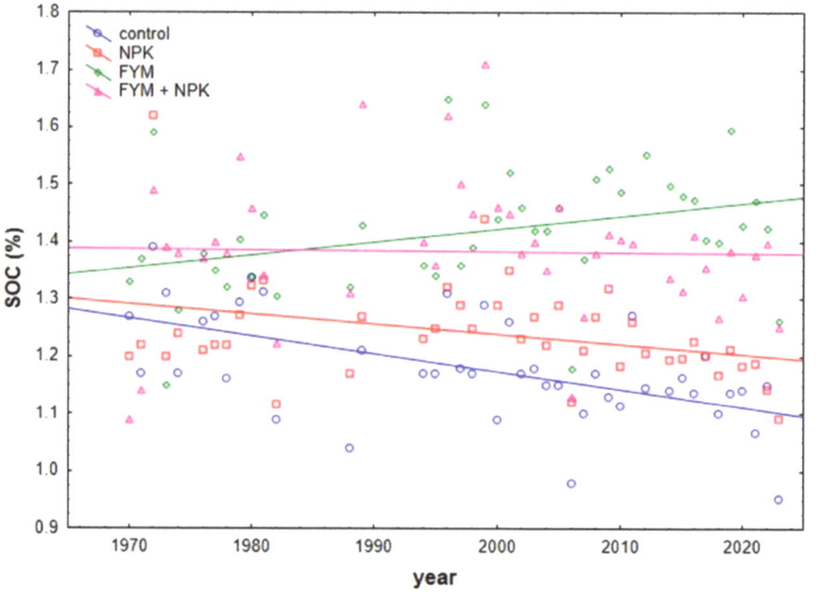

Figure 3. Dynamics of soil organic C (SOC) content in topsoil of SCR for the period of 1970–2022.

Table 2. Summary of statistical analysis of soil organic C (SOC) content dynamics in topsoil of SCR for the period of 1970–2022.

Treatment	r^2	r	m	*p*-Value	CI − 95%	CI + 95%
Control	0.314	−0.561 **	−0.003	0.001	1.15	1.207
NPK	0.108	−0.329 *	−0.002	0.032	1.215	1.27
FYM	0.0116	0.341 *	0.002	0.027	1.384	1.451
FYM + NPK	0.001	−0.021 ns	−0.000	0.893	1.345	1.421

r^2 = Coefficient of determination; r = Pearson's correlation coefficient; CI = confidence intervals; m = regression slope from regression equation; * $p < 0.05$, ** $p < 0.01$, ns = not significant.

The trend of the NPK + FYM treatment is worth a closer look. Regarding the trend since the beginning of the trial, SOC showed stagnation (Figure 3). However, the NPK + FYM treatment showed a gradual significant increase in SOC contents to approximately 1.40–1.50% in 1990. The rate of increase was even higher than that for the FYM treatment (Figures 4 and S2A,B; Table 3). Modeling SOC at the SCR also suggested that C turnover favored accumulation of SOC between 1972 and 1994, where seven out of eight simulation models predicted increasing SOC for the NPK + FYM treatment [25].

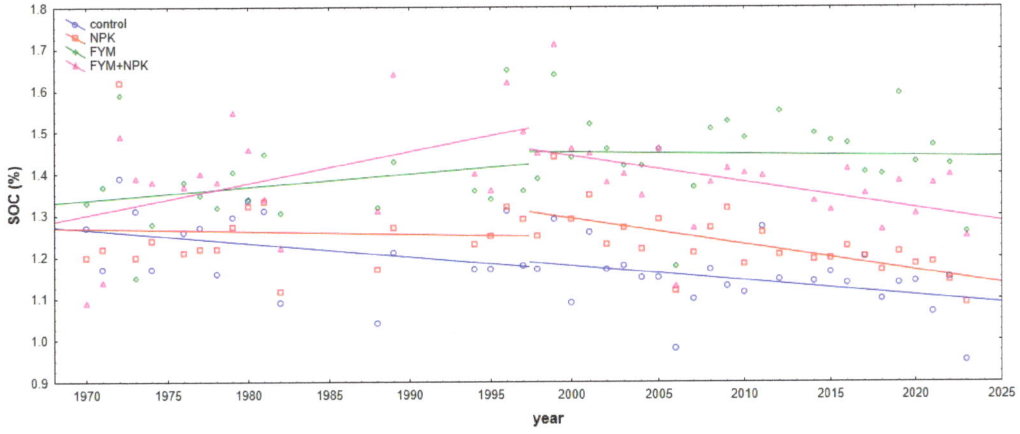

Figure 4. Dynamics of soil organic C (SOC) content in topsoil of SCR, and treatments FYM and FYM+NPK for the periods of 1970–1997 and 1997–2022.

Table 3. Summary of statistical analysis of soil organic C (SOC) content dynamics in topsoil of SCR for the periods of 1970–1997 (I.) and 1997–2022 (II.).

Statistic	Control		NPK		FYM		FYM + NPK	
	I.	II.	I.	II.	I.	II.	I.	II.
r^2	0.108	0.152	0.004	0.471	0.069	0.002	0.231	0.236
r	−0.328 ^{ns}	−0.39 *	−0.06 ^{ns}	−0.686 **	0.262 ^{ns}	−0.043 ^{ns}	0.481 *	−0.486 *
m	−0.003	−0.004	−0.001	−0.006	0.0032	−0.001	0.008	−0.006
p-value	0.184	0.049	0.813	0.000	0.293	0.84	0.043	0.014
CI − 95%	1.183	1.114	1.209	1.201	1.319	1.407	1.319	1.339
CI + 95%	1.274	1.174	1.314	1.261	1.429	1.487	1.463	1.426

r^2 = Coefficient of determination; r = Pearson's correlation coefficient; CI = confidence intervals; m = regression slope from regression equation; * $p < 0.05$, ** $p < 0.01$, ns = not significant.

In the last two decades, a turn in the SOC content in the organically fertilized treatments has been obvious (Figures 4 and S2A,B), demonstrated by a slight insignificant decrease in SOC content for the FYM treatment and a sharp (and statistically significant) decrease for the NPK + FYM treatment. These changes were also reflected in the total SOC stock in the plough layer. Within 1970–1997, the annual SOC stock increased by 97.3 kg C ha^{-1} and 253.1 kg C ha^{-1} for FYM and FYM + NPK, respectively. On the contrary, the annual increment of SOC stock for the FYM treatment reduced to 36.8 kg C ha^{-1} in the period 1997–2022. There was even a significant annual decrease in SOC stocks, on average by 177 kg C ha^{-1} for FYM + NPK. However, it must be considered that documented trends depend on the choice of the starting year, as observed, e.g., in [26].

The changes in the SOC dynamics can be linked to the gradual increase in average annual temperatures (Figure 1), which in the decade 2011–2020 was +2.3 °C compared to the decade 1970–1980. Trends demonstrated a decrease in soil organic C reserves, which were built up on organically fertilized treatments in the first cooler decades after the establishment of RFE. Contrary to our findings, Keel et al. [5] reported SOC losses at cooler sites in Swiss long-term experiments, with mean annual temperatures below 9 °C. However, they found that changes in air temperature did not prove to be a significant factor in SOC losses. Ross et al. [27] found the opposite trend of SOC content in a 12-year field experiment with rye monoculture. Although the mean temperature between September and July increased significantly in the second half of the experiment, SOC content evinced a sharp increase in the case of FYM + NPK and FYM + PK fertilization. However, control and NPK fertilization showed a slight increase in SOC content compared to the first six years, when the trend of SOC content was stagnant or decreasing. The authors also attributed the increasing SOC trend in the second half of the experiment to the change in rye cultivar.

The combination of high temperatures with higher precipitation could negatively affect SOC content [28]. Simulation models also showed that increasing temperatures negatively affect C storage in the soil. For example, a regional model simulating the future SOC dynamics in cropland and grassland soils of Bavaria in the 21st century showed a significant SOC decrease of 11–16% at an expected mean temperature increase of 3.3 °C, assuming unchanged C inputs [1]. Results indicated that C inputs must increase by 29% to maintain current SOC stocks in agricultural soils. On the other hand, the review simulating the likely changes in SOC with warming concluded that warming would have the effect of reducing SOC by stimulating decomposition rates [29]. In contrast, the effect of increasing CO_2 may simultaneously imply an increase in SOC through higher net primary production. However, any changes in SOC are also likely to be very slow.

3.2. Multi-Crop Rotation

Compared to the calculated SOC = 1.29% in 1955, no significant decrease in SOC was recorded on any of the treatments of multi-crop rotation, including forage crops (Figure 5; Table 4). Unlike SCR, even for the control treatment the SOC contents showed stagnation and reached 1.28% on average for the last nine-year rotation (Figure S3A). SOC contents of all fertilized treatments were higher in the last crop rotation than in 1955 (1.35% for NPK, 1.42% for FYM, and 1.54% for FYM + NPK). Annual increases in SOC stock ranged from 31 kg C ha^{-1} for NPK to 80 kg C ha^{-1} for FYM + NPK. SOC contents for treatments with manure application were significantly higher compared to those without manure application (Figure S3B–D). It was demonstrated that the long-term manure application significantly improved SOC compared to the unfertilized control and mineral N application, despite the increase in temperature [30]. Application of organic fertilizers was found to achieve long-term stable yields while maintaining soil at optimal quality, including higher quality of humus compared to mineral-only fertilization [21].

While SOC content was more-or-less stagnant under the NPK treatment in the period 1990–2022, a slight insignificant increase in SOC in control and FYM treatments was recorded (Figure S3B–D). The increase was significant for the FYM + NPK treatment. Trends of SOC contents appeared consistent throughout the monitoring period. Trend modeling showed periodic interannual changes according to the crop sequence, with SOC minima in the year of spring barley under sown by alfalfa and two maxima in the second year of alfalfa and the year of sugar beet cultivation [23]. Balík et al. [31] studied SOC stock in long-term field experiments (22–50 years in duration) on a Cambisol at four sites in the Czech Republic. Seven crops (clover, winter wheat, early potato, winter wheat, spring barley, potato, and spring barley with inter-seeded clover) were successively rotated in the sequence and organic and mineral fertilization was applied. The application of balanced mineral fertilizer in conjunction with farmyard manure resulted in an increased soil C sequestration, specifically by 22.9% in the FYM treatment and 45.6% in the FYM + NPK treatment.

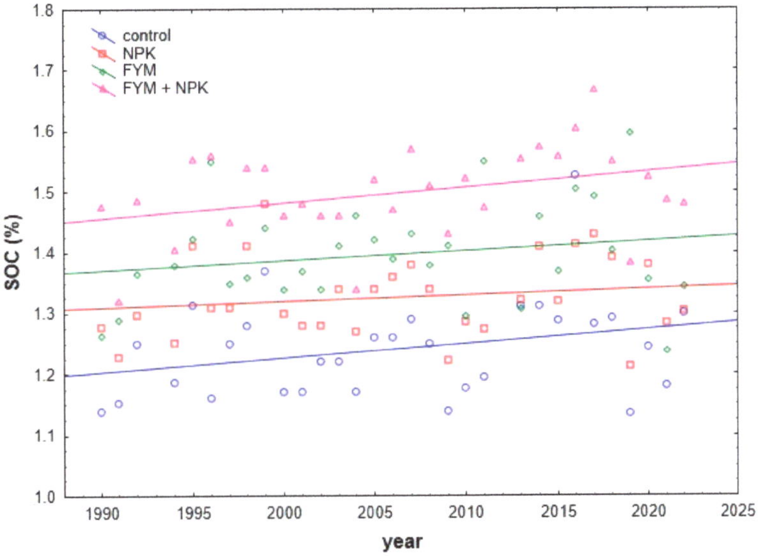

Figure 5. Dynamics of soil organic C (SOC) content in the topsoil of MCR for the period of 1990–2022.

Table 4. Summary of statistical analysis of soil organic C (SOC) content dynamics in topsoil of MCR for the period of 1990–2022.

Treatment	r^2	r	m	*p*-Value	CI − 95%	CI + 95%
Control	0.077	0.277 ^{ns}	0.002	0.131	1.213	1.271
NPK	0.022	0.147 ^{ns}	0.001	0.431	1.299	1.348
FYM	0.035	0.187 ^{ns}	0.002	0.315	1.364	1.424
FYM + NPK	0.11	0.332 ^{ns}	0.003	0.068	1.473	1.526

r^2 = Coefficient of determination; r = Pearson's correlation coefficient; CI = confidence intervals; m = regression slope from regression equation; ns = not significant.

3.3. *Effect of Crop Rotation Complexity*

Data from 1989 to 2022 were used to compare SOC contents at fields differing in crop rotations (Figures 5 and S3A–D). Intensive soil cultivation and unbalanced crop rotation on the SCR negatively affected SOC content compared to MCR. Significant differences between crop rotations were evident for all treatments, except for FYM. While SOC contents stagnated or slightly increased over time on the MCR in the case of the control, on NPK and FYM + NPK treatments, there was a significant decrease in SOC on the SCR. According to the result, the positive influence of the multi-crop rotation was evident for all treatments. Multi-crop rotation appears to be sustainable long term for stabilizing the level of soil organic C in a wide scale of fertilization scenarios thanks to the varied composition of cultivated crops, including forage crops. However, this conclusion might not hold for all combinations of soil, climate, and agronomy conditions. A positive effect of the diversified crop rotation with six different crops (including cover crops) on SOC stock, compared to a two-crop rotation with sugar beet and winter wheat, was not confirmed [32]. In contrast to our experiment, the remaining post-harvest sugar beet and wheat residues on the field played a positive role in SOC dynamics. Buysse et al. [33] did not attach as much importance to crop residues for C storage as to manure. While FYM caused significant SOC increases (10 g C year^{-1} over 50 years), treatment with residue restitution did not influence a significant SOC trend over the 50 years of the experiment. The authors justified this by noting the better C stabilization in manure than in crop residues, which is in agreement with the study from another Czech long-term field trial [34]. On the other hand, Koga et al. [35] concluded that not only manure application but also continuous C input to the soil through crop residue return is an essential practice for increasing C sequestration on an Andisol in northern Japan. Prudil et al. [36] simulated changes in SOC in the monoculture of spring barley and the Norfolk crop rotation during 1972–2100 using the RothC-263.3 model. Results showed that SOC stocks of Gleyic Fluvisol were mainly influenced by plant residue inputs and exogenous organic materials' application.

The negative impact of crop rotation on SCR was partly compensated by the farmyard manure application without mineral fertilizers. The expected positive effect of manure application on the soil C content was thus confirmed in both fields, which is in accordance with our previous results [17] or similar studies by other authors [16]. Similarly, the authors of [37] showed the greatest potential of nine-year crop rotation to increase the stable C content in the soil. Cultivation of crops with high C/N residues and the re-introduction of leguminous meadows can both represent successful agronomic management to maintain a good fertility level and to lower the carbon dioxide concentration in the atmosphere.

3.4. Carbon Sequestration

To demonstrate the impact of changes in SOC dynamics on the SOC stock balance in the plough layer, we calculated 10-year averages and rates of C sequestration (Tables 5 and 6). SOC stock in the plough layer was gradually depleted from the unfertilized plots on the SCR from −2.1 to −5.1 t C ha^{-1}. A negative balance was also observed in treatments with NPK application, where the C sequestration was negative, from −1.6 to −3.7 t C ha^{-1}. On the other hand, treatment with farmyard manure application showed a positive C balance, despite that the sequestration rate was negative in the last monitored period (as well as for other SCR treatments; Table 6). Organic amendments, particularly FYM application, played a pivotal role in promoting soil organic carbon (SOC) gain, as was shown in a synthesis study of German long-term field experiments [38]. These findings were also confirmed by research from Swiss long-term experiments [5], which indicated that fertilization with manure combined with higher doses of NPK fertilization had a positive effect on reducing SOC losses. Additionally, Ross et al. [27] observed the highest C sequestration rates of up to 0.5 t h^{-1} year^{-1} with a combination of mineral and FYM fertilization in a long-term field experiment with a winter rye monoculture.

On the contrary, in the present study, C sequestration decreased from 2.1 to 0.9 t C ha^{-1} in the case of NPK together with manure. This may be related to a higher decomposition of organic matter because of additional mineral N fertilization in the warmer period of 2000–2021. However, it should be considered that the SOC content in 1955 was used as a reference value.

Sequestration trends in treatments of MCR differed from those of SCR (Table 5). C sequestration on the unfertilized control was negative; however, it seems that over the course of the experiment, it tended toward neutral values (shifted from −1.8 to −0.3 t C ha^{-1}). A positive C balance was observed in fertilized treatments; however, significant differences existed between these treatments. While mineral NPK fertilization resulted in only a slight increase in C sequestration, from 1.0 to 1.5 t C ha^{-1} over three decades, both FYM and FYM + NPK applications increased C sequestration more intensively. Even in the case of manure-fertilized variants, it appears that the C sequestration trend has halted and stabilized in the last decade. In summary, C sequestration was significantly higher in crop rotations with a more diverse crop composition, with organically fertilized treatments increasing SOC stocks by an additional 88%.

The average SOC stock in Europe varies from 40 to 600 t ha^{-1}, depending on the soil type, texture, and land use [39]. However, in intensively managed soils, SOC stock tends to be in the tens of t/ha. Specifically, in Switzerland, SOC stock in arable soil exceeds 50 t ha^{-1} [40], in Belgium it is approximately 55 t ha^{-1} [41], and in Germany it is around 60 t ha^{-1} [39]. SOC stock is indeed highly influenced by management practices, which can result in significant differences, sometimes even doubling the SOC stock under different management regimes. For example, Prudil et al. [36] demonstrated that C sequestration for the monoculture of spring barley ranged from 40 to 66 t ha^{-1}. In contrast, the Norfolk crop rotation, which includes cereals, root crops, and forage crops, exhibited higher C sequestration levels, averaging approximately 70 to 100 t ha^{-1}. Therefore, even continuous changes on the order of a few t C ha^{-1} per decade are not insignificant or negligible.

Table 5. Carbon sequestration for individual treatments of the experiment for the period 1971–2020.

Crop Rotation	Period	C Sequestration Since 1955 (t C ha^{-1}, Average for the Decade)			
		Control	NPK	FYM	NPK + FYM
SCR	1971–1980	−2.10	−1.56	0.79	2.07
	1991–2000	−4.37	−1.57	3.14	4.71
	2011–2020	−5.11	−3.70	4.86	0.92
MCR	1991–2000	−1.82	1.01	2.84	5.58
	2011–2020	−0.33	1.55	4.67	7.55

Table 6. Carbon sequestration rates over decades, from 1971 to 2020.

Crop Rotation	C Sequestration Rate (t C ha^{-1} Year^{-1})					
	1971–1980		1991–2000		2011–2020	
	m	r^2	m	r^2	m	r^2
SCR						
Control	0.079	0.010	−0.014	0.001	−0.257	0.289
NPK	−0.277	0.046	0.371	0.341	−0.141	0.320
FYM	−0.086	0.006	0.465	0.161	−0.261	0.161
FYM + NPK	0.611	0.317	0.244	0.054	−0.191	0.174
MCR						
Control			0.249	0.117	−0.068	0.004
NPK			0.498	0.376	0.157	0.048
FYM			0.179	0.061	0.025	0.001
FYM + NPK			0.416	0.290	−0.032	0.002

m = Regression slope from regression equation; r^2 = coefficient of determination.

These results correspond to previously published data showing that diversified crop rotations contribute to higher C contents in the soil and thus to higher C sequestration. McDaniel et al. [42] found that adding one or more crops in rotation to a monoculture increased total soil C by 3.6% and total N by 5.3%. Grunwald et al. [32] documented significantly higher SOC stocks at a 0–20 cm soil depth in the sugar beet–winter wheat–winter wheat crop rotation compared to the sugar beet–winter wheat–silage maize one. Differences in SOC stocks were likely due to the different amounts and possibly the quality of crop residue C input. Management of post-harvest residues determines SOC stocks in the soil. Incorporation of aboveground sugar beet residues by tillage resulted in a SOC stock increase of 1.9 t C ha^{-1} for the regular amounts of residues compared to the removed residue treatment, while stocks increased a further 0.5 t C ha^{-1} for the doubled residue treatment compared to the regular residue amount [33]. In general, crop rotations providing more soil organic matter, e.g., through green manure or grass-clover cultivation, has many beneficial effects on soil properties and productivity, including C accumulation [43].

4. Conclusions

Evaluation of SOC dynamics in topsoil of the oldest Czech long-term field experiment identified multi-crop rotation with fodder as the key management strategy for increasing or at least maintaining SOC stocks long term. On the contrary, a simple crop rotation with only two crops appeared more sensitive to rising temperatures and showed higher SOC losses. These losses can be partially mitigated by FYM fertilization. Our results showed that a combination of appropriate management strategies, i.e., diversified crop rotation and FYM fertilization, can attenuate the negative impacts of environmental conditions on SOC stock or amplify the positive effects. These results are consistent with findings from other European long-term field experiments and need to be implemented into Good Agricultural Practices policy to prevent significant losses of SOC stocks in European agricultural soils.

Supplementary Materials: The following supporting information can be downloaded at: https://www.mdpi.com/article/10.3390/agriculture14060818/s1, Figure S1 (A). Dynamics of soil organic C content (SOC) in topsoil of SCR for the period of 1970–2022 for the control and NPK. Average values of SOC ± 95% confidence bands; (B). Dynamics of soil organic C content (SOC) in topsoil of SCR for the period of 1970–2022 for FYM and FYM + NPK. Average values of SOC ± 95% confidence bands; Figure S2 (A). Dynamics of soil organic C (SOC) content in topsoil of SCR, treatments FYM and FYM + NPK for the periods of 1970–1997. Average values of SOC ± 95% confidence bands; (B). Dynamics of soil organic C (SOC) content in topsoil of SCR, treatments FYM and FYM + NPK for the periods of 1997–2022. Average values of SOC ± 95% confidence bands; Figure S3 (A). Dynamics of soil organic C content (SOC) in the topsoil of SCR and MCR for the control in the period of 1990–2022. Average values of SOC ± 95% confidence bands; Figure S3 (B). Dynamics of soil organic C content (SOC) in the topsoil of SCR and MCR for NPK in the period of 1990–2022. Average values of SOC ± 95% confidence bands; Figure S3C. Dynamics of soil organic C content (SOC) in the topsoil of SCR and MCR for FYM in the period of 1990–2022. Average values of SOC ± 95% confidence bands; Figure S3D. Dynamics of soil organic C content (SOC) in the topsoil of SCR and MCR for FYM + NPK in the period of 1990–2022. Average values of SOC ± 95% confidence bands.

Author Contributions: Conceptualization, E.K. and T.Š.; methodology, E.K. and T.Š.; validation, T.Š. and M.M. (Mikuláš Madaras); formal analysis, M.M. (Markéta Mayerová) and T.Š.; investigation, T.Š.; resources, T.Š. and E.K.; data curation, T.Š. and M.M. (Markéta Mayerová); writing—original draft preparation, T.Š., M.M. (Markéta Mayerová) and M.M. (Mikuláš Madaras); writing—review and editing, M.M. (Markéta Mayerová), M.M. (Mikuláš Madaras) and T.Š.; supervision, M.M. (Mikuláš Madaras); project administration, E.K. All authors have read and agreed to the published version of the manuscript.

Funding: This research was funded by the Ministry of Agriculture of the Czech Republic, grant numbers QK21020155 and RO0423.

Institutional Review Board Statement: Not applicable.

Data Availability Statement: Some data are available upon reasonable request from the corresponding author.

Acknowledgments: We thank Michaela Friedlová for the precise analyses of organic carbon in soil samples.

Conflicts of Interest: The authors declare no conflicts of interest.

References

1. Wiesmeier, M.; Poeplau, C.; Sierra, C.A.; Maier, H.; Frühauf, C.; Hübner, R.; Kühnel, A.; Spörlein, P.; Geuß, U.; Hangen, E.; et al. Projected loss of soil organic carbon in temperate agricultural soils in the 21st century: Effects of climate change and carbon input trends. *Sci. Rep.* **2016**, *6*, 32525. [CrossRef]
2. Johnston, A.E.; Poulton, P.R.; Coleman, K. Soil organic matter: Its importance in sustainable agriculture and carbon dioxide fluxes. *Adv. Agron.* **2008**, *101*, 1–57.
3. Jones, A.; Panagos, P.; Barcelo, S.; Bouraoui, F.; Bosco, C.; Dewitte, O.; Gardi, C.; Erhard, M.; Hervás, J.; Hiederer, R.; et al. *The State of Soil in Europe: A Contribution of the JRC to the European Environment Agency's Environment State and Outlook Report-SOER 2010*; EUR 25186 EN; Publications Office of the European Union: Luxembourg, 2012; JRC68418.
4. Gubler, A.; Wächter, D.; Schwab, P.; Müller, M.; Keller, A. Twenty-five years of observations of soil organic carbon in Swiss croplands showing stability overall but with some divergent trends. *Environ. Monit. Assess.* **2019**, *191*, 277. [CrossRef]

5. Keel, S.G.; Anken, T.; Büchi, L.; Chervet, A.; Fliessbach, A.; Flisch, R.; Huguenin-Elie, O.; Mäder, P.; Mayer, J.; Sinaj, S.; et al. Loss of soil organic carbon in Swiss long-term agricultural experiments over a wide range of management practices. *Agric. Ecosyst. Environ.* **2019**, *286*, 106654. [CrossRef]
6. Bogusz, P.; Zimnoch, U.; Brodowska, M.S.; Michalak, J. The trend of changes in soil organic carbon content in Poland over recent years. *Arch. Environ. Prot.* **2024**, *50*, 35–44. [CrossRef]
7. Szatmári, G.; Pirkó, B.; Koós, S.; Laborczi, A.; Bakacsi, S.; Szabó, J.; Pásztor, L. Spatio-temporal assessment of topsoil organic carbon stock change in Hungary. *Soil Till. Res.* **2019**, *195*, 104410. [CrossRef]
8. Gonzalez-Sanchez, E.J.; Ordonez-Fernandez, R.; Carbonell-Bojollo, R.; VerozGonzalez, O.; Gil-Ribes, J.A. Meta-analysis on atmospheric carbon capture in Spain through the use of conservation agriculture. *Soil Till. Res.* **2012**, *122*, 52–60. [CrossRef]
9. West, T.O.; Post, W.M. Soil organic carbon sequestration rates by tillage and crop rotation. *Soil Sci. Soc. Am. J.* **2002**, *66*, 1930–1946. [CrossRef]
10. Dimassi, B.; Mary, B.; Wylleman, R.; Labreuche, J.; Couture, D.; Piraux, F.; Cohan, J.-P. Long-term effect of contrasted tillage and crop management on soil carbon dynamics during 41 years. *Agric. Ecosyst. Environ.* **2014**, *188*, 134–146. [CrossRef]
11. Powlson, D.S.; Stirling, C.M.; Jat, M.L.; Gerard, B.G.; Palm, C.A.; Sanchez, P.A.; Cassman, K.G. Limited potential of no-till agriculture for climate change mitigation. *Nat. Clim. Chang.* **2014**, *4*, 678–683. [CrossRef]
12. Shen, X.; Wang, L.; Yang, Q.; Xiu, W.; Li, G.; Zhao, J.; Zhang, G. Dynamics of soil organic carbon and labile carbon fractions in soil aggregates affected by different tillage managements. *Sustainability* **2021**, *13*, 1541. [CrossRef]
13. Haddaway, N.R.; Hedlund, K.; Jackson, L.E.; Kätterer, T.; Lugato, E.; Thomsen, I.K.; Jørgensen, H.B.; Isberg, P.-E. How does tillage intensity affect soil organic carbon? A systematic review. *Environ. Evid.* **2017**, *6*, 30. [CrossRef]
14. Herzfeld, T.; Heinke, J.; Rolinski, S.; Müller, C. Soil organic carbon dynamics from agricultural management practices under climate change. *Earth Syst. Dyn.* **2021**, *12*, 1037–1055. [CrossRef]
15. Bruni, E.; Guenet, B.; Clivot, H.; Kätterer, T.; Martin, M.; Virto, I.; Chenu, C. Defining quantitative targets for topsoil organic carbon stock increase in European croplands: Case studies with exogenous organic matter inputs. *Front. Environ. Sci.* **2022**, *10*, 824724. [CrossRef]
16. Balík, J.; Černý, J.; Kulhánek, M.; Sedlář, O. Soil carbon transformation in long-term field experiments with different fertilization treatments. *Plant Soil Environ.* **2018**, *64*, 578–586. [CrossRef]
17. Šimon, T.; Madaras, M. Chemical and spectroscopic parameters are equally sensitive in describing soil organic matter changes after decades of different fertilization. *Agriculture* **2020**, *10*, 422. [CrossRef]
18. Chatterjee, A.; Cooper, K.; Klaustermeier, A.; Awale, R.; Cihacek, L.J. Does crop species diversity influence soil carbon and nitrogen pools? *Agron. J.* **2016**, *108*, 427–432. [CrossRef]
19. Liu, X.; Tan, S.; Song, X.; Wu, X.; Zhao, G.; Li, S.; Liang, G. Response of soil organic carbon content to crop rotation and its controls: A global synthesis. *Agric. Ecosyst. Environ.* **2022**, *335*, 108017. [CrossRef]
20. Hlisnikovský, L.; Menšík, L.; Kunzová, E. The effect of soil-climate conditions, farmyard manure and mineral fertilizers on potato yield and soil chemical parameters. *Plants* **2021**, *10*, 2473. [CrossRef] [PubMed]
21. Menšík, L.; Hlisnikovský, L.; Pospíšilová, L.; Kunzová, E. The effect of application of organic manures and mineral fertilizers on the state of soil organic matter and nutrients in the long-term field experiment. *J. Soils Sediments* **2018**, *18*, 2813–2822. [CrossRef]
22. Ellert, B.H.; Bettany, J.R. Calculation of organic matter and nutrients stored in soils under contrasting management regimes. *Can. J. Soil Sci.* **1995**, *75*, 529–538. [CrossRef]
23. Żelazny, W.R. *Derivation of Total Soil Carbon Contents in 1955 (1.0)*; Zenodo: Geneva, Switzerland, 2024. [CrossRef]
24. Hlisnikovský, L.; Menšík, L.; Kunzová, E. Development and the effect of weather and mineral fertilization on grain yield and stability of winter wheat following alfaalfa—Analysis of long-term field trial. *Plants* **2023**, *12*, 1392. [CrossRef] [PubMed]
25. Smith, P.; Smith, J.U.; Powlson, D.S.; McGill, W.B.; Arah, J.R.M.; Chertov, O.G.; Coleman, K.; Franko, U.; Frolking, S.; Jenkinson, D.S.; et al. A comparison of the performance of nine soil organic matter models using datasets from seven long-term experiments. *Geoderma* **1997**, *81*, 153–225. [CrossRef]
26. Nerger, R.; Klüver, K.; Cordsen, E.; Fohrer, N. Intensive long-term monitoring of soil organic carbon and nutrients in Northern German. *Nutr. Cycl. Agrosyst.* **2020**, *116*, 57–69. [CrossRef]
27. Roβ, C.L.; Baumecker, M.; Ellmer, F.; Kautz, T. Organic manure increases carbon sequestration Far beyond the "4 per 1000 initiative" goal on a sandy soil in the thyrow long-term field experiment DIV.2. *Agriculture* **2022**, *12*, 170. [CrossRef]
28. Fantappiè, M.; L'Abate, G.; Costantini, E.A.C. The influence of climate change on the soil organic carbon content in Italy from 1961 to 2008. *Geomorphology* **2011**, *135*, 343–352. [CrossRef]
29. Kirschbaum, M.U.F. Will changes in soil organic carbon act as a positive or negative feedback on global warming? *Biogeochemistry* **2000**, *48*, 21–51. [CrossRef]
30. Ruimin, Q.; Juan, L.; Zhian, L.; Zhijie, L.; Yanting, L.; Xiangdong, Y.; Jianjun, Z.; Binggiang, Z. Temperature effects on soil organic carbon, soil labile organic carbon fractions, and soil enzyme activities under long-term fertilization regimes. *Appl. Soil Ecol.* **2013**, *102*, 36–45.
31. Balík, J.; Suran, P.; Sedlář, O.; Černý, J.; Kulhánek, M.; Procházková, S.; Asrade, D.A.; Smatanová, M. The effect of long-term farmyard manure and mineral fertilizer application on the increase in soil organic matter quality of cambisols. *Agronomy* **2023**, *13*, 2960. [CrossRef]
32. Grunwald, D.; Götze, P.; Koch, H.J. Soil organic carbon stocks in sugar beet rotations differing in residue management and associated rotational crop species. *J. Plant Nutr. Soil Sci.* **2021**, *184*, 556–561. [CrossRef]

33. Buysse, P.; Roisin, C.; Aubinet, M. Fifty years of contrasted residue management of an agricultural crop: Impacts on the soil carbon budget and on soil heterotrophic respiration. *Agric. Ecosyst. Environ.* **2013**, *167*, 52–59. [CrossRef]
34. Lipavský, J.; Kubát, J.; Zobač, J. Long-term effects of straw and farmyard manure on crop yields and soil properties. *Arch. Agron. Soil Sci.* **2008**, *54*, 369–379. [CrossRef]
35. Koga, N.; Tsuji, H. Effects of reduced tillage, crop residue management and manure application practices on crop yields and soil carbon sequestration on an Andisol in northern Japan. *Soil Sci. Plant Nutr.* **2009**, *55*, 546–557. [CrossRef]
36. Prudil, J.; Pospíšilová, L.; Dryšlová, T.; Barančíková, G.; Smutný, V.; Sedlák, L.; Ryant, P.; Hlavinka, P.; Trnka, M.; Halas, J.; et al. Assessment of carbon sequestration as affected by different management practices using the RothC model. *Plant Soil Environ.* **2023**, *69*, 532–544. [CrossRef]
37. Triberti, L.; Nastri, A.; Baldoni, D. Long-term effects of crop rotation, manure and mineral fertilisation on carbon sequestration and soil fertility. *Eur. J. Agron.* **2016**, *74*, 47–55. [CrossRef]
38. Gocke, M.I.; Gaigue, J.; Bauke, S.L.; Barkusky, D.; Baumecker, M.; Berns, A.E.; Hobley, E.; Honermeier, B.; Kögel-Knabner, I.; Koszinski, S.; et al. Interactive effects of agricultural management on soil organic carbon accrual: A synthesis of long-term field experiment in Germany. *Geoderma* **2023**, *438*, 116616. [CrossRef]
39. Dechow, R.; Franko, U.; Kätterer, T.; Kolbe, H. Evaluation of the RothC model as a prognostic tool for the prediction of SOC trends in response to management practices on arable land. *Geoderma* **2019**, *337*, 463–478. [CrossRef]
40. Wust-Galey, C.; Keel, C.G.; Leifeld, J. A model based carbon inventory for Switzerland's mineral agricultural soils using RothC. *Agroscope Sci.* **2020**, *105*, 115.
41. Chartin, C.; Stevens, A.; Goidts, E.; Krüger, I.; Carnol, M.; van Wesemael, B. Mapping soil organic carbon stocks and estimating uncertainties at the regional scale following a legacy sampling strategy (Southern Belgium, Wallonia). *Geoderma* **2017**, *9*, 73–86. [CrossRef]
42. McDaniel, M.D.; Tiemann, L.K.; Grandy, A.S. Does agricultural crop diversity enhance soil microbial biomass and organic matter dynamics? A meta-analysis. *Ecol. Appl.* **2014**, *24*, 560–570. [CrossRef]
43. Martyniuk, S.; Pikuła, D.; Kozieł, M. Soil properties and productivity in two long-term crop rotations differing with respect to organic matter management on an Albic Luvisol. *Sci. Rep.* **2019**, *9*, 1878. [CrossRef] [PubMed]

agriculture

MDPI

Article

Effects of Soil Quality Decline on Soil-Dwelling Mesofaunal Communities in Agricultural Lands of the Mollisols Region, China

Chen Ma, Xin Yao and Guoming Du *

School of Public Administration and Law, Northeast Agricultural University, Harbin 150030, China; mac580@neau.edu.cn (C.M.); s221202039@neau.edu.cn (X.Y.)
* Correspondence: duguoming@neau.edu.cn

Abstract: Soil quality decline can adversely affect ecosystem health and land productivity, with soil-dwelling mesofauna considered to potentially fulfill vital functions in accurately predicting these outcomes. However, the current state of research reveals a gap concerning the relationships between soil quality decline and soil-dwelling mesofauna in the Mollisols Region. For a more profound understanding of this issue, we conducted a comprehensive investigation of soil-dwelling mesofaunal communities in the different agricultural lands of the Mollisols Region. In this study, soil-dwelling mesofauna were collected, and 11 soil properties were determined following standard procedures, with soil quality levels quantified by utilizing soil quality index (SQI). Our results revealed that there was a gradient of soil quality across the different agricultural lands, which were divided into five levels, including very strong, strong, medium, weak, and very weak. Subsequently, this investigation provided empirical evidence that the decline in soil quality had implications for soil-dwelling mesofaunal communities in agricultural lands of the Mollisols region. A consistent decrease in the density of soil-dwelling mesofauna was observed with the decline of soil quality. In contrast, a greater richness was observed in areas with relatively weaker soil quality, suggesting that the consequences of soil quality decline on soil-dwelling mesofauna were not exclusively negative. Various taxa of soil-dwelling mesofauna exhibited varying degrees of response to the decline in soil quality. Oribatida was overwhelmingly dominant in the sampling fields with medium soil quality, and most Entomobryidae were found in agricultural lands with very weak soil quality. During soil quality decline, soil nutrients were observed to correlate positively with the density of soil-dwelling mesofauna. Overall, the outcomes of this investigation carry significance for comprehending how soil quality decline relates to soil-dwelling mesofauna, and can provide valuable ecological insights for formulating biodiversity guidelines targeted at preserving soil resources in the Mollisols region.

Keywords: soil quality index; soil-dwelling mesofauna; community characteristics; belowground ecosystems; Mollisols resources

Citation: Ma, C.; Yao, X.; Du, G. Effects of Soil Quality Decline on Soil-Dwelling Mesofaunal Communities in Agricultural Lands of the Mollisols Region, China. *Agriculture* **2024**, *14*, 766. https://doi.org/10.3390/agriculture14050766

Academic Editors: Anetta Siwik-Ziomek and Anna Figas

Received: 26 March 2024
Revised: 11 May 2024
Accepted: 14 May 2024
Published: 16 May 2024

1. Introduction

Soil quality decline is attributed to a synergistic effect of both natural forces and human activities [1]. It can cause reductions in agricultural productivity, decreases in land use efficiency, and the deterioration of belowground ecosystems [2]. Mollisols, a soil order in the United States Soil Taxonomy, forms beneath temperate and cold–temperate grassland or steppe vegetation and is characterized by a surface layer containing dark humified organic material [3]. Although Mollisols cover only 7% of the global land area, their richness in organic content and exceptional fertility contribute significantly to global agricultural production [4]. Nevertheless, Mollisols worldwide have been experiencing different levels of quality decline after cultivation [5]. Soil quality decline in Mollisols regions seems certain to drive changes in belowground ecosystems, ultimately altering soil environments [6].

As a result, these declines have been recognized as a serious issue closely correlated with sustainable agriculture worldwide and the balance of belowground ecosystems [7].

The adverse effects of soil quality decline on ecosystem health and land productivity are considered to be pervasive and systemic. Accordingly, numerous research endeavors have focused on how soil quality decline impacts belowground systems, especially regarding the relationships between soil physical properties, nutrients, or microbial communities, and the decline in soil quality, yielding definitive findings [8–10]. All things considered, soil quality decline results in soil becoming thinner, less fertile, and more compacted, thereby restricting ecological functions in belowground systems [11]. Consequently, the impacts of soil quality decline on belowground systems have recently gained increased attention. Among these, the links between soil quality decline and soil faunal communities have been increasingly explored. For instance, Yan et al. illustrated that the abundance and functional traits of soil fauna, based on the mixed taxonomic resolution, could be used to assess soil quality [12]. Analysis of Megascolecidae and Lumbricidae could exhaustively describe the physical and chemical features as well as the biological properties associated with changes in soil quality [13,14]. Following decline in soil quality, communities of spiders, beetles and ants underwent significant simplification, but they demonstrated potential for recovery within only four years of land restoration in Northeast Brazil [15]. In addition to the studies on soil macrofauna communities mentioned above, Martin et al. employed factor analysis strategies to integrate the community structure and function of soil nematodes into the current framework of soil health indicators [16]. Du Preez et al. demonstrated that the metabolic footprints of soil nematodes could serve as indirect measures of soil quality [15]. Thus, these studies have provided valuable insights into the intricate connections between soil faunal communities and soil quality decline, building upon earlier observations. Soil-dwelling mesofauna constitute integral components of soil biological communities, and thus the implications of declining soil quality for them cannot be underestimated.

Soil-dwelling mesofauna are invertebrates with body sizes ranging from 0.1 mm to 2 mm, typically extracted by the Tullgren funnel extraction method [17]. Soil-dwelling mesofauna, constituting the bulk of soil organisms, are instrumental in various processes essential for soil formation, development, and improvement [16]. Most soil-dwelling mesofauna, such as mites and springtails, possess minuscule body sizes, fragile body structures, and limited migratory abilities, making them extremely sensitive to changes in environmental conditions [18]. In environmental gradient processes in particular, soil-dwelling mesofauna appear to directly indicate variations in ecosystem deterioration [19]. In this regard, soil-dwelling mesofauna are considered to potentially fulfill vital functions in the accurate prediction of soil health, and some previous studies have focused on this issue. From the perspective of mesofaunal community statistical analysis, simple counting of soil-dwelling mesofauna could effectively reflect changes in soil environment, surpassing the analysis of soil microbial properties [20]; mesofaunal community characteristics (abundance, richness, diversity) were closely associated with soil properties, but the additional nitrogen might have a general negative impact on the community [21]. Considering various taxonomic groups of mesofauna, epedaphic and euedaphic Collembolans played relatively important roles in assessment of land degradation [22]; abundance, richness and diversity of mesostigmatid mite communities could increase with declining nitrate-nitrogen levels in European ash stands [23]. Nevertheless, studies examining the relationships between soil-dwelling mesofauna and soil quality, particularly compared to physical, chemical, and other biological features, have been relatively lacking [24].

Considering this background, soil quality levels were quantified by utilizing soil quality index (SQI). Soil-dwelling mesofauna were collected from the agricultural lands under investigation. Here, we hypothesize (H1) that the community characteristics of medium-sized soil invertebrates are negatively affected by soil quality decline in the Mollisols region and (H2) that soil quality decline impacts different taxa of soil-dwelling mesofauna to varying extents.

2. Materials and Methods

2.1. Study Area and Sampling Design

To gain deeper insights into this knowledge gap, we set up an investigation in agricultural lands of the Mollisols Region, which is located on the Songnen Plain of Northeast China. This investigation was carried out in the southeast region of the Songnen Plain, Harbin, Heilongjiang Province, China (45°43′–45°46′ N, 126°53′–126°58′ S), and this region is situated on the south bank of the Songhua River, at the intersection of an alluvial plain and low hills (Figure 1). It features a temperate continental monsoonal climate and experiences an average annual temperature of 5.6 °C and an average annual rainfall of 550 mm. Mollisols are the dominant soil type in the study area, with pH levels that are slightly acidic to neutral and a rich organic material content. The predominant crop cultivated in the area is maize (*Zea mays*).

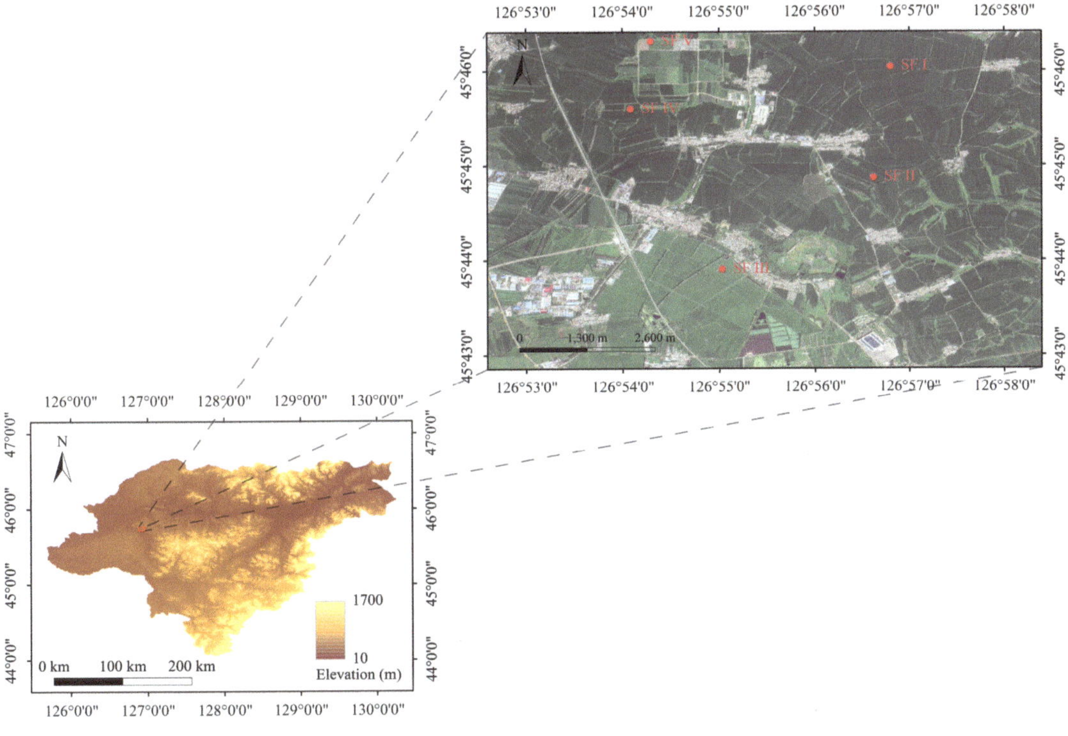

Figure 1. Location of the sampling fields.

The most conspicuous sign of soil quality decline is decrease in crop yield [25]. Consequently, we searched for continuously cropping fields with different levels of average annual maize yields in this study area. The average annual maize yield over three years in these fields was estimated by consulting local farmers after the final harvest in 2021, and five levels of average annual yields were selected for this study (Table 1). Then, we roughly estimated soil nutrients using an IN-HT300 rapid soil nutrient tester (Comecause, China) to select fields within a gradient of soil quality decline. Finally, five sampling fields were selected in this study (Figure 1). In this study, we requested that the field administrators maintain agricultural management methods at the same level for maize cultivation. The maize variety was Hongxu 78. Conventional fertilizer application and tillage practices commonly used in Northeast China were applied. Encapsulated urea, triple superphosphate and potassium sulphate were used as fertilizer, with application rates of N 180 kg/hm^2, P_2O_5 120 kg/hm^2, K_2SO_4 100 kg/hm^2, respectively. To avoid impacting soil-dwelling mesofauna, insecticides were not used in this study. Soil samples were collected in August 2022 (R3 stage of maize), corresponding to the active period of soil-dwelling mesofauna, while minimizing the disturbance from tillage. Three separate sampling stands (200 m × 200 m) were selected from each sampling field in the study area. Randomly, five replicate sampling plots (1 m × 1 m) were set up, one in each sampling field. In each sampling plot, a soil sample (cross-section: 100 cm^2; depth: 20 cm) was taken to extract soil-dwelling mesofauna. In this investigation, we collected 75 fresh samples (5 levels × 3 replicated stands × 5 replicated plots) from different fields. Concomitantly, we obtained other soil samples from each plot (depth: 20 cm) using soil augers to determine soil properties. The weather during the collection of these samples was sunny and rain-free.

Table 1. Maize yield levels and average annual yields in different sampling fields.

Sampling Fields	Yield Levels	Average Annual Yields (kg/hm^2)
SF I	Very high yields	12,000
SF II	High yields	10,000
SF III	Moderate yields	8000
SF IV	Low yields	6000
SF V	Very low yields	4000

2.2. Processing of Samples

The soil samples (used for extracting soil-dwelling mesofauna) were brought back to the laboratory and then placed in Tullgren funnel extractors. Standard bulbs of 25 W output provided both light and heat. The mesh size of Tullgren funnel extractors was 10 mesh sieves (2 mm). The soil samples were extracted for 48 h at 40 °C. After extraction, soil-dwelling mesofaunal specimens were all stored in 75% ethanol. Counting of specimens and identification of their taxonomic ranks up to the familial (subordinal) levels were conducted under a Nikon SMZ745T stereoscopic microscope [26].

The soil samples (used for determining soil properties) were naturally air-dried, with the removal of litter, roots, and gravels. Determination of soil properties was carried out following standard procedures [27–30]. Briefly, the Walkley-Black method was employed for measuring soil organic matter (SOM); a Seal AA3 continuous flow auto-analyzer was applied to determine soil total nitrogen (TN), phosphorus (TP), available nitrogen (AN), phosphorus (AP); a Thermo atomic absorption spectrophotometer was applied to determine soil total potassium (TK), available potassium (AK), exchangeable calcium (ExCa), magnesium (ExMg), and manganese (ExMn); and a cylinder method was utilized for quantifying soil bulk density (SBD).

2.3. Assessment of Soil Quality

Soil quality levels were quantified by utilizing soil quality index (SQI) in different sampling fields. Firstly, the minimum dataset (MDS) was performed to select soil properties indicators for calculating SQI [31]. The MDS was established by combining normalized values based on the method of data reduction using principal component analysis (PCA). All soil properties from the entire sample set were included in the PCA as descriptor indicators. In PCA, the variability was examined by eigenvalues of different principal components (PCs), and the PC with an eigenvalue of 1 or more was considered [32]. To prevent some important indicators from being left out, the norm values of soil properties indicators' weight loadings were calculated [33]. Subsequently, the soil properties indicators were chosen for the MDS if their norm values were at least 90% of the maximum value. The following equations were utilized to calculate the norm value:

$$N_{ik} = \sqrt{\sum_{1}^{k}(u_{ik}^{2}\lambda_{k})}$$

where N_{ik} represents the norm value of soil properties indicators i on first k-th PCs, u_{ik} is loading of soil properties indicators i on PC k, and λ_k is the eigenvalue of the PC k.

After that, soil properties indicators were normalized to combinable scores within the range of 0 to 1 using the linear scoring (LS) and non-linear scoring (NLS) systems [34]. Two categories of indicators' scoring functions were assigned to soil properties indicators based on their impact on soil productivity [35]. Briefly, the SBD was classified as "the lower, the better", whereas other indicators were categorized as "the more, the better" [36]. The following two equations were utilized to calculate the combinable scores in the LS system [37]:

$$LS = \frac{X}{H} \quad (1)$$

$$LS = \frac{L}{X} \quad (2)$$

where LS represents the score of a soil properties indicator based on linear scoring (LS) systems, Equation (1) is the scoring function which follows "the lower, the better" category, Equation (2) is the scoring function which follows "the more, the better" category, X represents the original value of an indicator, and L and H represent the lowest and highest values of an indicator among all the samples, respectively.

In the NLS system, a sigmoidal curve was applied using the following equation to calculate the combinable scores [38]:

$$NLS = \frac{1}{\left(1 + \frac{X}{X_0}\right)^b}$$

where NLS represents the score of a soil properties indicator based on non-linear scoring (NLS) systems, X represents the value of an indicator, X_0 represents average value of an indicator among all the samples, b indicates slopes assumed as 2.5 for the category of "the lower, the better" and -2.5 for the category of "the more, the better" [38].

Next, the weight value of a soil properties indicator was calculated based on its communality in PCA. The weight value (W_i) was calculated as the proportion of the communality of a soil properties indicator to the summation of all indicators' communalities evaluated in the PCA [39]. Upon scoring and weighing the indicators, the soil quality index (SQI) was carried out based on the following equation [40]. Finally, the SQIs were divided into several levels based on statistical differences of SQI values between each sampling field [41].

$$SQI = \sum_{i=1}^{n} W_i N_i$$

where SQI represents the value of soil quality index, N_i represents the combinable score of soil properties of the i-th indicator, W_i represents weight value of the i-th indicator, and n represents the number of indicators chosen based on the MDS.

2.4. Statistical Analysis

Principal component analysis (PCA) was executed to choose the most appropriate indicators for establishing MDS and calculating SQI. The normality of the soil-dwelling mesofauna and soil property data were examined by a Shapiro–Wilk test in each sampling field. Tukey's HSD test was applied to compare the differences in soil-dwelling mesofaunal density (individuals/m^2), soil properties, and SQIs among each sampling field. The "stats" R 4.3.2 package was used to perform the aforementioned statistical analyses [42].

For the identification of unique and shared taxa across the different levels of soil quality, endemic taxa underwent manual screening. Then, the "VennDiagram" R 4.3.2 package was used to illustrate a Venn diagram [43]. To demonstrate the richness of the soil-dwelling mesofaunal communities, rarefaction curves were created to compare the differences in richness using the "vegan" R 4.3.2 package [44]. The richness of the soil-dwelling mesofaunal communities was quantitatively represented by the taxon number. An unweighted pair group method with arithmetical averages (UPGMA) was implemented to demonstrate community structure of soil-dwelling mesofauna within different levels of soil quality via the "stats" R 4.3.2 package [42]. Then, the result of hierarchical clustering was visualized via a heatmap using the "pheatmap" R 4.3.2 package [45].

Data matrices of the soil-dwelling mesofauna (density, richness), soil physical properties (SBD), total nutrients (SOM, TN, TP, and TK), available nutrients (AN, AP, and AK), and mineral nutrients (ExCa, ExMg, and ExMn) were set up to explore the effects of soil quality decline on soil-dwelling mesofaunal communities. Then, a Mantel test was conducted to examine relationships between different taxa of soil-dwelling mesofauna and each soil property via the "vegan" R 4.3.2 package [44]. Subsequently, to assess the correlations within different data matrices and to estimate the links of soil-dwelling mesofauna to soil quality decline, multiple-factor analysis (MFA) was implemented via the "FactoMineR" R 4.3.2 package [18,46].

3. Results

3.1. Soil Quality

In this investigation, 11 soil properties across the different sampling fields are presented in Table 2. Soil bulk density (SBD) ranged from 1.20 to 1.37 g/cm^3, peaking in Sampling Field V (SF V) and bottoming out in SF I. In contrast, all the remaining 10 soil properties reached their minimum values in SF V. These soil properties in SF V, with the exception of soil total nitrogen (TN), potassium (TK), and exchangeable magnesium (ExMg), were significantly lower compared to other sampling fields ($p < 0.05$). At the same time, the difference in TK among different sampling fields was not significant ($p > 0.05$).

Table 2. Summary of descriptive statistics of the 11 soil properties across the different sampling fields (mean ± SE).

Soil Properties	Sampling Fields				
	SF I	**SF II**	**SF III**	**SF IV**	**SF V**
Soil bulk density (g/cm^3)	1.20 ± 0.01 c	1.21 ± 0.02 c	1.26 ± 0.01 b	1.29 ± 0.01 b	1.37 ± 0.02 a
Soil organic matter (g/kg)	18.45 ± 0.01 a	13.60 ± 0.14 b	10.92 ± 0.08 c	8.29 ± 0.04 d	7.24 ± 0.07 e
TN (g/kg)	2.23 ± 0.01 a	2.11 ± 0.03 b	1.65 ± 0.01 c	1.35 ± 0.01 d	1.3 ± 0.01 d
TP (g/kg)	0.63 ± 0.01 a	0.54 ± 0.01 b	0.46 ± 0.01 cd	0.47 ± 0.01 c	0.44 ± 0.01 d
TK (g/kg)	14.54 ± 0.27 a	14.40 ± 0.27 a	14.26 ± 0.01 a	14.04 ± 0.08 a	14.00 ± 0.38 a
AN (mg/kg)	225.37 ± 1.12 a	223.84 ± 2.15 a	210.36 ± 1.44 b	181.90 ± 1.12 c	173.75 ± 1.09 d
AP (mg/kg)	17.03 ± 0.24 a	15.38 ± 0.32 b	12.69 ± 0.08 c	12.24 ± 0.04 c	11.1 ± 0.26 d
AK (mg/kg)	181.73 ± 2.35 a	143.97 ± 1.78 b	137.26 ± 1.47 c	133.79 ± 2.04 c	106.28 ± 1.94 d
ExCa (g/kg)	2.52 ± 0.03 a	2.08 ± 0.04 b	1.59 ± 0.01 c	1.53 ± 0.01 c	1.12 ± 0.01 d
ExMg (g/kg)	0.82 ± 0.01 a	0.59 ± 0.02 b	0.56 ± 0.01 b	0.57 ± 0.01 b	0.56 ± 0.01 b
ExMn (mg/kg)	63.29 ± 2.00 a	60.9 ± 1.03 ab	57.74 ± 0.36 b	51.77 ± 0.36 c	38.20 ± 0.25 d

Note: different letters indicate a significant difference between each sampling field at $p < 0.05$ based on the statistical method of Tukey's HSD test.

Soil quality levels were quantified based on a statistical method utilizing soil quality index (SQI), and a minimum dataset (MDS) was performed to select soil properties indicators for calculation of SQI using principal component analysis (PCA). Results derived from PCA indicated that only the eigenvalue of the first principal component (PC1) was higher than 1, accounting for 81.74% of the cumulative percentage of total variation (Supplementary Table S1). Subsequently, the norm value was calculated to select soil properties indicators if the value was at least 90% of the maximum value. The norm values showed that TK (Norm: 1.52), ExMg (Norm: 2.43), and ExMn (Norm: 2.66) were less than 90% of the maximum norm value, and thus they were excluded from the MDS (Supplementary Table S1). Consequently, SBD, SOM, TN, TP, AN, AP, AK, and ExCa were included in MDS to estimate the SQI.

The SQIs from the different sampling fields were illustrated in Figure 2. Whether the linear transformation (LS) or the non-linear transformation (NLS) was performed, the SQIs exhibited the same trend in this investigation. Briefly, the SQIs from SF I were greater than those from other sampling fields; it was obvious that SF V had the lowest value; the rest of the sampling fields were in intermediate positions, respectively. Additionally, the difference in SQI among different sampling fields was significant ($p < 0.05$), and these SOIs were classified into five levels following Zeraatpisheh's method [36], including very strong, strong, medium, weak, and very weak. Therefore, to more clearly demonstrate how soil quality decline affects soil-dwelling mesofaunal communities, we designated "SF I" as "very strong soil quality" (VSSQ); "SF II" as "strong soil quality" (SSQ); "SF III" as "medium soil quality" (MSQ); "SF IV" as "weak soil quality" (WSQ); and "SF V" as "very weak soil quality" (VWSQ) in the following text.

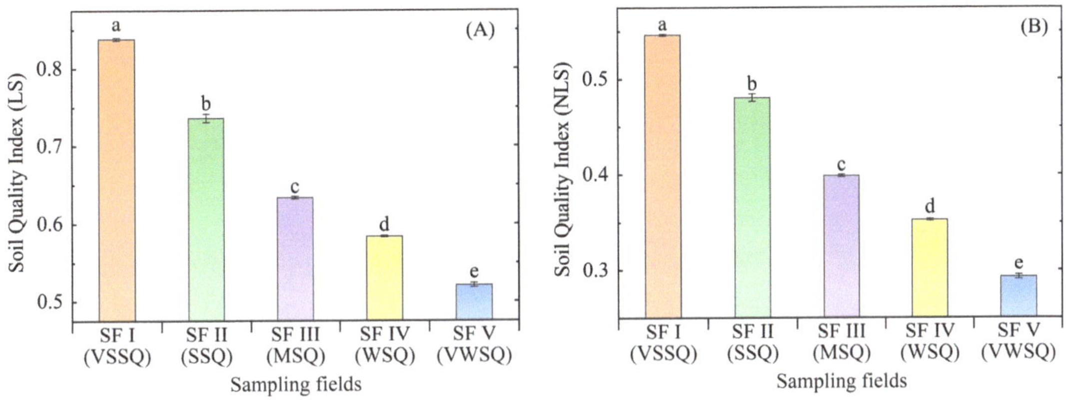

Figure 2. Soil quality indexes (SQIs) in agricultural lands of the Mollisols region. (**A**) Soil properties indicators were normalized to combinable scores using linear scoring (LS). (**B**) Soil properties indicators were normalized to combinable scores using non-linear scoring (NLS) systems. Different letters indicate a significant difference between each sampling field at $p < 0.05$ based on the statistical method of Tukey's HSD test. VSSQ, very strong soil quality; SSQ, strong soil quality; MSQ, medium soil quality; WSQ, weak soil quality; VWSQ, very weak soil quality.

3.2. *Soil-Dwelling Mesofauna*

We collected 4815 individuals from all sampling fields, and these soil-dwelling mesofauna belonged to 36 taxa (families/suborders), 20 orders, 6 classes, and 2 phyla. The taxonomic composition of soil-dwelling mesofauna in the different levels of soil quality are displayed in Figure 3. We found that soil quality declines resulted in variations in the taxonomic compositions of the soil-dwelling mesofauna (Figure 3A). In the sampling fields with very strong soil quality (VSSQ), Isotomidae (34.17%), Actinedida (34.17%), and Elateridae (10.79%) were the dominant taxa. Actinedida (23.32%) and Isotomidae (18.50%) occupied dominant positions in the sampling fields with strong soil quality (SSQ). Oribatida (64.68%) was only absolutely dominant in the sampling fields with medium soil quality (MSQ). Gamasida (31.94%) and Actinedida (13.09%) were the dominant taxa in the sampling fields with weak soil quality (WSQ). Entomobryidae (59.63%) and Actinedida (11.93%) were the dominant taxa in the sampling fields with very weak soil quality (VWSQ), in which Entomobryidae accounted for more than half of the individuals.

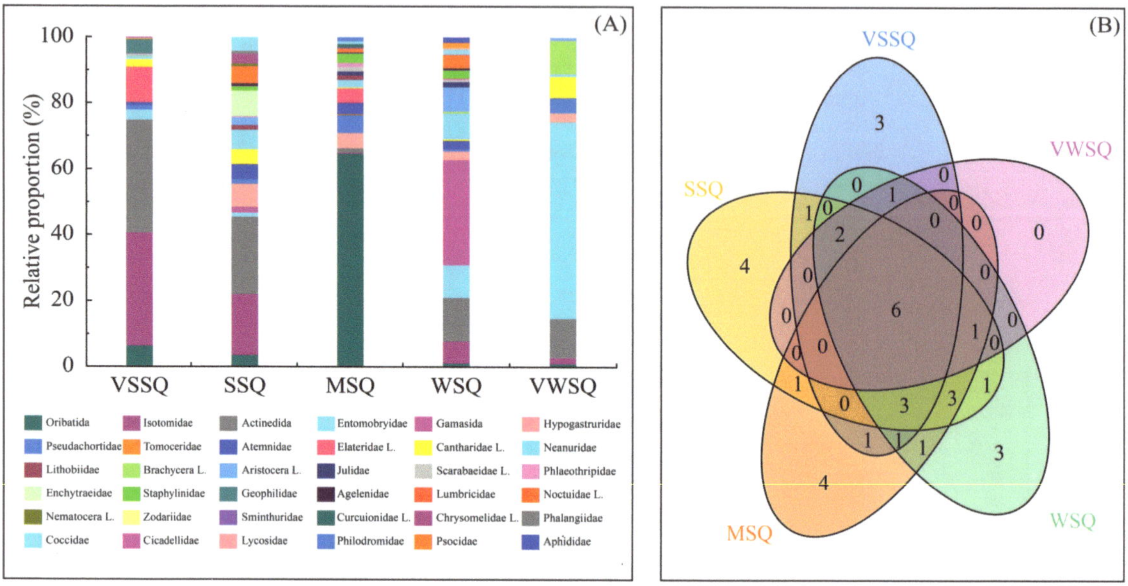

Figure 3. Soil-dwelling mesofaunal community structures in agricultural lands of the Mollisols region. (**A**) Taxonomic composition (%) of soil-dwelling mesofauna in the different soil quality levels. (**B**) The overlap and distinctiveness of soil-dwelling mesofaunal taxa in the different soil quality levels. VSSQ, very strong soil quality; SSQ, strong soil quality; MSQ, medium soil quality; WSQ, weak soil quality; VWSQ, very weak soil quality.

The unique and shared taxa of soil-dwelling mesofaunal communities are summarized in Figure 3B. Oribatida, Isotomidae, Actinedida, Pseudachortidae, Cantharidae, and Neanuridae were found in all of the sampling fields and contributed 27.27–60.00% of the soil-dwelling mesofaunal taxa. Geophilidae, Zodariidae, and Cicadellidae were unique taxa in the VSSQ fields; Enchytraeidae, Lumbricidae, Nematocera, and Phalangiidae were unique taxa in the SSQ fields; Tomoceridae, Sminthuridae, Curcuionidae, and Philodromidae were the only unique taxa in the MSQ fields; and Lycosidae, Psocidae, and Aphididae were unique taxa in the WSQ fields. However, the VWSQ fields exhibited no unique taxa.

The density of soil-dwelling mesofauna varied across different levels of soil quality (Figure 4A). Regarding the box and whisker plot, specifically, VSSQ fields exhibited significantly higher density in these agricultural lands ($p < 0.05$); both SSQ and MSQ fields showed significantly greater density than WSQ and VWSQ fields ($p < 0.05$), with no significant distinction in this respect between SSQ and MSQ fields ($p > 0.05$); the density in WSQ fields significantly surpassed that in VWSQ fields ($p < 0.05$); and the minimum value of density was recorded in the VWSQ fields.

The rarefaction curves revealed variations in the richness among these plots (Figure 4B). Plateau formations were observed in the curves for all locations, indicating the comprehensive detection of the majority of the soil-dwelling mesofaunal taxa in these agricultural lands. Among these curves, the richness of soil-dwelling mesofauna was consistently at the lowest level in the VWSQ fields, whereas it was consistently higher in WSQ fields than those in other levels of soil quality. However, the richness in the VSSQ fields was relatively low compared to the other levels of soil quality.

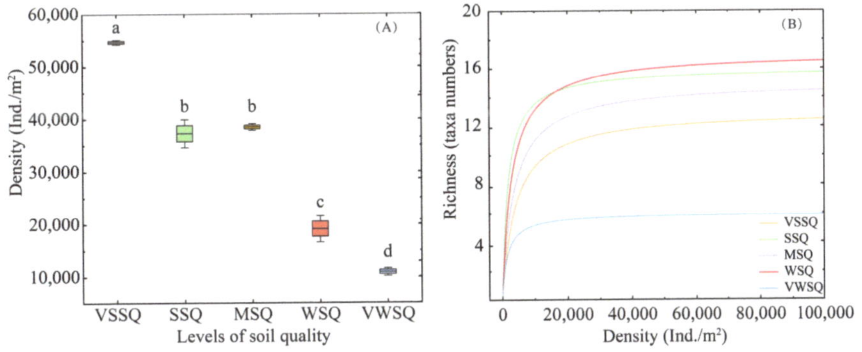

Figure 4. Density and richness of soil-dwelling mesofauna in agricultural lands of the Mollisols region. (**A**) Box and whisker plots depicting density (Ind/m^2) of soil-dwelling mesofauna across different soil quality levels, with line in the box indicating mean values of density. (**B**) Rarefaction curves illustrating richness (taxon number) for density (Ind/m^2) of soil-dwelling mesofauna across different soil quality levels. Matching letters indicate no significant difference between each level of soil quality level at $p < 0.05$. VSSQ, very strong soil quality; SSQ, strong soil quality; MSQ, medium soil quality; WSQ, weak soil quality; VWSQ, very weak soil quality.

A heatmap demonstrated that the five sampling fields were clustered into four clusters (Figure 5). It revealed that substantial similarity in the community structure of soil-dwelling mesofauna could be found in the WSQ and VWSQ fields, and there were some differences among the other sampling fields. At the same time, a high level of similarity could be observed among the remaining sampling fields. The soil-dwelling mesofaunal communities were also clustered into three clusters. Briefly, Oribatida exclusively constituted the first cluster; Isotomidae and Actinedida were the second cluster; and other species constituted another cluster. The VSSQ fields exhibited a higher density of Isotomidae and Actinedida, and a large count of Oribatida was also observed in the MSQ fields. Additionally, a large portion of soil-dwelling mesofauna was evenly distributed throughout these agricultural lands.

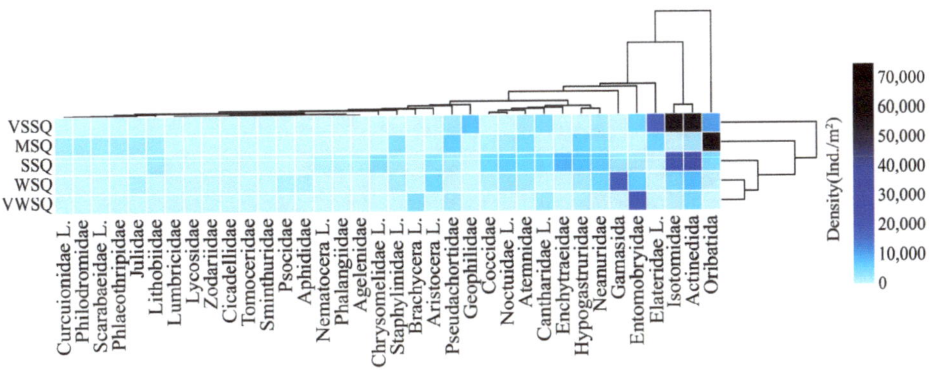

Figure 5. Distribution characteristics of soil-dwelling mesofauna in agricultural lands of the Mollisols region. Colors symbolize soil-dwelling mesofaunal density (Ind/m^2). The right dendrogram represents the clustering of different soil quality levels, while the upper dendrogram shows the clustering of different soil-dwelling mesofauna. These clusters are formed using an unweighted pair group method with arithmetical averages (UPGMA). VSSQ, very strong soil quality; SSQ, strong soil quality; MSQ, medium soil quality; WSQ, weak soil quality; VWSQ, very weak soil quality.

3.3. Relationship between Soil-Dwelling Mesofauna and Soil Quality

To effectively evaluate the effects of soil quality decline on soil-dwelling mesofaunal communities, a comprehensive evaluation was meticulously conducted to delineate the relationship between soil-dwelling mesofaunal communities and soil quality decline utilizing a Mantel test and an MFA (Figures 6 and 7). We found that the effects of soil physical properties, soil total nutrients, soil available nutrients, and soil mineral nutrients on soil-dwelling mesofaunal communities were primarily characterized by positive influences, and none of the negative influences were significant ($p > 0.05$). Among these soil properties, the effects of soil mineral nutrients were lower compared to the other factors. At the same time, different soil properties correlated with each taxon of soil-dwelling mesofauna in various ways. For instance, inconspicuous correlations were observed between Oribatida and all of the soil properties; Isotomidae, Actinedida, Elateridae, Cantharidae, Brachycera, Zodariidae, and Cicadellidae were susceptible to the influence of soil properties; and TK exhibited an effect that did not reach statistical significance on the majority of soil-dwelling mesofauna ($p > 0.05$).

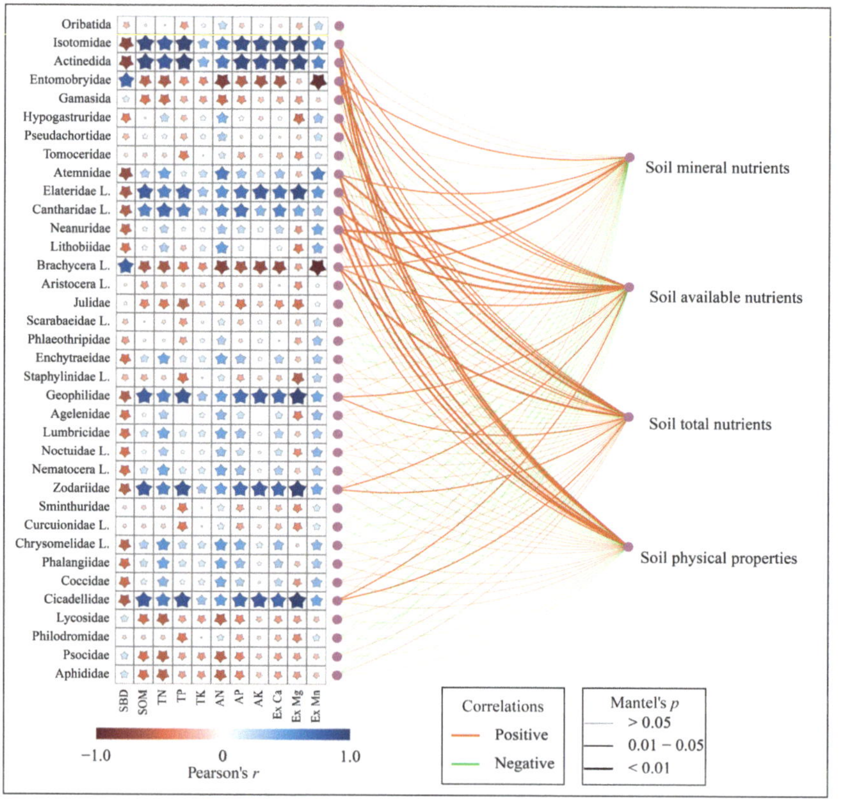

Figure 6. Correlations between different taxa of soil-dwelling mesofauna and soil properties. Stars correspond to correlation coefficients between different soil-dwelling mesofauna and each soil property via a Pearson correlation method. Star sizes represent the correlation coefficient's absolute value. Star color indicates the correlation coefficient. Pairwise comparisons of predictors (soil physical properties, total nutrients, available nutrients, and mineral nutrients) are shown on the right. Edge colors denote correlations via a Mantel test. Edge width represents the correlation coefficient's absolute value via the Mantel test.

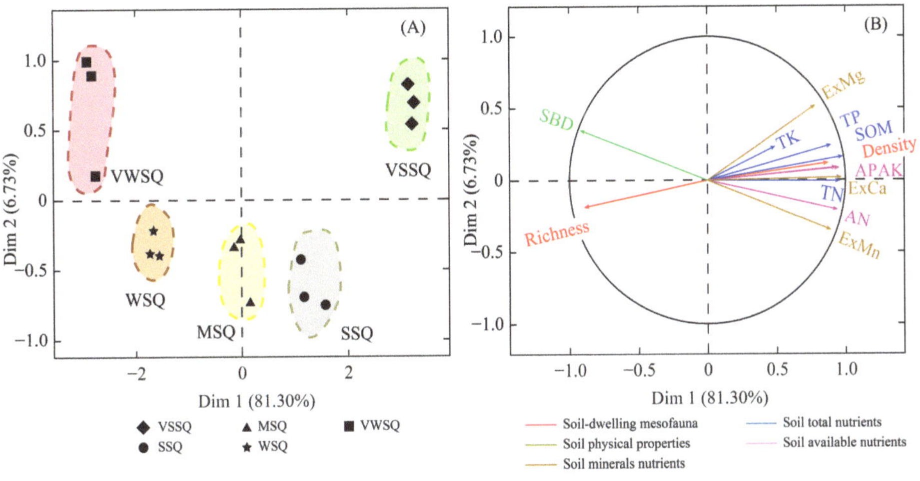

Figure 7. Relationships between soil-dwelling mesofauna and soil quality decline. (**A**) Two-dimensional scatter graph illustrating the variation among the different sampling fields. (**B**) Two-dimensional ordination diagram representing the relationships based on the entire dataset, including soil-dwelling mesofauna, soil physical properties, total nutrients, available nutrients, and mineral nutrient data. The two-dimensional plots are generated using a multiple-factor analysis (MFA). VSSQ, very strong soil quality; SSQ, strong soil quality; MSQ, medium soil quality; WSQ, weak soil quality; VWSQ, very weak soil quality. SBD, soil bulk density; SOM, soil organic matter; TN, total N; TP, total P; TK, total K; AN, available N; AP, available P; AK, available K; ExCa, exchangeable Ca; ExMg, exchangeable Mg; ExMn, exchangeable Mn.

Multiple-factor analysis (MFA) was implemented for estimating the links between soil-dwelling mesofauna and soil quality decline. The results of MFA were demonstrated by a two-dimensional diagram (Figure 7). The two-dimensional space defined by the MFA explained 88.03% of the total variability. Specifically, the axis of dimension 1 (Dim 1) explained 81.30% of the variance. Due to the fact that the VWSQ, WSQ, MSQ, SSQ, and VSSQ fields were positioned along Dim 1 from the negative to the positive direction, we deemed that Dim 1 indicated the gradient of soil quality decline (Figure 7A). This concretely showed that the positive direction indicated a greater level of soil quality, whereas the negative direction indicated a lower level of soil quality.

In the two-dimensional ordination diagram (Figure 7B), the vector representing the density of soil-dwelling mesofauna exhibited an evident positive correlation with Dim 1's positive orientation. This indicated that soil quality decline decreased the density of soil-dwelling mesofauna. At the same time, the two-dimensional ordination diagram displayed that the vectors representing richness pointed towards WSQ fields and showed negative correlations with density. In addition, the density of soil-dwelling mesofauna was negatively related to soil temperature and soil bulk density.

4. Discussion

4.1. Community Characteristics of Soil-Dwelling Mesofauna under Soil Quality Decline

Soil quality levels in this study were quantified by utilizing soil quality index (SQI), and a gradient of soil quality decline existed among the different sampling fields (Figure 2). In practice, frequent plowing and cultivation during reclamation and planting are primary contributors to soil quality decline in the Mollisols region [47]. The sampling fields in the study area were managed by different local farmers, resulting in varying cultivation histories and levels of land-use intensity, which contributed to a gradient of soil quality decline in this study.

In this study, it was evident that community structures and distribution characteristics of soil-dwelling mesofauna varied across five levels of soil quality. Firstly, we found variations in the taxonomic compositions of soil-dwelling mesofauna across different levels of soil quality (Figure 3A). Compared to other fields with different soil quality levels, Oribatida experienced absolute dominance in the fields with medium soil quality (MSQ), accounting for 64.68% of the total. Oribatida is among the most abundant taxa of soil mites, which live in a variety of habitats where they feed as scavengers on bacteria and fungi [48]. A previous study has revealed that enhanced species diversity and more intricate co-occurrence networks of bacterial communities emerged during the mid-term of soil quality decline [49]. Consequently, this phenomenon might have led to Oribatida being absolutely predominant in the MSQ fields. Meanwhile, Entomobryidae accounted for 59.63% of the total taxa, emerging as the absolute dominant taxa in the sampling fields with very weak soil quality (VWSQ). Entomobryidae, a family of Entomobryomorpha within the Collembola, is commonly consumed by various predators of soil macrofauna, such as Reduviidae and Coccinellidae [50]. Rousseau et al. revealed that soil quality decline could negatively affect soil macrofauna in agroecosystems [51]. Consequently, the very weak soil quality led to the absence of natural predators for Entomobryidae, thereby contributing to their higher proportion in the taxonomic compositions. Subsequently, we observed that the VWSQ fields exhibited no unique taxa; rather, all observed taxa were common within this study area (Figure 3B). This indicated that soil quality decline could reduce the probability of soil-dwelling rare mesofaunal taxa appearing. As substantiated by previous studies, rare taxa of soil fauna were vulnerable to the effects of soil environmental deterioration [52,53].

In the context of soil quality decline, a consistent decrease in the density of soil-dwelling mesofauna was noted (Figure 4A). This finding partially confirms our hypothesis (H1) that the community characteristics of soil-dwelling mesofauna were negatively affected by soil quality decline in the Mollisols region. In general, substantial input of organic matter is an essential element in the maintenance of soil quality in the Mollisols region [54]. Some studies revealed that one of the primary factors contributing to decline in soil quality was decrease in soil organic matter content [55,56], and our investigation also agreed with these findings, as depicted in Table 2. Guidi et al. revealed that a substantial quantity of organic matter can promote good living conditions and food resources for soil-dwelling mesofauna [57]. Therefore, the lack of organic matter led to a shortage of food resources for soil-dwelling mesofauna, resulting in a decrease in their density in areas with weaker soil quality. Concurrently, it was observed that in the sampling fields categorized as weak soil quality (WSQ), medium soil quality (MSQ), and strong soil quality (SSQ), the richness of the soil-dwelling mesofauna was higher than those in the sampling fields with very strong soil quality (VSSQ) (Figure 4B). This was not consistent with our hypothesis (H1). In the VSSQ sampling fields, soil-dwelling mesofaunal density was significantly higher than that in the other sampling fields ($p < 0.05$) (Figure 4A). A greater number of dominant individuals suggests that the majority of ecological niches have been filled, posing challenges for subordinate individuals in locating suitable habitats and resources, thus limiting their richness [58]. Consequently, a relatively lower richness was found in the VSSQ sampling fields.

Additionally, the heatmap in this study demonstrated that substantial similarity in the community structure of soil-dwelling mesofauna could be found in the WSQ and VWSQ fields, and there were some differences among the other sampling fields (Figure 5). It indicated that the effects of soil quality decline might be more pronounced on the community structure of soil-dwelling mesofauna in environments with relatively stronger soil quality. The competitive exclusion principle implies that when resources are limited, competition will lead to the exclusion or reduction in numbers of organisms with weaker competitive abilities [59]. There is relatively greater resource availability in environments with higher soil quality, leading to relatively less competition among soil organisms [60]. Moreover, we observed notable density variations of Isotomidae from VSSQ to MSQ, with no discernible differences noted in the WSQ (Figure 5). Isotomidae, a family of long-bodied springtails within the Collembola order, is characterized by its minuscule body size, weak physical constitution, and limited capacity for migration [61]. These characteristics might render them particularly vulnerable to competition. Consequently, as soil quality declined in this study, resources became more limited, intensifying competition, pronouncedly altering the community structure of soil-dwelling mesofauna in environments with relatively stronger soil quality. Conversely, a reduction in resources might result in relatively reduced competition among soil-dwelling mesofauna. Therefore, despite a decrease in soil quality, a drastic competitive exclusion effect might not occur, allowing for a relatively similar community structure in the WSQ and VWSQ fields.

4.2. Exploring the Relationship between Soil-Dwelling Mesofaunal Communities and Soil Quality Decline

Our results revealed that 11 soil properties exhibited varied correlations with different taxa of soil-dwelling mesofauna during soil quality decline (Figure 6). This finding confirmed our hypothesis (H2) that soil quality decline impacts different taxa of soil-dwelling mesofauna to varying extents. It is known that significant differences in diets, nutritional requirements, life histories, and life forms are evident among the majority of soil-dwelling mesofaunal taxa [62,63]. In this study, a total of 94.41% of the individuals isolated belonged to the following taxa: Actinedida (19.75%), Oribatida (18.69%), Isotomidae (16.64%), Entomobryidae (6.60%), Elateridae (4.74%), Gamasida (4.24%), Hypogastruridae (3.24%), Neanuridae (3.24%), Atemnidae (2.49%), Pseudachortidae (2.43%), Cantharidae (2.43%), Noctuidae (2.06%), Enchytraeidae (1.81%), Aristocera (1.56%), Geophilidae (1.50%), Coccidae (1.37%), and Staphylinidae (1.31%), and the majority of them were saprophagous or phytophagous soil-dwelling mesofauna, while a minority belonged to predatory fauna. In general, saprophagous and phytophagous soil-dwelling mesofauna depend more on organic matter and inorganic nutrients in the soil for their survival and reproduction [64]. This results in soil properties exerting positive influences on saprophagous and phytophagous fauna. Subsequently, an increase in phytophagous and saprophagous fauna might draw in more predatory fauna to form colonies [65]. Consequently, the effects of soil properties on soil-dwelling mesofauna were primarily characterized by positive influences in this study. At the same time, we observed negative impacts on certain taxa of soil-dwelling mesofauna. However, Mantel's test indicated that the negative influences of soil properties on soil-dwelling mesofaunal communities were not predominant. While soil properties had negative impacts on partial taxa of soil-dwelling mesofauna, other factors such as food resource availability, competition relationships, and migration patterns, might play more significant roles in the formation and maintenance of mesofaunal communities [66,67]. In this study, these negatively correlated taxa (e.g., Entomobryidae, Brachycera larvae) were mainly distributed in the VWSQ fields. Therefore, contents of soil nutrients might not be the primary factor influencing community composition in environments with very low soil quality.

In addition, our results were congruent with prior research, indicating that monitoring the characteristics of soil-dwelling mesofaunal communities could provide crucial information about soil environmental changes, aiding in maintaining ecosystem health [68]. The two-dimensional ordination plot constructed based on the MFA provided further support for our hypothesis (H2), and it revealed that the density of soil-dwelling mesofauna could signal soil quality decline in agricultural lands of the Mollisols region (Figure 7). A reduction in density could be considered an ecosystem response to degradation, reflecting challenges for soil-dwelling mesofauna in terms of adaptation and survival [69]. Concurrently, soil total nitrogen (TN), phosphorus (TP), and potassium (TK); available nitrogen (AN), phosphorus (AP), and potassium (AK); and exchangeable calcium (ExCa), magnesium (ExMg), and manganese (ExMn) were positively related to the density of soil-dwelling mesofauna. These soil nutrients can promote plant growth, thereby improving the contents of organic matter entering the belowground environment as root exudates and finally increasing the abundance of soil-dwelling mesofauna [70]. Therefore, this finding could suggest that the community characteristics of soil-dwelling mesofauna could reflect soil quality decline to some extent in Mollisols regions.

5. Conclusions

To summarize, changes occurred in the communities of soil-dwelling mesofauna as soil quality declined in agricultural lands of the Mollisols region. The investigation results demonstrated a consistent decrease in the density of soil-dwelling mesofauna with declining soil quality, indicating that density might signal soil quality decline in agricultural lands of the Mollisols region. Amidst soil quality decline, the effects on soil-dwelling mesofaunal communities were not exclusively negative, as greater richness was observed in areas with relatively weaker soil quality. Soil quality decline impacted different taxa of soil-dwelling mesofauna to varying extents. Oribatida was absolutely predominant in agricultural lands with medium soil quality, and most Entomobryidae were found in agricultural lands with very weak soil quality. Additionally, the changes in soil properties during soil quality decline had various effects on the communities of soil-dwelling mesofauna. Soil nutrients, in particular, exhibited a positive correlation with the abundance of soil-dwelling mesofauna in agricultural lands of the Mollisols region, but they might not constitute the primary factor influencing community composition in environments with very weak soil quality. The outcomes of this study carry significance for comprehending how soil quality decline relates to soil-dwelling mesofauna, and can provide valuable ecological insights for formulating biodiversity guidelines targeted at preserving soil resources in the Mollisols region.

Supplementary Materials: The following supporting information can be downloaded at: https://www.mdpi.com/article/10.3390/agriculture14050766/s1, Table S1: Results of principal component analysis (PCA) of 11 soil properties and their norm values for constructing evaluation indicators for the minimum data set (MSD).

Author Contributions: Conceptualization, C.M. and G.D.; methodology, C.M.; software, X.Y.; validation, C.M., X.Y. and G.D.; formal analysis, X.Y.; investigation, X.Y.; resources, C.M.; data curation, C.M.; writing—original draft preparation, C.M.; writing—review and editing, X.Y.; visualization, C.M.; supervision, C.M.; project administration, C.M.; funding acquisition, C.M. and G.D. All authors have read and agreed to the published version of the manuscript.

Funding: This research was funded by the National Key R&D Program of China, grant No.2021YFD1500101; National Natural Science Foundation of China, grant No. 42307413; Heilongjiang Provincial Natural Science Foundation of China, grant No. LH2023D002, and the Youth Talent Project of the Northeast Agricultural University of China, grant No. 19QC35.

Institutional Review Board Statement: Not applicable.

Data Availability Statement: Data will be made available upon request.

Conflicts of Interest: The authors declare no conflicts of interest.

References

1. Lehmann, J.; Bossio, D.A.; Kögel-Knabner, I.; Rillig, M.C. The concept and future prospects of soil health. *Nat. Rev. Earth Environ.* **2020**, *1*, 544–553. [CrossRef] [PubMed]
2. IPCC. *IPCC Special Report: Climate Change and Land*; IPCC: Geneva, Switzerland, 2019.
3. Simonson, R.W. Soil Classification in the United States. *Science* **1962**, *137*, 1027–1034. [CrossRef] [PubMed]
4. Kassam, A.; Friedrich, T.; Derpsch, R. Global spread of conservation agriculture. *Int. J. Environ. Stud.* **2019**, *76*, 29–51. [CrossRef]
5. Kravchenko, Y.S.; Chen, Q.; Liu, X.; Herbert, S.J.; Zhang, X. Conservation practices and management in Ukrainian mollisols. *J. Agric. Sci. Technol.* **2016**, *18*, 845–854.
6. Ouyang, W.; Wu, Y.; Hao, Z.; Zhang, Q.; Bu, Q.; Gao, X. Combined impacts of land use and soil property changes on soil erosion in a mollisol area under long-term agricultural development. *Sci. Total Environ.* **2018**, *613–614*, 798–809. [CrossRef] [PubMed]
7. Wang, S.; Wang, Z.; Heinonsalo, J.; Zhang, Y.; Liu, G. Soil organic carbon stocks and dynamics in a mollisol region: A 1980s–2010s study. *Sci. Total Environ.* **2022**, *807*, 150910. [CrossRef] [PubMed]
8. Adetunji, A.T.; Ncube, B.; Mulidzi, R.; Lewu, F.B. Management impact and benefit of cover crops on soil quality: A review. *Soil Till. Res.* **2020**, *204*, 104717. [CrossRef]

9. Zhan, T.; Zhang, Z.; Sun, J.; Liu, M.; Zhang, X.; Peng, F.; Tsunekawa, A.; Zhou, H.; Gou, X.; Fu, S. Meta-Analysis demonstrating that moderate grazing can improve the soil quality across China's grassland ecosystems. *Appl. Soil Ecol.* **2020**, *147*, 103438. [CrossRef]
10. Hermans, S.M.; Buckley, H.L.; Case, B.S.; Curran-Cournane, F.; Taylor, M.; Lear, G. Using soil bacterial communities to predict physico-chemical variables and soil quality. *Microbiome* **2020**, *8*, 79. [CrossRef]
11. Leul, Y.; Assen, M.; Damene, S.; Legass, A. Effects of land use types on soil quality dynamics in a tropical sub-humid ecosystem, Western Ethiopia. *Ecol. Indic.* **2023**, *147*, 110024. [CrossRef]
12. Yan, S.; Singh, A.N.; Fu, S.; Liao, C.; Wang, S.; Li, Y.; Cui, Y.; Hu, L. A soil fauna index for assessing soil quality. *Soil Biol. Biochem.* **2012**, *47*, 158–165. [CrossRef]
13. Li, Y.; Wang, J.; Shao, M. Assessment of earthworms as an indicator of soil degradation: A case-study on Loess Soils. *Land Degrad. Dev.* **2021**, *32*, 2606–2617. [CrossRef]
14. Kamau, S.; Barrios, E.; Karanja, N.K.; Ayuke, F.O.; Lehmann, J. Dominant tree species and earthworms affect soil aggregation and carbon content along a soil degradation gradient in an agricultural landscape. *Geoderma* **2020**, *359*, 113983. [CrossRef]
15. De Araújo, A.S.F.; Eisenhauer, N.; Nunes, L.A.P.L.; Leite, L.F.C.; Cesarz, S. Soil surface-active fauna in degraded and restored lands of Northeast Brazil. *Land Degrad. Dev.* **2013**, *26*, 1–8. [CrossRef]
16. Martin, T.; Wade, J.; Singh, P.; Sprunger, C.D. The integration of nematode communities into the soil biological health framework by factor analysis. *Ecol. Indic.* **2022**, *136*, 108676. [CrossRef]
17. Bradford, M.A.; Jones, T.H.; Bardgett, R.D.; Black, H.I.J.; Boag, B.; Bonkowski, M.; Cook, R.; Eggers, T.; Gange, A.C.; Grayston, S.J.; et al. Impacts of soil faunal community composition on model grassland ecosystems. *Science* **2002**, *298*, 615–618. [CrossRef] [PubMed]
18. Ma, C.; Yin, X.; Xu, H.; Tao, Y. Responses of soil Collembolans to vegetation restoration in temperate coniferous and broad-leaved mixed forests. *J. For. Res.* **2020**, *31*, 2333–2345. [CrossRef]
19. Beaumelle, L.; Thouvenot, L.; Hines, J.; Jochum, M.; Eisenhauer, N.; Phillips, H.R.P. Soil fauna diversity and chemical stressors: A review of knowledge gaps and roadmap for future research. *Ecography* **2021**, *44*, 845–859. [CrossRef]
20. Arboláez, H.P.H.; Hu, J.; Orozco, Y.N.; Gebremikael, M.T.; Alcantara, E.A.; Sleutel, S.; Höfte, M.; De Neve, S. Mesofauna as effective indicators of soil quality differences in the agricultural systems of Central Cuba. *Appl. Soil Ecol.* **2023**, *182*, 104688. [CrossRef]
21. Ren, H.; Lv, H.; Xu, Q.; Yao, Z.; Yao, P.; Zhao, N.; Wang, Z.; Huang, D.; Cao, W.; Gao, Y.; et al. Green manure provides growth benefits for soil mesofauna by promoting soil fertility in agroecosystems. *Soil Tillage Res.* **2024**, *238*, 106006. [CrossRef]
22. Ma, C.; Nie, R.; Du, G. Responses of soil collembolans to land degradation in a black soil region in China. *Int. J. Environ. Res. Public Health* **2023**, *20*, 4820. [CrossRef] [PubMed]
23. Urbanowski, C.K.; Turczański, K.; Andrzejewska, A.; Kamczyc, J.; Jagodziński, A.M. Which soil properties affect soil mite (Acari, Mesostigmata) communities in stands with various shares of European ash (*Fraxinus Excelsior* L.)? *Appl. Soil Ecol.* **2022**, *180*, 104633. [CrossRef]
24. Mantoni, C.; Pellegrini, M.; Dapporto, L.; Del Gallo, M.; Pace, L.; Silveri, D.; Fattorini, S. Comparison of soil biology quality in organically and conventionally managed agro-ecosystems using microarthropods. *Agriculture* **2021**, *11*, 1022. [CrossRef]
25. D'Hose, T.; Cougnon, M.; De Vliegher, A.; Vandecasteele, B.; Viaene, N.; Cornelis, W.; Van Bockstaele, E.; Reheul, D. The positive relationship between soil quality and crop production: A case study on the effect of farm compost application. *Appl. Soil Ecol.* **2014**, *75*, 189–198. [CrossRef]
26. Yin, W.Y. *Soil Animals of China*; Science Press: Beijing, China, 2000.
27. Walkley, A.; Black, I.A. An examination of the degtjareff method for determining soil organic matter, and a proposed modification of the chromic acid titration method. *Soil Sci.* **1934**, *37*, 29–38. [CrossRef]
28. Bao, S. *Soil Agro-Chemistrical Analysis*; China Agricultural Press: Beijing, China, 2000.
29. Klute, A. *Methods of Soil Analysis: Physical and Mineralogical Methods*; American Society of Agronomy: Madison, WI, USA, 1986.
30. Page, A.L. *Methods of Soil Analysis: Chemical and Microbiological Properties*; American Society of Agronomy: Madison, WI, USA, 1982.
31. Nakajima, T.; Lal, R.; Jiang, S. Soil quality index of a Crosby silt loam in Central Ohio. *Soil Till. Res.* **2015**, *146*, 323–328. [CrossRef]
32. Brejda, J.J.; Moorman, T.B.; Karlen, D.L.; Dao, T.H. Identification of regional soil quality factors and indicators I. Central and Southern High Plains. *Soil Sci. Soc. Am. J.* **2000**, *64*, 2115–2124. [CrossRef]
33. Zhang, G.; Bai, J.; Xi, M.; Zhao, Q.; Lu, Q.; Jia, J. Soil quality assessment of coastal wetlands in the Yellow River Delta of China Based on the minimum data set. *Ecol. Indic.* **2016**, *66*, 458–466. [CrossRef]
34. Nabiollahi, K.; Golmohamadi, F.; Taghizadeh-Mehrjardi, R.; Kerry, R.; Davari, M. Assessing the effects of slope gradient and land use change on soil quality degradation through digital mapping of soil quality indices and soil loss rate. *Geoderma* **2018**, *318*, 16–28. [CrossRef]
35. Andrews, S.S.; Karlen, D.L.; Cambardella, C.A. The soil management assessment framework. *Soil Sci. Soc. Am. J.* **2004**, *68*, 1945–1962. [CrossRef]
36. Zeraatpisheh, M.; Bakhshandeh, E.; Hosseini, M.; Alavi, S.M. Assessing the effects of deforestation and intensive agriculture on the soil quality through digital soil mapping. *Geoderma* **2020**, *363*, 114139. [CrossRef]

37. Raiesi, F. A minimum data set and soil quality index to quantify the effect of land use conversion on soil quality and degradation in native rangelands of upland arid and semiarid regions. *Ecol. Indic.* **2017**, *75*, 307–320. [CrossRef]
38. Bastida, F.; Luis Moreno, J.; Hernández, T.; García, C. Microbiological degradation index of soils in a semiarid climate. *Soil Biol. Biochem.* **2006**, *38*, 3463–3473. [CrossRef]
39. Sheidai Karkaj, E.; Sepehry, A.; Barani, H.; Motamedi, J.; Shahbazi, F. Establishing a suitable soil quality index for semi-arid rangeland ecosystems in Northwest of Iran. *J. Soil Sci. Plant Nut.* **2019**, *19*, 648–658. [CrossRef]
40. Andrews, S.S.; Karlen, D.L.; Mitchell, J.P. A Comparison of soil quality indexing methods for vegetable production systems in Northern California. *Agr. Ecosyst. Environ.* **2002**, *90*, 25–45. [CrossRef]
41. Qi, Y.; Darilek, J.L.; Huang, B.; Zhao, Y.; Sun, W.; Gu, Z. Evaluating soil quality indices in an agricultural region of Jiangsu Province, China. *Geoderma* **2009**, *149*, 325–334. [CrossRef]
42. R Core Team. R: A Language and Environment for Statistical Computing. Available online: https://www.R-project.org/ (accessed on 31 October 2023).
43. Chen, H. VennDiagram: Generate High-Resolution Venn and Euler Plots. R Package Version 1.7.3. Available online: https://CRAN.R-project.org/package=VennDiagram (accessed on 12 April 2022).
44. Oksanen, J.; Simpson, G.; Blanchet, F.; Kindt, R.; Legendre, P.; Minchin, P.R.; O'Hara, R.B.; Solymos, P.; Stevens, M.H.H.; Szoecs, E.; et al. Vegan: Community Ecology Package_R Package Version 2.6-4. 2022. Available online: https://CRAN.R-project.org/package=vegan (accessed on 11 October 2022).
45. Kolde, R. Pheatmap: Pretty Heatmaps. R Package, Version 1.0.12. Available online: https://CRAN.R-project.org/package=pheatmap (accessed on 4 January 2019).
46. Le, S.; Josse, J.; Husson, F. FactoMineR: An R package for multivariate analysis. *J. Stat. Softw.* **2008**, *25*, 1–18. [CrossRef]
47. Peng, J.; Yang, Q.; Zhang, C.; Ni, S.; Wang, J.; Cai, C. Aggregate pore structure, stability characteristics, and biochemical properties induced by different cultivation durations in the Mollisol region of Northeast China. *Soil Tillage Res.* **2023**, *233*, 105797. [CrossRef]
48. Oszust, M.; Klimaszyk, P. Soil conditions under cormorant colonies favor for mites excepting Oribatida. *Acarologia* **2022**, *62*, 974–988. [CrossRef]
49. Xiao, H.; Xu, M.; Wang, Z.; Wang, L.; Wang, J.; Xiao, Y.; Shi, Z. Role of autotrophic microbes in organic matter accumulation in soils degraded by erosion. *Land Degrad. Dev.* **2022**, *33*, 2092–2102. [CrossRef]
50. Bernard, E.C. Soil Arthropods: Underfoot and all around. In *Encyclopedia of Soils in the Environment*; Goss, M.J., Oliver, M., Eds.; Academic Press: Cambridge, MA, USA, 2023.
51. Rousseau, L.; Fonte, S.J.; Téllez, O.; van der Hoek, R.; Lavelle, P. Soil macrofauna as indicators of soil quality and land use impacts in smallholder agroecosystems of Western Nicaragua. *Ecol. Indic.* **2013**, *27*, 71–82. [CrossRef]
52. Li, Y.; Ma, L.; Wang, J.; Shao, M.; Zhang, J. Soil faunal community composition alters nitrogen distribution in different land use types in the Loess Plateau, China. *Appl. Soil Ecol.* **2021**, *163*, 103910. [CrossRef]
53. Xu, G.; Zhang, Y.; Zhang, S.; Ma, K. Biodiversity associations of soil fauna and plants depend on plant life form and are accounted for by rare taxa along an elevational gradient. *Soil Biol. Biochem.* **2020**, *140*, 107640. [CrossRef]
54. Wen, Y.; Kasielke, T.; Li, H.; Zhang, B.; Zepp, H. May agricultural terraces induce gully erosion? A case study from the Black Soil Region of Northeast China. *Sci. Total Environ.* **2021**, *750*, 141715. [CrossRef]
55. Li, H.; Zhu, H.; Qiu, L.; Wei, X.; Liu, B.; Shao, M. Response of soil OC, N and P to land-use change and erosion in the black soil region of the Northeast China. *Agric. Ecosyst. Environ.* **2020**, *302*, 107081. [CrossRef]
56. Jensen, J.L.; Schjønning, P.; Watts, C.W.; Christensen, B.T.; Obour, P.B.; Munkholm, L.J. Soil degradation and recovery–Changes in organic matter fractions and structural stability. *Geoderma* **2020**, *364*, 114181. [CrossRef]
57. Guidi, C.; Frey, B.; Brunner, I.; Meusburger, K.; Vogel, M.E.; Chen, X.; Stucky, T.; Gwiazdowicz, D.J.; Skubała, P.; Bose, A.K.; et al. Soil fauna drives vertical redistribution of soil organic carbon in a long-term irrigated dry pine forest. *Glob. Change Biol.* **2022**, *28*, 3145–3160. [CrossRef] [PubMed]
58. Wiens, J.J.; Ackerly, D.D.; Allen, A.P.; Anacker, B.L.; Buckley, L.B.; Cornell, H.V.; Damschen, E.I.; Jonathan Davies, T.; Grytnes, J.; Harrison, S.P.; et al. Niche Conservatism as an Emerging Principle in Ecology and Conservation Biology. *Ecol. Lett.* **2010**, *13*, 1310–1324. [CrossRef]
59. Hardin, G. The competitive exclusion principle: An idea that took a century to be born has implications in ecology, economics, and genetics. *Science* **1960**, *131*, 1292–1297. [CrossRef]
60. Wardle, D.A.; Bardgett, R.D.; Klironomos, J.N.; Setälä, H.; van der Putten, W.H.; Wall, D.H. Ecological linkages between aboveground and belowground biota. *Science* **2004**, *304*, 1629–1633. [CrossRef]
61. Rusek, J. Biodiversity of Collembola and their functional role in the ecosystem. *Biodivers. Conserv.* **1998**, *7*, 1207–1219. [CrossRef]
62. Betancur-Corredor, B.; Lang, B.; Russell, D.J. Organic nitrogen fertilization benefits selected soil fauna in global agroecosystems. *Biol. Fertil. Soils* **2023**, *59*, 1–16. [CrossRef]
63. Hedde, M.; Blight, O.; Briones, M.J.; Bonfanti, J.; Brauman, A.; Brondani, M.; Sanou, I.C.; Clause, J.; Conti, E.; Cortet, J.; et al. A common framework for developing robust soil fauna classifications. *Geoderma* **2022**, *426*, 116073. [CrossRef]
64. Maraun, M.; Martens, H.; Migge, S.; Theenhaus, A.; Scheu, S. Adding to 'the enigma of soil animal diversity': Fungal feeders and saprophagous soil invertebrates prefer similar food substrates. *Eur. J. Soil Biol.* **2003**, *39*, 85–95. [CrossRef]

65. Woodcock, B.A.; Potts, S.G.; Westbury, D.B.; Ramsay, A.J.; Lambert, M.; Harris, S.J.; Brown, V.K. The importance of sward architectural complexity in structuring predatory and phytophagous invertebrate assemblages. *Ecol. Entomol.* **2007**, *32*, 302–311. [CrossRef]
66. Lindo, Z.; Winchester, N.N. Spatial and Environmental Factors Contributing to Patterns in Arboreal and Terrestrial Oribatid Mite Diversity across Spatial Scales. *Oecologia* **2009**, *160*, 817–825. [CrossRef] [PubMed]
67. Kardol, P.; Reynolds, W.N.; Norby, R.J.; Classen, A.T. Climate Change Effects on Soil Microarthropod Abundance and Community Structure. *Appl. Soil Ecol.* **2011**, *47*, 37–44. [CrossRef]
68. Malheiro, C.; Cardoso, D.N.; Loureiro, S.; González-Alcaraz, M.N. Effects of climate conditions on the avoidance behavior of *Folsomia candida* and *Enchytraeus crypticus* towards metal (loid)-contaminated soils. *Sci. Total Environ.* **2020**, *741*, 140368. [CrossRef]
69. Kou, X.; Tao, Y.; Wang, S.; Wu, Z.; Wu, H. Soil meso-fauna community composition predicts ecosystem multifunctionality along a coastal-inland gradient of the Bohai Bay. *Land Degrad. Dev.* **2021**, *32*, 4574–4582. [CrossRef]
70. Zhang, Y.; Peng, S.; Chen, X.; Chen, H.Y. Plant diversity increases the abundance and diversity of soil fauna: A meta-analysis. *Geoderma* **2022**, *411*, 115694. [CrossRef]

agriculture

MDPI

Article

Enhancing Sustainable Agriculture in China: A Meta-Analysis of the Impact of Straw and Manure on Crop Yield and Soil Fertility

Zhe Zhao [1,2,3], Yali Yang [1,3,*], Hongtu Xie [1,3], Yixin Zhang [2], Hongbo He [1,3], Xudong Zhang [1,3] and Shijun Sun [2,*]

[1] Institute of Applied Ecology, Chinese Academy of Sciences, Shenyang 110016, China; 2023200015@stu.syau.edu.cn (Z.Z.)
[2] College of Water Conservancy, Shenyang Agricultural University, Shenyang 110866, China
[3] Key Lab of Conservation Tillage and Ecological Agriculture, Liaoning Province, Shenyang 110016, China
* Correspondence: yali_yang@iae.ac.cn (Y.Y.); sunshijun2000@syau.edu.cn (S.S.)

Abstract: As the main organic materials, straw and manure play a critical role in soil organic carbon (SOC) sequestration and crop yield in China. This meta-analysis evaluated the impact of straw and manure amendments, both individually and combined, on crop yield, SOC, and soil nutrients in China by collecting 173 studies. The findings of this study revealed that straw return and manure application increased crop yields by 14.4% and 70.4%, respectively, overall. Combined straw and manure application gained a better improvement effect than straw alone but was less effective than manure alone. Regarding the straw return results, rice straw and a 3000–6000 kg ha^{-1} returning quantity improved crop yield, SOC, available phosphorus (AP), available potassium (AK), and total nitrogen (TN) the most; regarding the straw return form, straw incorporated into soil and biochar increased crop yield and SOC more, respectively; and <5 years and ≥5 years of straw return treatment increased crop yield and TN more, respectively. Regarding manure application, pig and chicken manure increased crop yield and TN more, respectively; a 50–80% substitution ratio and 10–20 years of duration were best for improving crop yield, SOC, AP, AK, and TN. This study highlights the importance of optimal organic amendment through straw or manure applications to achieve a win–win between crop yield and soil fertility under the requirement of sustainable agriculture.

Keywords: meta-analysis; organic amendment; China; carbon sequestration; management strategy

Citation: Zhao, Z.; Yang, Y.; Xie, H.; Zhang, Y.; He, H.; Zhang, X.; Sun, S. Enhancing Sustainable Agriculture in China: A Meta-Analysis of the Impact of Straw and Manure on Crop Yield and Soil Fertility. *Agriculture* **2024**, *14*, 480. https://doi.org/10.3390/agriculture14030480

Academic Editors: Anetta Siwik-Ziomek and Anna Figas

Received: 31 January 2024
Revised: 10 March 2024
Accepted: 14 March 2024
Published: 16 March 2024

1. Introduction

In the earth system, soil organic carbon (SOC) constitutes up to 75% of the terrestrial carbon pool, representing the largest carbon pool, which is approximately three times greater than the amount of carbon stored in the atmosphere or vegetation [1–3]. Findings have demonstrated that the sequestration of SOC could play an irreplaceable role in significantly reducing carbon emissions [4,5]. However, in China, long-term traditional agriculture management has induced soil deterioration from carbon sink to carbon source [6,7]. Meanwhile, the decrease in soil organic matter (SOM) and nutrients seriously threatens food security [8,9]. Consequently, the application of suitable management practices to increase SOC content and enhance crop yield is very important.

In agriculture, the practice of returning straw to the field is a feasible and promising strategy to sequestrate SOC without reducing yield [10–13]. A large number of studies have shown that returning straw to the field has a positive impact on crop yield, SOC, and soil nutrients [1,3,14,15]. China, as one of the world's largest crop straw producers, produces 800 to 1000 MT of crop straw every year, accounting for about 20% to 40% of the global aggregate (Table S1) [16–18]. In recent years, China has gradually shifted from incinerating crop straw to directly returning it to the field. The quantity of straw directly returned can reach more than 40% of the total crop straw [19,20]. However, there are still many obstacles to directly returning straw to the field. For example, the difficult decomposition of straw and the fact that incompletely decomposed straw can impede crop growth have resulted in farmers preferring to dispose of the straw by burning it [21–24]. Moreover, there is a multitude of factors that can impact the effects of returning straw to the field (e.g., its duration, quantity, form, and category), which makes it difficult for farmers to ensure a proper strategy of straw return [21,25,26]. Due to these reasons, as well as the huge quantity of straw itself and its great potential in restoring soil and mitigating carbon emissions, it is, therefore, essential to apply appropriate straw return management methods in actual production practices to foster green development in China. Current meta-analyses related to returning straw to the field are limited in their investigation of the effects of different categories and forms of straw on SOC, crop yield, and soil nutrient content. Additionally, the categorization rules for research factors vary among different meta-analyses. Given the continuous updates to meta-analyses, a newly integrative and comprehensive meta-analysis should be conducted.

Apart from returning straw to the field, the strategy of replacing chemical fertilizers with organic ones can also effectively reverse the decrease in SOC caused by the excessive use of chemical fertilizers [27–30]. Currently, China has the largest production of manure, similar to crop residues, in the world, which is over twice the combined total of the United States and the European Union (Table S2) [31]. So far, except for increasing SOC content, a myriad of experiments have testified to the important role of manure in restoring soil nutrients and improving crop yield [32–36]. However, the excessive and sole use of manure can diminish its advantages compared with the proper combination of manure and chemical fertilizer, even leading to counterproductive outcomes such as lower yields than those achieved with chemical fertilizer due to its slow decomposition process [35]. Moreover, the difficulty and high labor cost of collecting manure are not seen as worthwhile or appropriate by farmers, thereby causing reluctance and resistance from farmers toward propaganda and promotion concerning the replacement of chemical fertilizer with organic fertilizer [37]. Therefore, it is both inevitable and crucial to analyze an effective management strategy for manure application. Based on our meta-analysis, the effects of manure application and its management strategies can be explained further. Likewise, the result can be regarded as a reference to manure application. In order to explore the impact of different manure management modes on the effectiveness of manure, we divided three groups: the source, substitution ratio, and duration of manure for the meta-analysis.

In addition to the development of a survey on returning straw and applying manure, the effects of combining straw and manure were explored to determine the specific role of integrating these two organic amendments.

In this context, this study carried out a meta-analysis including 2647 data derived from 173 eligible articles. The objectives were (1) to analyze the different effects of straw return management on crop yield, soil nutrients, and SOC; (2) to analyze the different effects of manure application management on crop yield, soil nutrients, and SOC; (3) to analyze the effects of manure application plus straw return on crop yield, soil nutrients, and SOC; and (4) to provide valid references for acquiring win–win management between crop yield and SOC.

2. Materials and Methods

2.1. Data Collection

In the rudimentary stage, we conducted a search and identified a total of 9320 relevant articles using the following keywords: "straw", "manure", "straw plus manure", "crop yield", "soil nutrient", "soil organic carbon", etc. Relevant articles published from 2013 to 2023 were gathered from the China National Knowledge Infrastructure (CNKI) and the Web of Science. In detail, the articles were collected without review and communication, the language of the articles is English or Chinese, and specific search strings were used (e.g., TS = soil organic carbon or SOC or soil nutrient or yield or crop production or crop productivity and TS = crop residue or straw or stover or stalk or manure or organic fertilizer or slurry, etc.). The entire article search process is illustrated in a flowchart (Figures S1–S3). The data were sieved by conforming to the following rules: (1) The field experiment was conducted in China. (2) The article includes at least one of the following relevant indicators: crop yield, SOC, total nitrogen (TN), available phosphorus (AP), and available potassium (AK); to avoid confusion about the soil stock, the soil variables must be recorded as soil content, and all indicators must be observed or calculated. (3) The replication number must be ≥ 2 to ensure the function of the meta-analysis. (4) The treatments contained a control, namely no straw, manure, or chemical fertilizer application. (5) For crop data collection, only rice, wheat, and maize were considered. (6) SOC and soil nutrients were collected only from the 0–20 cm soil layer, regardless of crop category. (7) No other variables are present that may interfere with the analysis except for those necessary. For different amendments, the collection principles were distinct (Table S3). The details of these groupings are provided in the Supplementary Materials or the annotation under the following Figures.

An aggregate of 173 articles passed the standard screening process. We extracted the experiment location, soil and crop category, and replication number; straw return category, form, quantity, and duration; manure source, substitution ratio, and duration; nitrogen, phosphorus, and potassium fertilizer amount; and the mean and standard deviation (SD) of crop yield, SOC, TN, AP, and AK. GetData Graph Digitizer 2.26 was used to extract data in figures. We also used the function "refine category" of MetaWin 2.1 to deal with the missing data.

The collection of means and SDs was assisted by Formula (1) and Engauge Digitizer, Origin 2021. For the data that did not include the SD, if the standard error (SE) could be found, Formula (1) was used to obtain the SD; if the SD and SE could not be found in the article, we estimated the SD by calculating the 10% mean [38–40]. When only SOM could be found in the data, we used the method of multiplying SOM values with 0.58, a correction factor, to obtain the SOC values [41].

$$\mathrm{SD} = \mathrm{SE} \times \sqrt{\mathrm{n}} \tag{1}$$

In Formula (1), n means replication number.

The collected data were classified to analyze the effect on crop yield, SOC, and soil nutrients of TN, AP, and AK contents after the classification of straw, manure, and a combination of straw with manure application using assorted management strategies, respectively (Table S4).

2.2. Meta-Analysis

The method used to calculate effect size and sampling variance was the natural logarithm of the response ratio (lnRR) by MetaWin 2.1 for this study. The calculation process was operated automatically by the intrinsic program of MetaWin 2.1, which was appropriate for the form of the data being extracted. Formulas (2) and (3) were used to calculate the effect size and sampling variance, respectively, in the lnRR.

$$\ln \mathrm{R} = \ln\,(\mathrm{m_e}/\mathrm{m_c}) \tag{2}$$

In Formula (2), m_e and m_c are the sample means for the experimental and control groups.

$$\sigma_{\ln \mathrm{R}}{}^2 = \mathrm{s_e}^2/\mathrm{n_e}\mathrm{m_e}{}^2 + \mathrm{s_c}^2/\mathrm{n_c}\mathrm{m_c}{}^2 \tag{3}$$

In Formula (3), n_e and n_c are the sample sizes, s_e^2 and s_c^2 are sample variances for the two groups, and m_e and m_c are the same as above.

Next, the random effect model was applied, which can simultaneously consider the within-case and cross-case variations in the effect size. Summary analysis was conducted by imposing acquired effect size and sampling variance, and categorical analysis was used for cumulative effect value and its 95% confidence interval (CI) in this process. If the 95% CIs overlapped with 0, straw, manure, or straw plus manure application did not significantly influence the research indexes (crop yield, SOC, TN, AP, and AK); if the 95% CIs did not involve 0, the foregoing amendments had a positive (>0) or negative (<0) effect [38,42]. If the 95% CIs overlapped with the effect sizes at different levels of the same group in categorical analysis, there was no significant difference ($p < 0.05$) [43]. In order to ensure the reliability of the results, each category level can only be regarded as valid if the data volume is not less than 10 or if it comes from three different independent articles [44]. For results that meet the requirements, the sample size and effect value shown in brackets are presented in black on the plot; otherwise, the data are shown in red. Under this principle, we ensured the credence of these black-colored results; the red-colored results may possess little relevance but should still be further explored once sufficient samples are generated in the future. The effect value was present in the form of a percentage, which represented the improvement effects of different amendments; the percentage consisted of effect size and 95% CI. In crop yield, we used the following four different indicators: total yield, maize yield, wheat yield, and rice yield; the total yield means the average increase in crop yield without the consideration of crop type. Additionally, Q values for each group were displayed in each forest plot to demonstrate heterogeneity [45].

The forest plots and correlation were analyzed using GraphPad Prism 9.5 and SPSS Statistics 23. ArcGIS 10.8.1 was used to designate the experimental locations (Figure 1). Rosenberg's fail-safe number was used for testing the publication bias (Table S5) [46].

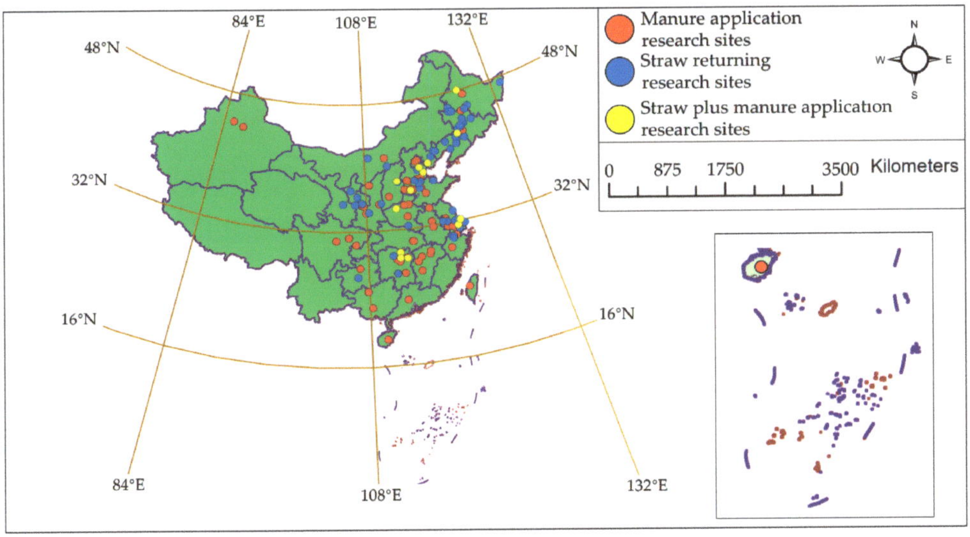

Figure 1. The research sites in China.

3. Results

3.1. Overall Effects of Straw Return on Crop Yield, SOC, and Soil Nutrients

3.1.1. Straw Return Effects on Crop Yield

In Figure 2, the forest plots showed the overall effect on improving crop yield, duration effect, quantity effect, form effect, and category effect of straw return in turn. Overall, straw return increased the crop yield by 14.4%. The rice, wheat, and maize yield significantly increased by 24.5%, 10.1%, and 12.1%, respectively. Among them, the response degree of rice was significantly higher than maize and wheat (Figure 2a).

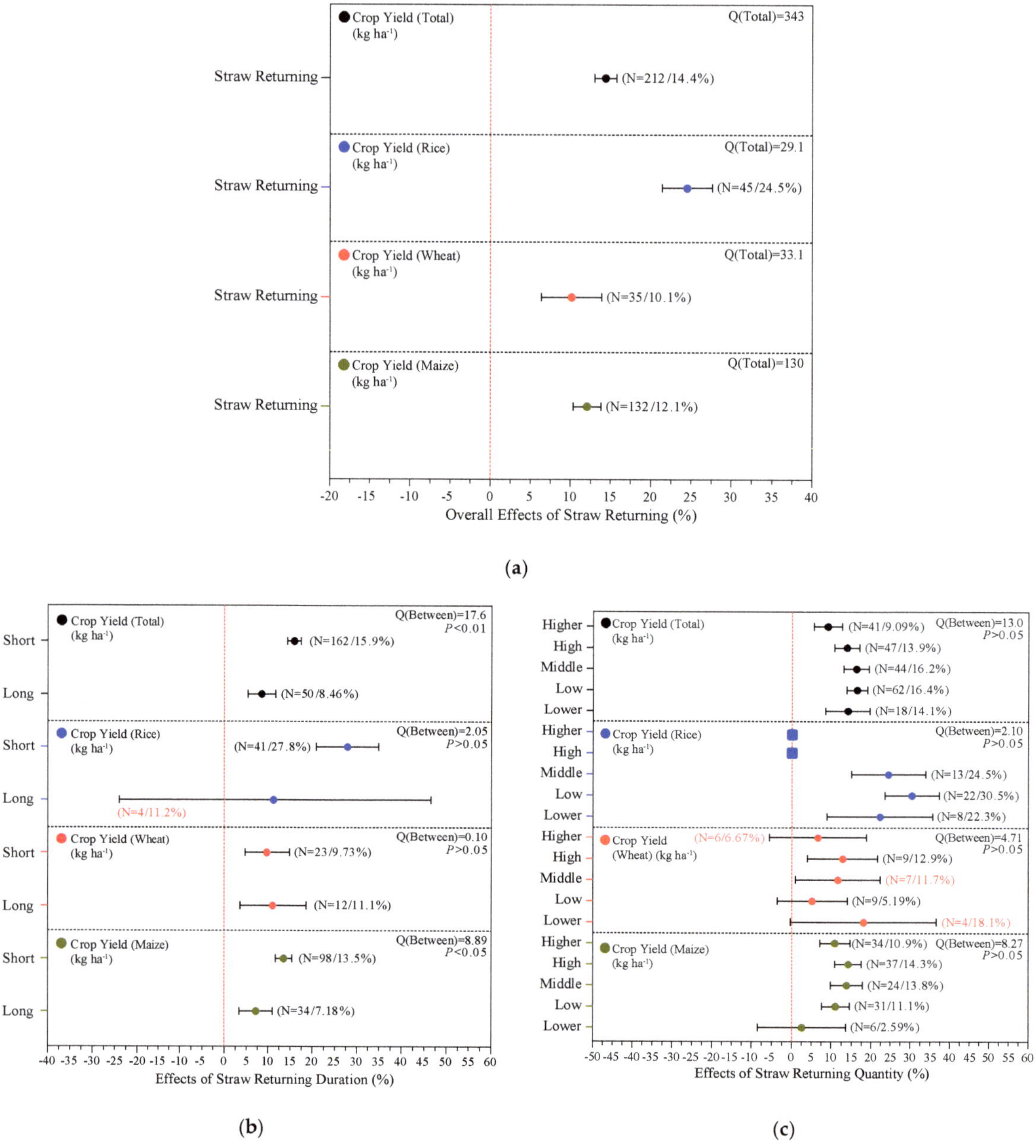

Figure 2. *Cont.*

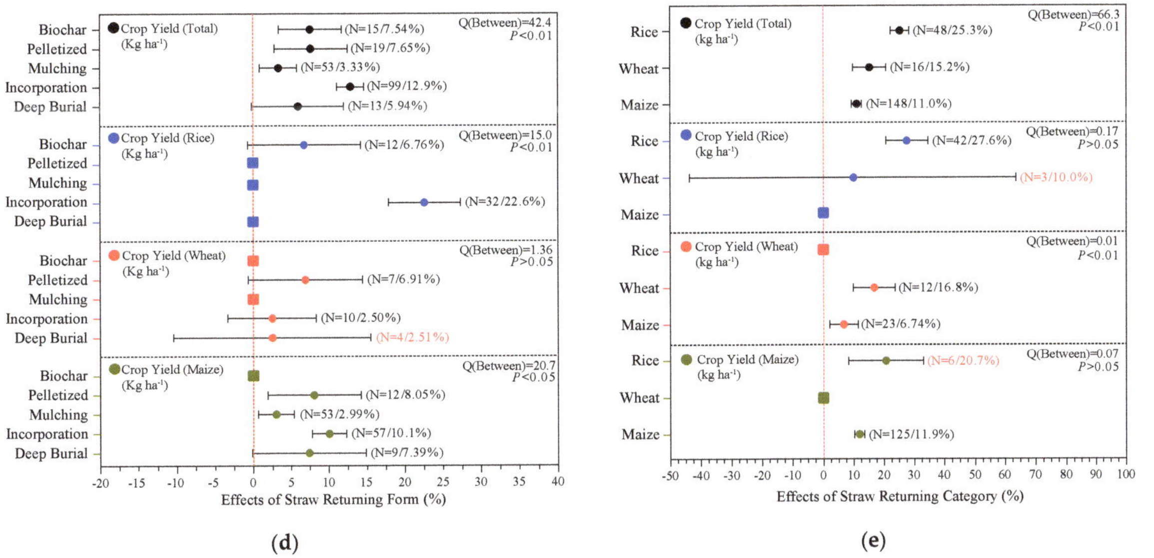

Figure 2. The forest plots of the effects of straw returning on yield. (**a**) Straw return overall effects on yield; (**b**) duration on yield; (**c**) quantity on yield; (**d**) form on yield; and (**e**) category on yield. In (**b**), long means ≥ 5 years, and short means < 5 years. In (**c**), higher, high, middle, low, and lower means ≥ 12,000 kg ha^{-1}, 9000–12,000 kg ha^{-1}, 6000–9000 kg ha^{-1}, 3000–6000 kg ha^{-1}, and <3000 kg ha^{-1}, respectively. Dot means effects values; black line means 95% CIs; square means void value; N means sample number; (%) means effect value; and all clusters were divided by different colors. Q (total) is the total heterogeneity, and Q (between) is the between-group heterogeneity; $p < 0.05$ indicates a significant difference in the group; $p < 0.01$ indicates a particularly significant difference in the group; and $p > 0.05$ indicates no significant difference in the group.

For the effect of straw return duration, except for the duration of ≥5 years, on increasing rice yield, all durations significantly increased crop yield. The duration of <5 years improved crop yield significantly more than ≥5 years overall. Compared with ≥5 years, <5 years significantly increased maize yield by 5.90% (Figure 2b).

All the quantity degrees of straw return possessed significantly positive impacts on total crop yield. Compared with the straw return quantity ≥12,000 kg ha^{-1}, 3000–6000 kg ha^{-1} and 6000–9000 kg ha^{-1} significantly increased crop yield by 6.70% and 6.52%, respectively. For individual crops, the quantity of <9000 kg ha^{-1} significantly increased rice yield; 6000–12,000 kg ha^{-1} significantly increased wheat yield; and ≥3000 kg ha^{-1} significantly increased maize yield (Figure 2c).

For the effects of form on crop yield, in the total crop yield, all forms significantly increased crop yield except deep burial of straw, and straw incorporation significantly enhanced crop yield by 9.26% compared with straw mulching. In terms of rice yield, straw incorporation significantly increased rice yield by 22.6%. Maize yield was significantly increased by pelletized straw, straw mulching, and straw incorporation; among them, straw incorporation significantly promoted maize yield by 6.90% compared with straw mulching (Figure 2d).

In terms of the relation of straw category to total crop yield, all straw categories significantly increased total yield, and rice straw significantly enhanced yield by 8.77% and 13.9%, respectively, compared with wheat straw and maize straw. Depending on the cultivation crop to classify, rice yield was significantly increased by rice straw; wheat yield was significantly increased by wheat and maize straw; and maize yield was significantly increased by rice and maize straw (Figure 2e).

3.1.2. Straw Return Effects on SOC and Soil Nutrients

In Figure 3, the forest plots showed the overall effect on improving the SOC and soil nutrients, duration effect, quantity effect, form effect, and category effect of straw return in turn. Overall, the contents of AK, AP, TN, and SOC were significantly increased by straw return. Among them, the degree of response of TN content to straw return was significantly less than other elements; SOC response was significantly higher than TN but lower than AP and AK (Figure 3a).

Figure 3. *Cont.*

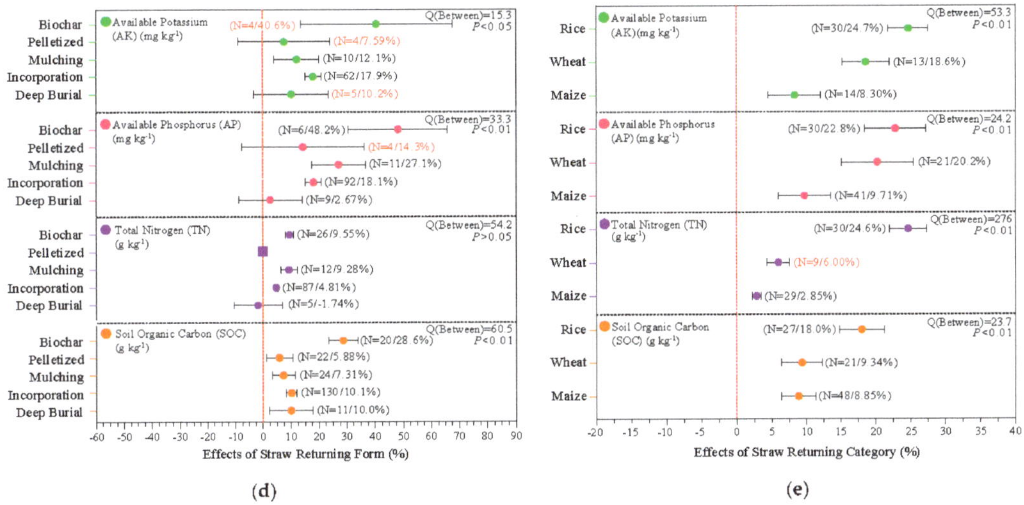

Figure 3. The forest plots of the effects of straw return on SOC and soil nutrients. (**a**) Straw return overall effects on SOC and soil nutrients; (**b**) duration on SOC and soil nutrients; (**c**) quantity on SOC and soil nutrients; (**d**) form on SOC and soil nutrients; and (**e**) category on SOC and soil nutrients. The rest of the notes are the same as in Figure 2 and were omitted for clarity.

Regarding the straw duration role in the content of SOC and soil nutrients, compared with the control, all durations significantly and positively affected the contents of AK, AP, TN, and SOC. Regarding increases in TN, compared with the straw return duration <5 years, ≥5 years significantly promoted TN by 7.64% (Figure 3b).

In terms of the effects regarding different straw return quantities, the quantity of 3000–12,000 kg ha^{-1} significantly increased the contents of AK, AP, TN, and SOC. Compared with the quantities of 6000–9000 kg ha^{-1} and ≥12,000 kg ha^{-1}, TN was increased by 3000–6000 kg ha^{-1} and 9000–12,000 kg ha^{-1} significantly (Figure 3c).

Among different straw return forms, straw biochar, straw incorporation, and straw mulching significantly increased the contents of AK, AP, and TN; all forms significantly enhanced SOC. Regarding the contents of AP, TN, and SOC, straw biochar significantly and positively affected AP more than straw incorporation and deep burial of straw; straw incorporation significantly decreased TN compared with straw mulching and straw biochar; and straw biochar significantly increased SOC compared with other forms. Namely, under equal conditions, these proper forms can increase the foregoing indicators with more efficiency compared with other forms (Figure 3d).

Regarding the impact of the straw category on SOC and soil nutrients, all categories significantly increased the contents of AK, AP, TN, and SOC. Rice and wheat straw significantly enhanced AK, AP, and TN compared with maize straw; rice straw significantly increased TN by 17.5% and 21.1% compared with wheat and maize straw, respectively; and wheat and maize straw significantly increased SOC less than rice straw (Figure 3e).

3.2. Overall Effects of Manure Application on Crop Yield, SOC, and Soil Nutrients

3.2.1. Manure Application Effects on Crop Yield

In Figure 4, the forest plots show the overall effect on improving crop yield, source effect, substitution ratio effect, and duration effect of manure application in turn. Overall, manure application increased the crop yield by 70.4%. The rice, wheat, and maize yield significantly increased by 47.0%, 78.0%, and 76.9%, respectively. Among them, the response degree of rice was significantly lower than maize and wheat (Figure 4a).

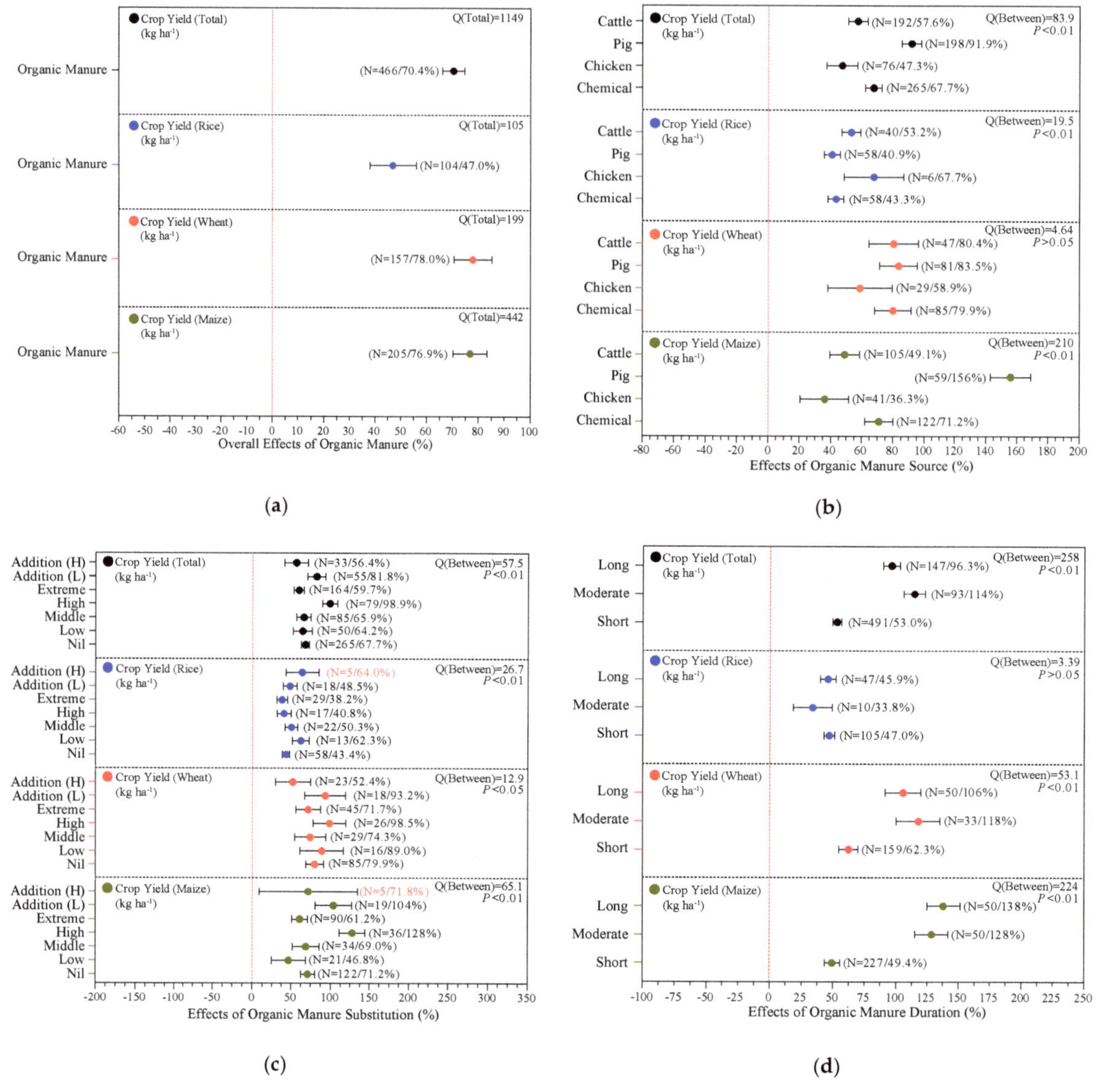

Figure 4. The forest plots of the effects of organic manure on yield. (**a**) Manure application overall effects on yield; (**b**) source on yield; (**c**) substitution ratio or additional replenishment on yield; and (**d**) duration on yield. In (**c**), nil, low, middle, high, extreme, addition (L), and addition (H) mean 0%, 0–30%, 30–50%, 50–80%, 80–100%, <30 t ha^{-2}, and ≥30 t ha^{-2}, respectively. In (**d**), long means ≥ 20 years, moderate means 10–20 years, and short means < 10 years. The rest of the notes are the same as in Figure 2 and were omitted for clarity.

For the effects of different manure sources, all sources significantly increased crop yield. In the total crop yield, compared with chemical fertilizer, pig manure significantly increased crop yield by 14.4%, but chicken manure significantly decreased crop yield by 12.2%. For individual crops, pig manure significantly increased rice yield less than cattle and chicken manure, but pig manure significantly increased maize yield more than other sources; moreover, chemical fertilizer significantly increased maize yield compared with cattle and chicken manure (Figure 4b).

Regarding increases in crop yield, the forest plot was made to determine the impacts of substitute proportion change. Compared with the control, all substitution ratios significantly and positively increased crop yield; except in the case where <30 t ha^{-2}, the substitution ratio 50–80% significantly increased total crop yield compared with other ratios. Under crop categorization, the substitution ratio <30% significantly increased rice yield more than the 0% substitution ratio; the wheat yield was significantly increased by the ratio of 50–80% compared with ≥30 t ha^{-2}; and the substitution ratio 50–80% significantly promoted maize yield by 40.5% and 41.4% compared with <50% and >80%, respectively (Figure 4c).

Finally, the impacts of different durations of manure application on yield were determined, and all durations significantly increased crop yield. The duration of 10–20 years significantly increased total crop yield by 9.02% and 39.9% compared with ≥20 years and <10 years, respectively; the duration of ≥20 years also significantly increased total crop yield by 28.3% compared with <10 years. In terms of wheat and maize yield, the duration of <10 years decreased yield compared with 10–20 years and ≥20 years significantly (Figure 4d).

3.2.2. Manure Application Effects on SOC and Soil Nutrients

In Figure 5, the forest plots showed the overall effect on improving the SOC and soil nutrients, source effect, substitution ratio effect, and duration effect of manure application in turn. Overall, the contents of AK, AP, TN, and SOC were significantly increased by manure application. Among them, the response degree of AP content to manure application was significantly higher than other elements; AK responded significantly more than TN and SOC but less than AP (Figure 5a).

In terms of the effects of manure sources on SOC and soil nutrients, all sources significantly increased the contents of AK, AP, TN, and SOC. Regarding increases in AP, pig manure significantly increased AP by 59.7% and 63.7% compared with cattle manure and chemical fertilizer, respectively. Chicken manure significantly increased TN content compared with other sources; cattle and pig manure significantly and positively affected TN more than chemical fertilizer but less than chicken manure. All manure sources significantly increased SOC content compared with chemical fertilizers (Figure 5b).

Regarding the effects of the substitution ratio on SOC and soil nutrients, AK was significantly enhanced by the ratios except <50%; all ratios significantly increased the contents of AP, TN, and SOC. For individual elements, the ratio of 50–80% significantly increased AK by 54.0% and 36.2% compared with <30% and >80%, respectively; AP was significantly promoted by 68.6% and 52.4% through the ratio of 50–80% compared with <50% and >80%, respectively; and compared with 0%, except when <30%, all ratios significantly increased TN and SOC (Figure 5c).

Under the impact of duration variation, all durations significantly increased the contents of AK, AP, TN, and SOC. Compared with the duration of <10 years, 10–20 years and ≥20 years significantly increased the contents of AK, AP, and SOC; regarding TN, the duration of ≥20 years significantly enhanced TN by 11.4% (Figure 5d).

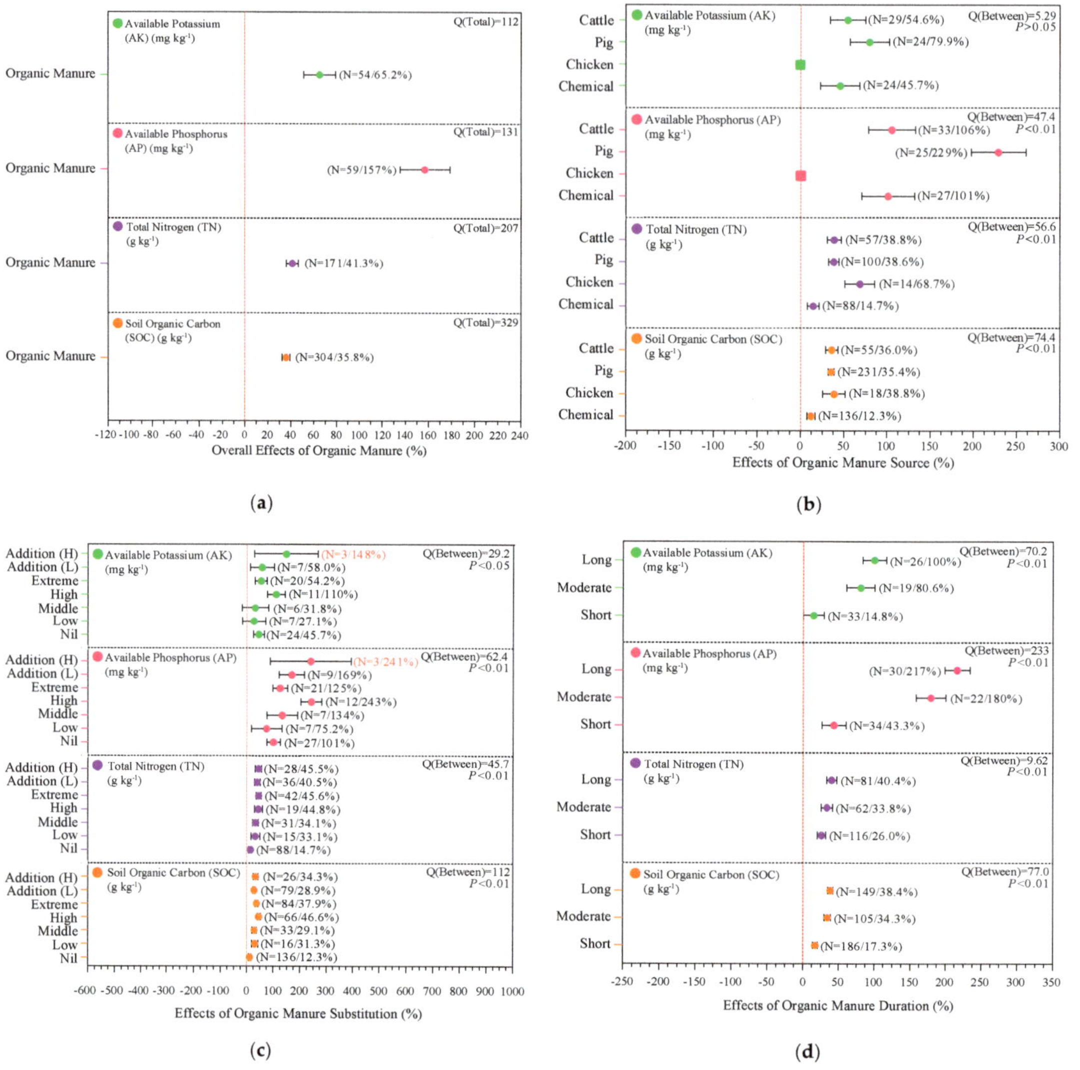

Figure 5. The forest plots of organic manure on SOC and soil nutrients. (**a**) Manure application overall effects on SOC and soil nutrients; (**b**) source on SOC and soil nutrients; (**c**) substitution ratio on SOC and soil nutrients; and (**d**) duration on SOC and soil nutrients. The rest of the notes are the same as in Figure 4 and were omitted for clarity.

3.3. Overall Effects of Straw Plus Manure on Crop Yield, SOC, and Soil Nutrients

Compared with the control, straw returning plus manure application significantly increased the crop yield of total, rice, wheat, and maize by 35.0%, 40.8%, 37.2%, and 27.9%, respectively (Figure 6a).

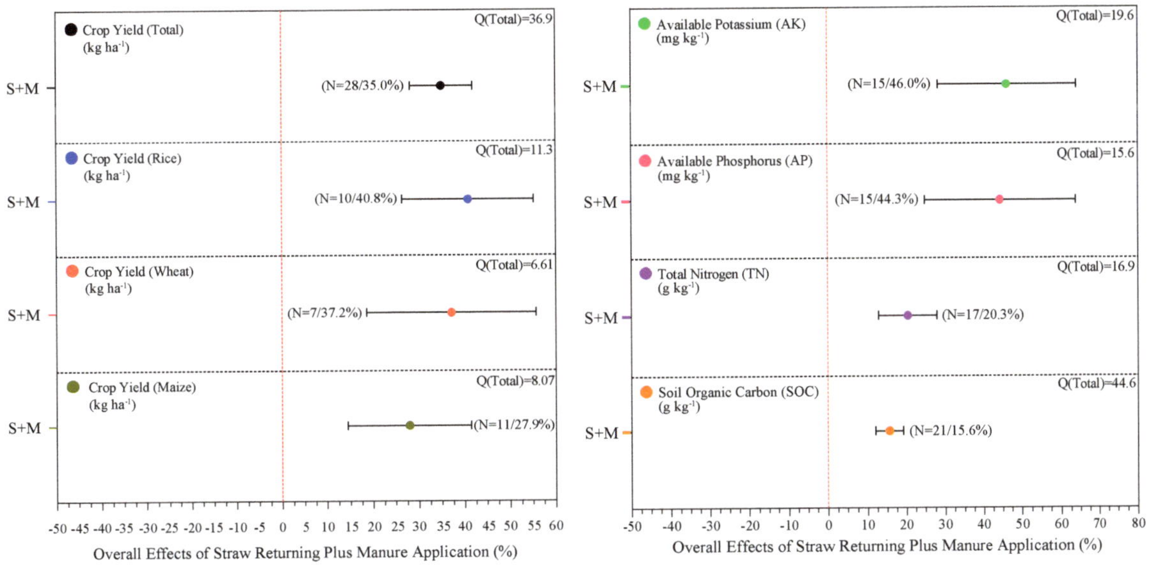

Figure 6. (**a**) The forest plot of straw returning plus manure application to yield; (**b**) SOC and soil nutrients. S + M means straw plus manure. The rest of the notes are the same as in Figure 2 and were omitted for clarity.

Comprehensively, compared with the control, straw returning plus manure application significantly increased the contents of AK, AP, TN, and SOC by 46.0%, 44.3%, 20.3%, and 15.6%, respectively. The response degree of SOC content to straw returning plus manure application was significantly lower than for AK and AP; moreover, AK significantly responded more than TN (Figure 6b).

3.4. Correlation Analysis

The correlation coefficients between the net increase in crop yield and net increase in SOC, TN, C/N, AP, and AK were calculated and divided into straw return (Figure S4) and manure application groups (Figure S5).

In the straw return group, SOC and soil nutrients had significant and positive effects on the net increase in crop yield. The correlation coefficient of net-increased SOC, TN, C/N, and AK were more than 0.42, while AP was the lowest (0.22) (Figure 7a).

In terms of manure application, the net-increased C/N was irrelevant to crop yield. The correlation coefficients of the net-increased SOC and TN both were significantly positive to the net-increased crop yield, which reached 0.35 and 0.33, respectively. The biggest correlation was 0.40 of the net-increased AK (Figure 7b).

Overall, the significance and coefficients of different factors regarding crop yield under straw return or manure application were determined. Under the straw return treatment, the increase in SOC and soil nutrients was more closely related to yield.

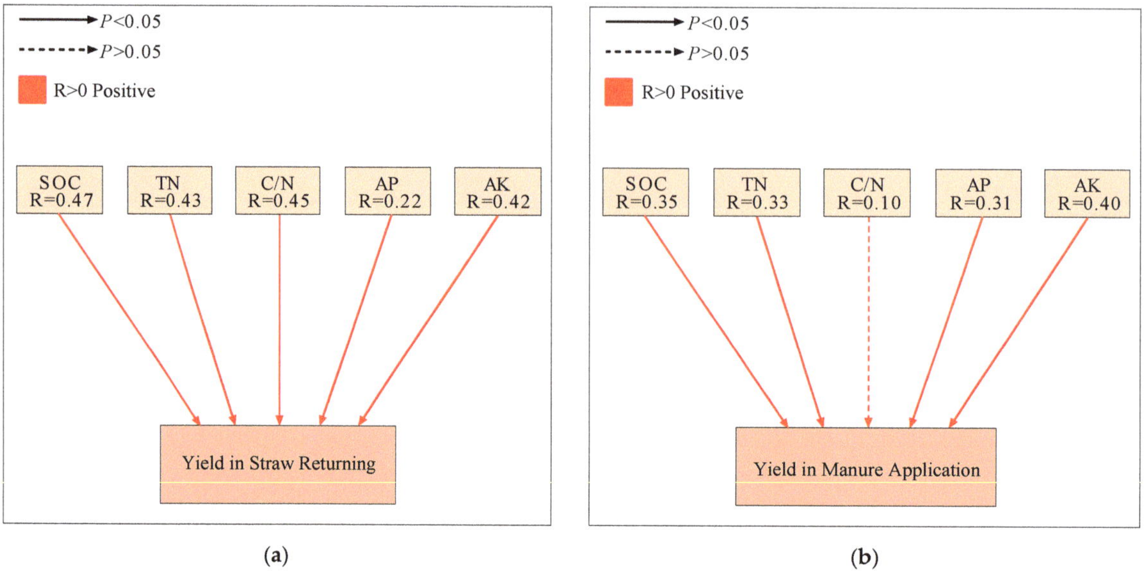

Figure 7. (**a**) The correlation between different factors and crop yield under the straw return application and (**b**) under the manure application. Full line and dashed line, respectively, mean the correlation was significant ($p < 0.05$) and not significant; the red color means positive effect; and R means correlation coefficient.

4. Discussion

4.1. Overall Effects of Straw Return on Crop Yield, SOC, and Soil Nutrients

4.1.1. Straw Return Overall Effects

Straw return can significantly improve crop yield, SOC, and soil nutrients [14,47–50]. Through our meta-analysis, the ranks of the overall effect of straw return on improving the yield of different crops and the content of different soil indicators were as follows: rice (24.5%) > maize (12.1%), wheat (10.1%); AK (18.1%), AP (16.6%) > SOC (11.3%) > TN (4.48%). It is important to note that the ranking pertains to the magnitude of improvement attributed to straw return across various parameters. The ranking of straw return's overall effect on enhancing the yield of different crops and the levels of various soil indicators is based on significant differences.

The yield-increasing effect of rice was significantly greater than maize and wheat after straw return, which possibly resulted from the cultivation conditions. The pH of paddy soil pH is 5.5–6.5, presenting slight acidity. The increased yield effect of straw return in acidic soils may be higher because crop straw is an alkaline material, and its return to the field reduces soil acidity, enhancing the retention capacity of soil to fertility [48]. Moreover, straw return increased maize yield slightly more than wheat yield, which was mainly affected by different growth seasons. Wheat is cultivated in winter or spring, which possess lower precipitation and temperature compared with the growth season of maize—summer. Lower precipitation and temperature are not beneficial to the acceleration of straw decomposition, thus impacting the release of straw nutrients [39]. AK and AP increased more compared with the SOC under straw return, which may be attributed to the quickly released minerals contained within straw; the release of available nutrients mainly occurs through chemical decomposition and microbial activity, and the processing is comparatively faster than SOC accumulation. In terms of the difference between SOC and TN, the high C/N of straw could be one major reason. When straw is added to soil, microorganisms use up available nitrogen to decompose the straw, temporarily slowing the short-term increase in TN [51]. This shows that the improvement effect of straw return on different indicators is influenced by their characteristics. However, the improvement effect is also affected by changes in environmental and management factors. For instance, the effectiveness of straw return on increasing SOC is significantly influenced by the initial SOC levels, with soils having lower initial SOC showing greater response ratios [52]. Proper use of nitrogen fertilizer can optimize soil C/N and boost microbial activity, thereby facilitating the nutrient release of straw [38]. Moreover, compared with no-tillage practices, tillage could mix straw into the deeper soil layer, making soil nutrients well distributed and improving the straw decomposition rate [53]. In practical production, these factors should also be considered under straw return.

4.1.2. Straw Duration

Our meta-analysis found the straw return of <5 years significantly increased crop yields compared with ≥5 years. This was mainly because of the high C/N of straw [17]. Continuous straw return results in relatively more exogenous carbon and less nitrogen inputs [38]. In order to maintain growth, the soil microbe will begin to compete with crops for nitrogen, which impacts crop yields [7]. Regarding soil nutrients, ≥5 years was more effective in enhancing soil nutrient levels, particularly TN. The net-increased TN was possibly generated from the remaining straw after long-term straw return treatment. In terms of SOC, there was no significant difference between <5 years and ≥5 years in improving its content. This result for SOC could be attributed to the soil gradually reaching a carbon threshold with extended periods of straw return, which may limit further increases in SOC [38]. Moreover, long-term straw return can enhance the potential for carbon decomposition. In contrast, only a small proportion of carbon remains in the soil as stable carbon, reducing the carbon sequestration effect [20].

From our meta-analysis, a universal conclusion could be found. Namely, the duration < 5 years increased crop yield more, and the duration ≥ 5 years increased soil nutrients more. However, the determination of the optimal duration always needs to consider the local environment. Many studies have waged an exploration based on their local environment [25,38,39]. Under long-term straw return, Goran et al. and Wang et al. discovered the content of TN significantly increased [14,25], which was similar to our outcome. One study suggested that long-term straw returning over 20 years could considerably increase SOC content by two or three times compared with short-term straw returning [52]. However, the SOC will decrease rapidly once the continuous straw return stops due to the introduction of numerous labile carbon compounds and the acceleration of "priming effects" [41]. In a rain-fed upland, the practice of long-term rice straw return was used to increase maize productivity; although compared with the control, straw return significantly increased maize yield, a year-on-year comparison within the straw return group showed a decline over the years [51]. Jiang et al. and Wang et al. found that straw return significantly increased crop yield after continuous return for 5 years [54,55]. Altogether, and based on other experiments, the regulation of straw returning duration needs to also consider the local conditions of soil properties and meteorology to determine the most suitable duration.

4.1.3. Straw Quantity

As an important index and controllable variable of weighing crop residue return effects, the quantity of straw is one of the most popular topics in the subject of returning straw to the field. The change in returning straw quantity can always significantly affect crop yield, SOC, and soil nutrient content [14,20,56,57]. Many studies have revealed that a higher quantity of straw return has an improvement role [7,49,56]. However, theoretically, there is no linear relationship between crop yield, SOC, soil nutrients, and straw returning quantity [15,47]. Moreover, in terms of the economic dimension, an excessive returning straw quantity can decrease the potential economic benefit of farmers. Therefore, the most vital crux is to identify a proper quantitative level for straw return based on the premise of balancing improvement and economy.

Based on the data that we compiled, the alteration of straw return quantity significantly affected TN content and crop yield. Initially, the increase in straw return quantity could significantly improve TN content and crop yield. After arriving at a limitation, there was no significant increment; furthermore, there was even a reduction in the increasing rate with superfluous straw quantity. In the meta-analysis, the return of 3000–6000 kg ha^{-1} of straw to the field significantly increased the total crop yield and TN content compared with the higher returning quantity. Regarding the improvement in other indicators, no significant difference among different quantities was found. Similar to the continuous straw return, when an excessive amount of straw is returned to the field, it induces competition between microbes and crops for available nitrogen [7,38]. This may be the main reason why the returned straw amount of 3000–6000 kg ha^{-1} could best increase the total crop yield. In conclusion, following the collection of an enormous amount of data, a universal returning quantity was gained from our meta-analysis. The quantity of 3000–6000 kg ha^{-1} could effectively increase crop yield and TN while arriving at the same effect and saving straw resources compared with other quantities on the improvement for all indicators. Therefore, the quantity of 3000–6000 kg ha^{-1} of straw return could be recommended based on the premise of balancing improvement and economy.

However, in specifically local conditions, the best straw quantity always varies with the cultivation environment. As the temperature decreased and SOC content increased in the experimental area, the applicable quantity of straw return also decreased. For example, research results from Ustalfs, an area in a warm temperate zone with low SOC content, revealed that the highest maize straw return of 13,500 kg ha^{-1} resulted in the most marked increase in TN and SOM [14]. In the cultivation system of a calcareous, subtropical zone, returned maize straw treatments of 5000 kg ha^{-1} and wheat straw treatments of 6000 kg ha^{-1} enhanced the content of SOC more than other lower quantitative treatments [56]. In the results of research conducted in Chinese semiarid areas, warm temperate zones, and low SOC content areas, treatment with 13,500 kg ha^{-1} of maize stalk incorporation accomplished the most effective boost in the contents of SOC and TN under the maize–millet rotation system [57]. The conditions of maize monoculture, cold monsoon climate, and treatment with the 2500 kg ha^{-1} of maize straw on farmland can significantly enhance the relative abundance of fungi under the no-tillage condition, which benefits the increase and immobilization of SOC [15]. In terms of saline–alkali paddy soil in a cold monsoon climate, the treatment of 7300–7500 kg ha^{-1} of rice straw decreased crop yield compared with the 5475–5625 kg ha^{-1} treatment [47]. Ultimately, because of China's complicated terrain and climate conditions, soil type varies in China [58]. The effect of increasing straw return quantity should also incorporate a range of factors, including soil physicochemical characteristics and meteorological and management conditions [38].

4.1.4. Straw Form

A change in returning form can considerably affect the improvement in straw return on crop yield, SOC, and soil nutrients [21,42,59–63]. Different straw return forms each have their unique advantages. Compared with other forms, straw mulching has shown superiority in reducing water evaporation and increasing soil moisture, thereby enhancing crop yield and water use efficiency [50]. Furthermore, straw mulching has been observed to lower soil temperature, which is beneficial for ensuring crop growth and soil moisture during the high temperatures of the summer [51]. In contrast to mulching, straw incorporation increases the contact area between straw and soil microorganisms when the straw is buried at the soil layer of 0–20 cm, which accelerates the decomposition rate of straw [7]. One study has suggested that due to its water conservation capabilities, straw mulching may be more suitable for areas with insufficient precipitation, while straw incorporation might be better suited for regions with sufficient precipitation [38]. Deep straw burial enhances subsurface microbial activity, boosting microbial metabolism and SOM formation compared with straw mulching; comparatively, straw mulching leaves straw on the surface in a semi-dry status, resulting in more carbon and nitrogen loss through gas during decomposition [64]. Straw biochar significantly enhances soil nutrient effectiveness through its high adsorption capacity, engaging in nutrient cycling by adsorbing key nutrients (nitrogen, phosphorus, and potassium) and facilitating ion exchange [21]. Pelletized straw disrupts the original structure and cuticle of straw, enhancing the contact area with soil and improving decomposition rates by increasing microbial biomass [65,66]. Moreover, after straw is processed into pellets, its release of carbon and nitrogen from the straw is enhanced due to the pelletization process breaking down the original adhesive structure of the straw [67]. It also helps preserve fine straw particles by forming soil aggregates, reducing carbon loss; the pelletized straw boosts SOC more effectively than unprocessed straw [68]. However, although one study discovered that the pelletized straw was observed to significantly increase grain yield in the short term more than incorporation, straw incorporation showed a long-term, slow-release benefit in enhancing SOC levels over the years [69].

Based on the results of our meta-analysis on the increase in total crop yield and maize yield, straw incorporation significantly and positively affected the yield compared with straw mulching. The possible reason for this difference is that straw mulching decomposes slower than straw incorporation, making the straw less available to crops; this may result in nutrient deficiencies and crop yield decrease [50,70,71]. The straw biochar and the pelletized straw possessed a similar effect value on each other; the effect of straw biochar on improving crop yield is contingent on a lot of external factors such as the category of soil and environmental conditions [72]. In SOC and soil nutrients, straw biochar had the biggest effect value on SOC, but there was no significant difference with straw mulching on TN, AP, and AK. Straw biochar can significantly uplift the contents of available soil nutrients and SOC sequestration due to its physicochemical characteristics [72,73]. Compared with straw incorporation, straw mulching significantly increased TN, but there was no significant difference between straw incorporation and straw mulching in improving AP, AK, and SOC. Different forms have varying impacts, each with its own best improvement in different indicators (SOC, AP, AK, TN, maize yield, wheat yield, and rice yield). Through a comparison, our findings show that straw incorporation boosted crop yield more effectively, while straw biochar enhanced SOC more efficiently compared with other forms.

However, the choice of straw return form should also be based on different objectives and environmental conditions. For instance, under cultivation conditions with limited soil moisture, opting for straw mulching might be more beneficial for ensuring crop growth compared with straw incorporation, pelletized straw, or deep burial of straw. Moreover, many experiments also explored the best form based on their local conditions. For example, straw mulching has been found to enhance SOC levels, leading to improved soil fertility and nutrient availability, particularly in dryland regions and under no-till systems [21,74,75]. In the rice–wheat system, straw incorporation can lead to the retention of SOC from farmyard manure or crop straw, limiting nutrient loss and increasing SOC sequestration [61]. The outcome of a study on Argiudolls suggested that deep straw burial can increase SOC and TN contents compared with other straw incorporation methods [64]. In the paddy soils of northeast China, biochar treatment was superior to straw in terms of SOC accumulation and increasing soil fertility [76]. Except for the foregoing forms, pelletized residue return was salutary to the rapid increment in soil nutrients, SOM, and crop yield of cultivated hibernal wheat [68]. In a warm temperate continental monsoon climate, under the winter wheat–summer maize rotation system or monoculture maize cultivation, pelletized straw significantly increased crop yield and SOC content [69,77].

4.1.5. Straw Category

As three of the most predominant crops in China, rice, maize, and wheat constitute more than three-fourths of China's total crop residues [38], which is why these crops are the major subjects of straw returning research [37,78,79]. The effects of straw return vary with the change in straw category. Under the rice–wheat rotation system, a 10-year experiment indicated rice straw was better than wheat straw in improving SOC and other nutrient properties of soil [80]. In the wheat–maize rotation system, the combination of wheat and maize straw return diversified the abundance of soil bacteria and fungi before wheat cultivation; compared with the combination, the sole application of maize straw led to the decrease, which was regarded as unfavorable toward the sequestration of SOC [81]. The disposal of blending maize and wheat stalk return could reinforce the formation of soil aggregate, finally boosting internal SOC storage and crop yield under the integrated application of controlled-release nitrogen fertilizer [1].

In our meta-analysis, the improvement effect of the three categories of straw presented a tendency: rice straw > wheat straw > maize straw overall. Upon comparing the obtained results, it was found that the rice straw possessed the optimal effect for improving crop yield, SOC, and soil nutrient content. This distinction could possibly be attributed to their different physicochemical characteristics. In the results of the nutrients of crop residues, it was found that the rice, wheat, and maize straw were similar in carbon content but significantly different in the content of nitrogen, phosphorus, and potassium. Many articles pointed out that rice straw contained more potassium and phosphorus compared with wheat and maize straw [17,82]. Crops could ingest and assimilate the nutrients generated in the decaying process of rice straw better; rice straw could also foster the activity of microbes, releasing more nitrogen than wheat straw [80]. Moreover, the decomposition of rice straw was quick, and its nutrients were liable to be mineralized easily; thus, the employment of rice straw was conducive to the recovery of soil fertility [47]. The maize straw represented the lowest effect value in our study, which may be due to the slow decomposition. According to previous research, maize straw decomposed slower than wheat straw [83]. A study found that maize straw decreased the diversity of soil fungi compared with wheat straw; theoretically, fungi tend to decompose complex organic matter [81]. In the physical structure, the internal structure and pore characteristics of crop residues such as rice straw, wheat straw, and maize straw differ significantly. This leads to variations in their specific surface area, pore volume, average pore size, cumulative pore volume, total pore area, porosity, etc. These characteristics also influence their compactness, which is reflected in different bulk densities. The bulk density of maize straw is greater than that of wheat straw, which means greater compactness; however, greater compactness may not be conducive to decomposition [17].

4.2. Overall Effects of Manure Application on Crop Yield, SOC, and Soil Nutrients

4.2.1. Manure Application Overall Effects

The employment of manure could significantly increase soil fertility, crop yield, and SOC sequestration [84–86]. Based on the results of our meta-analysis, the ranks of the overall effect of manure application on improving the yield of different crops and the contents of different soil indicators were as follows: wheat (78.0%), maize (76.9%) > rice (47.0%); AP (157%) > AK (65.2%) > TN (41.3%), SOC (35.8%). Similar to the straw return, the ranking is based on the improvement role of manure and significant differences.

The results convey that the wheat yield increased more than the maize yield, and the possible reason could be the difference in growing season and fertilization management. The response of rice yield to manure application was lower than that of maize and wheat, which was similar to the results of a previous study [87]. This result implies that the rice yield increased under the treatment of organic manure plus chemical fertilizer but decreased under the full employment of manure. The high requirement of soil nutrients in rice growth and the sluggish process of manure nutrient release were inferred as the possible causes. In terms of the content of SOC and soil nutrients, the results of our meta-analysis were similar to the results of the previous study [88]. The difference among various elements may be because of their different characters and different manure constitutions.

4.2.2. Manure Source

Different types of organic manure can have varying nutrient contents and physical properties, which can affect their impact on soil and crops. In acidic paddy soil, the additional replenishment of cattle manure on chemical fertilizer significantly enhanced the rice yield, TN, AP, AK, and SOC storage, which was conducive to increasing soil fertility and SOC sequestration [89]. Under rice–wheat rotation cultivation, pig manure plus chemical fertilizer could significantly increase SOC content and AP more than chemical fertilizer alone, but no significant difference in TN and AK was found [90].

In our meta-analysis, compared with the chemical fertilizer, pig manure had significantly greater effects on increasing the crop yield of total and maize indicators and the contents of AP, TN, and SOC; cattle and chicken manure significantly enhanced TN and SOC. The group of cattle, pig, and chicken manure contained treatments that partially substituted the chemical fertilizer via the corresponding manure source, which was the main reason why the pig source acquired a better effect value than the chemical fertilizer. The effect of pig manure plus chemical fertilizer was also similar to the results of the precedent study [91], which indicated that manure plus chemical fertilizer achieved a better effect value. Since the precedent study lacked classification of the manure category, it reported different effect values compared with our study. The difference in effect values among different manure categories could possibly be attributed to the difference in physical and chemical characteristics regarding cattle, chicken, and pig manure.

According to the results of the present study, pig manure could be regarded as the suitable manure source for the crop yield of total and maize indicators and the content of AP; however, chicken manure was effective in enhancing the content of TN compared with other manure sources. Since the organic manure treatments included different replacement degrees of chemical fertilization, partial manure treatment attained more remarkable effects than chemical fertilizer in terms of improving yield. However, the analysis results convey that it is not apt to recommend the total substitution of chemical fertilizers with manure; instead, the content of chemical fertilizer could be reasonably reduced after optimization and then substituted by manure to increase the crop yield. Namely, the use of a manure source could be better than using chemical fertilizer provided a proper substitute is used; therefore, the results of our meta-analysis of the use of manure sources demonstrate that pig, cattle, and chicken manure are not conclusively better than chemical fertilizer without the consideration of proportion between manure and chemical fertilizer.

4.2.3. Manure Substitute Ratio of Chemical Fertilizer

Depending on a multitude of data, the partial substitution of organic fertilizer with chemical fertilizer could acquire a higher crop yield and considerably increase SOC content and sequestration and soil fertility [35,36,92]. For example, in the rice inter-row planting experiment, partial use of organic fertilizer instead of chemical fertilizer significantly improved soil quality and accelerated soil fertility recovery [32]. In addition, under the maize cultivation system, organic fertilizer replaced 50% of chemical fertilizer, which significantly improved soil fertility and crop yield [92]. The ramifications from northern China clarified that the combination of 50% organic fertilizer and 50% chemical fertilizer can uplift the labile organic and mineral nitrogen pool while decreasing the risk of nitrogen loss, which increases the stability of soil fertility under intensive cultivation conditions [33].

The results from our meta-analysis displayed that the substitution ratio of 50–80% could significantly increase the crop yield of total and maize indicators and the contents of AK, AP, TN, and SOC compared with the ratio of 0%, namely chemical fertilizer application alone. In contrast to other substitution ratios, the substitution ratio of 50–80% significantly increased most of the indicators compared with chemical fertilizer. However, the ratio of 50–80% performed worse than the ratio <30% in the group of rice yield, and it did not possess a significant increasing role compared with other ratios in the group of TN, which showed similarity with the earlier results [93,94]. The results from the wheat–maize rotation system showed that the 25-year substitution ratio of 50–80% of organic manure to chemical fertilizer increased AP, AK, and wheat yield by 23.2%, 186.5%, and 19.0%, respectively, and decreased SOC and TN by 6.84% and 31.3%, respectively [93]. In acidic paddy soil, the substitution ratio of 50–80% increased the foregoing indicators more when compared with the ratio of <30% and 30–50% [94]. As seen from the correlation analysis, the net increases in AK, AP, TN, and SOC were significantly relevant to the net increase in crop yield. The ratio of 50–80% significantly increased AK, AP, TN, and SOC, which could be positive for crop yield increase. Thus, although the ratio of 50–80% cannot increase rice yield better than the ratio <30%, it increases other indicators more than the rest of the ratios compared with chemical fertilizer. Therefore, a ratio of 50–80% is recommended as the suitable ratio. The specific ratio used should be based on the soil and crop types.

4.2.4. Manure Duration

As one of the critical ingredients in soil amendments, the duration of manure application could influence the effects of manure, which further affects crop yield, SOC sequestration, and soil nutrients [95]. So far, many studies have demonstrated the effects of long-term manure application on crop yield and soil [88,90,95,96]. However, a study discovered that inordinately long-term manure application decreased the effect of manure application on the improvement in SOC content compared with shorter application durations [91]. Therefore, the determination of manure application duration is very important to the optimal performance of organic amendments.

Compared with the duration < 10 years of manure application, the results from our meta-analysis revealed that the durations ≥ 20 years and 10–20 years significantly increased the crop yield of total, wheat, and maize indicators, but there was no significant difference in the rice yield. In terms of SOC and soil nutrients, the duration ≥ 20 years significantly increased SOC, TN, AP, and AK; the duration of 10–20 years was insignificantly positive in promoting TN; and the result was similar to the previous study [88]. The increment performed as a regular tendency, which gradually increased with the advance of time. As seen in our correlation analysis, the increase in SOC and soil nutrients positively affected the increase in crop yield, which can explain the promotion of maize and rice yields with the rise in SOC and soil nutrient content [55]. The rice, wheat, and maize yield was not significantly increased by the duration ≥ 20 years compared with the duration of 10–20 years. The main possible reason is that after long-term cumulation, soil nutrients and SOC are less crucial than climate, water availability, etc. [7,55]. In terms of the results of this study and other relevant previous studies, since the duration 10–20 years significantly increased the yield of the total indicator compared with the duration ≥ 20 years and there was no other significant difference between these two durations, the duration 10–20 years could be suitable.

4.3. *Overall Effects of Straw plus Manure on Crop Yield, SOC, and Soil Nutrients*

As a kind of plentiful source of carbon, straw could be provided for stock-raising as fodder and then transformed into manure via livestock digestion [97]. Practically, some experiments have tried to compare the improvement effect between straw and manure [97–100]. In terms of the difference between straw and manure, straw has a higher C/N (approximately 55:1) and a compact structure that is recalcitrant to decomposition [17], whereas manure has a lower C/N (approximately 25:1), which facilitates the decomposition of exogenous carbon by soil microbes, thereby accelerating the mineralization speed of soil nutrients [101]. Most manure undergoes composting and fermentation before use, a process that not only increases the content of lignin and polyphenols in the manure but also enhances the SOC content [102]. A study comparing the effects of straw return and manure application on crop yield found that manure application significantly increased crop yield by 49%, while straw return only resulted in an 8% increase [103]. This indicates that due to the significant yield-increasing effect of manure, more crop roots remain in the soil, indirectly increasing the input of exogenous carbon [104]. Therefore, compared with straw return, manure application more effectively improves soil nutrients because of its characteristics. With the increase in soil nutrients, crop yield and SOC increased mutually.

Based on their characteristics, we found that straw return can sustain soil moisture and regulate soil temperature [50,51], and manure could provide more nutrients to crops [101]. By compromising their C/N, the combined application of straw and manure may offer a more effective method to leverage the complementary benefits of the two amendments. In terms of individual experiments, lots of studies have experimented with the combination of straw and manure. In the summer maize–winter wheat system, compared with no fertilizer application, the combined application of crop straw and manure increased SOM, soil nutrients, crop biomass, and yield [97]. Based on the results from the monoculture condition of maize, the unification of crop straw and farmyard manure could sharpen the circulation of soil nutrients and carbon storage without generating a negative impact on yield. Furthermore, compared with the chemical fertilizer, the organic fertilizer was able to immobilize surplus TN, preventing its loss [98]. In dryland farming, the combination of organic fertilizer with straw can boost soil microbial quantity, soil enzyme activity, and crop yield under maize cultivation [99]. In the results of a study on Andisols, the application of crop residues with manure could attain the biggest yield among respective treatments in both no-tillage and tillage systems [100]. However, there is no meta-analysis concentrated on the integrative effects of straw plus manure on crop yield, SOC, and soil nutrients; the specific effect of it is still not clarified well. The lack of data and the complexity of it may be regarded as the main reasons. Although the combination of straw and manure has been noticed by numerous scientists, the number of relevant articles on the subject is still obviously lower than the articles regarding the sole use of straw or manure, as the combination has always been designed as a secondary treatment rather than the primary subject. Moreover, the variables of the combination of straw and manure are excessive; for example, the quantity of straw and the substitution ratio of manure should be considered simultaneously. For these reasons, we made a meta-analysis to analyze its approximately increasing effect on crop yield, SOC, and soil nutrients that did not involve its variables.

From the results of our meta-analysis, the effect values of straw plus manure were higher than the effect values of straw return, while the values were lower than the effect values of manure application. Their combined application does not gain the anticipated synergistic benefit. The decrease in crop yield under the combination of straw plus manure application might be due to an excess of nitrogen in this treatment, which led to the production of ineffective tillers in crops [97,105]. Moreover, since a portion of manure (e.g., cattle manure) has a relatively high C/N ratio compared with others, combining them with straw might not effectively neutralize the C/N. The ratio of C/N and the amount of exogenous carbon input in the combination of straw plus manure are higher compared with the treatment with manure alone, which can affect the mineralization activity of microbes. Generally, a C/N ratio of 25:1 is considered suitable for the decomposition rate of organic matter; when the C/N ratio exceeds 25:1, the decomposition rate of organic matter decreases [106]. At the same time, most of the manure used in current experiments or production practices is composted in high-temperature fermentation tanks, resulting in a product that contains few microbes after composting, which has a negligible effect on enhancing the decomposition rate of straw.

However, considering the scarcity of present research articles on the aspect of combined application between straw return and organic manure, its specific impacts on the foregoing indexes cannot be concluded well; therefore, further research still should be conducted. Notwithstanding, economically, it is predictable that utilizing straw for livestock breeding to alleviate stock-raising costs, meanwhile producing manure through animal digestion, may become a gradually popular trend in the future. An attempt to combine straw with manure to notably enhance the efficiency of straw returning while vigorously developing manure application could potentially emerge as a major direction in sustainable agriculture practices.

4.4. *Vista of Straw Return and Manure Application in China*

By comparing the overall effects of straw return, manure application, and straw plus manure application, a hypothesis could be assumed. If we conduct these organic amendments to cover the major farmland of China completely, under the employment of straw, manure, or straw plus manure, how much SOC will be in sequestration? How much crop yield could be increased?

According to the data of the Chinese National Bureau of Statistics, in 2022, the sown areas of three staple crops were collected; the calculation results are shown in Table 1. The computation was based on the mean density of SOC in China [107], and the harvest yield of three major crops in China was taken from the dataset of the Chinese National Bureau of Statistics. We determined that the generalization of straw return and manure application will benefit the development of Chinese sustainable agriculture considerably.

Table 1. The mock computation of yearly increased SOC content and crop yield in China.

Crop	Sown Area (Hectare) In 2022	Expectedly Boosted SOC (MT)	Expectedly Boosted Yield (Kg)
Maize	43,070 × 10^3	Straw: 389 Manure: 1234 Straw plus manure: 537	Straw: 36,781,780 × 10^3 Manure: 23,434,387 × 10^4 Straw plus manure: 85,020,180 × 10^3
Wheat	23,520 × 10^3	Straw: 213 Manure: 674 Straw plus manure: 293	Straw: 13,970,880 × 10^3 Manure: 10,732,176 × 10^4 Straw plus manure: 51,132,480 × 10^3
Rice	29,450 × 10^3	Straw: 266 Manure: 843 Straw plus manure: 367	Straw: 51,066,300 × 10^3 Manure: 97,862,350 × 10^3 Straw plus manure: 85,051,600 × 10^3

4.5. Scientific Implication, Current Limitations, and Future Prospects

As a country with a large population, agriculture is the critical artery of China. In turn, the rational management of organic amendments is very important to agriculture, which impacts the development of the whole country indirectly. Under the meta-analysis based on a multitude of published articles, we confirmed the specific duration, quantity, form, and category of straw returning and the specific source, substitution ratio, and duration of manure application that possess the optimal improvement effects. This result is helpful to foster the proper utilization of organic matter returning and accelerate the achievement of the "Zero Increase Action of Chemical Fertilizer and Pesticide Usage" proposed by the Chinese government [108]. Meanwhile, precise and appropriate use of straw returning or organic manure can not only play a better role in improving soil but also reduce costs, which is conducive to reaching a harmonious balance among ecological, social, and economic benefits [109].

However, our meta-analysis did not consider the factors of original soil physicochemical characteristics (SOC, TN, soil bulk, soil moisture, etc.) and meteorological conditions (temperature, precipitation, etc.), which can affect the response of crop yield, SOC, and soil nutrients to straw returning and manure application. Moreover, the meta-analysis could be conducted under specific conditions (e.g., nitrogen use amount and temperature), which will further explore sustainable organic amendment management in the specific condition. As the impact of different management methods on straw returning and manure application improvement has been researched in this study, the suggested solution should combine this study's results and productive practice.

Some researchers presented results demonstrating that microbes impacted the improvement in straw returning and manure application significantly [15,35]. Thus, except to consider the foregoing limitations, further research is also needed on processes (biochemical, microbiological, and biogeochemical) that occur in the soil after straw and manure application.

5. Conclusions

This meta-analysis researched the effects of straw return, manure application, and their combination on crop yield, SOC, and soil nutrient content in China. It also evaluated the impact of different management methods for these amendments, including the category, form, quantity, and duration of straw, as well as the source, substitution ratio, and duration of manure. The main objective was to identify optimal strategies for straw and manure application. The key findings include the following:

(1) In terms of effect values, the rank of the three organic amendments could be described as straw return < the combination of straw and manure < manure application. All significantly increased the crop yield, SOC, and soil nutrient content; however, manure application resulted in the greatest increase.

(2) In straw return, the optimal duration varied: <5 years was beneficial to improve crop yield, while ≥5 years increased TN the most. The optimal quantity was the low quantity of 3000–6000 kg ha^{-1}. The optimal form varied: incorporation was beneficial to improve crop yield; biochar increased SOC more; and the optimal category was rice straw.

(3) In manure application, pig manure was beneficial in improving crop yield, while chicken manure increased TN the most. The optimal substitution ratio was the high ratio of 50–80%, and the optimal duration was 10–20 years.

Overall, this study is fit for the development policy of China. In productive practice, the research presented in this study allows farmers to apply straw or manure precisely and decrease waste. At present, the theory of green development is prevailing around the world; this study not only provides benefits to sustainable agriculture in China but also complies with the trend of global development. On a global scale, the comprehensive and proper use of straw and manure will positively solve the problems of food crisis, soil deterioration, and global warming. However, the further exploration of straw plus manure and the consideration of microbial effects in the returning process should be researched in the future.

Supplementary Materials: The following supporting information can be downloaded at: https://www.mdpi.com/article/10.3390/agriculture14030480/s1, Figure S1: A flowchart of article selection regarding straw returning; Figure S2: A flowchart of article selection regarding organic manure; Figure S3: A flowchart of article selection regarding straw returning plus organic manure; Figure S4: The correlation between yield and other variables of straw returning; Figure S5: The correlation between yield and other variables of manure application; Figure S6: Frequency distribution diagrams of different organic amendments to SOC; Table S1: Brief circumstance of annual crop residue production between China and global aggregate; Table S2: Succinct profile of annual livestock manure production in China; Table S3: The selection qualification of articles in this meta-analysis; Table S4: Data grouping of the management strategies in this meta-analysis; Table S5: Rosenberg's fail-safe number; Table S6: The correlation between crop yield and other variables in two different groups.

Author Contributions: Conceptualization, Y.Y. and S.S.; methodology, Z.Z. and Y.Y.; formal analysis, Z.Z.; investigation, Z.Z.; writing—original draft, Z.Z.; writing—review and editing, all authors; project administration, Y.Y., H.X. and S.S.; funding acquisition, Y.Y., H.X., S.S., H.H. and X.Z.; supervision, Y.Y., H.X., S.S., H.H. and X.Z.; validation, Y.Z. All authors have read and agreed to the published version of the manuscript.

Funding: This work was supported by the National Key Research and Development Program of China (No. 2023YFD1500302), Joint Funds of the National Natural Science Foundation of China (U22A20610), China Postdoctoral Science Foundation (2023M733673), National Key Technology Support Program for Foreign Experts (ZCNC2022001L), Innovation Team for Research, Development and Application of Conservation Tillage in Liaoning Province (XLYC2008015), and Liaoning Province Applied Basic Research Program Project (2023JH2/101300070 and 2022JH2/101300195).

Institutional Review Board Statement: Not applicable.

Informed Consent Statement: Not applicable.

Data Availability Statement: The datasets generated and/or analyzed during the current study are available from the corresponding author upon reasonable request.

Acknowledgments: We would like to thank all the reviewers and editors for their constructive comments and advice.

Conflicts of Interest: The authors declare no conflicts of interest.

References

1. Gao, Y.X.; Feng, H.J.; Zhang, M.; Shao, Y.Q.; Wang, J.Q.; Liu, Y.L.; Li, C.L. Straw returning combined with controlled-release nitrogen fertilizer affected organic carbon storage and crop yield by changing humic acid composition and aggregate distribution. *J. Clean. Prod.* **2023**, *415*, 137783. [CrossRef]
2. Paul, S.; Flessa, H.; Veldkamp, E.; López-Ulloa, M. Stabilization of recent soil carbon in the humid tropics following land use changes: Evidence from aggregate fractionation and stable isotope analyses. *Biogeochemistry* **2008**, *87*, 247–263. [CrossRef]
3. Zheng, L.; Wu, W.L.; Wei, Y.P.; Hu, K.L. Effects of straw return and regional factors on spatio-temporal variability of soil organic matter in a high-yielding area of Northern China. *Soil Till. Res.* **2015**, *145*, 78–86. [CrossRef]
4. Wu, L.P.; Zhang, S.R.; Ma, R.H.; Chen, M.M.; Wei, W.L.; Ding, X.D. Carbon sequestration under different organic amendments in saline-alkaline soils. *Catena* **2021**, *196*, 104882. [CrossRef]
5. Zuo, W.A.; Gu, B.X.; Zou, X.W.; Peng, K.; Shan, Y.L.; Yi, S.Q.; Shan, Y.H.; Gu, C.H.; Bai, Y.C. Soil organic carbon sequestration in croplands can make remarkable contributions to China's carbon neutrality. *J. Clean. Prod.* **2023**, *382*, 135268. [CrossRef]
6. Liu, X.; Herbert, S.J.; Hashemi, A.M.; Zhang, X.; Ding, G. Effects of agricultural management on soil organic matter and carbon transformation—A review. *Plant Soil Environ.* **2006**, *52*, 531–543. [CrossRef]
7. Wei, T.; Zhang, P.; Wang, K.; Ding, R.X.; Yang, B.P.; Nie, J.F.; Jia, Z.K.; Han, Q.F. Effects of wheat straw incorporation on the availability of soil nutrients and enzyme activities in semiarid areas. *PLoS ONE* **2015**, *10*, e0120994. [CrossRef] [PubMed]
8. Silver, W.L.; Perez, T.; Mayer, A.; Jones, A.R. The role of soil in the contribution of food and feed. *Philos. T. R. Soc. B* **2021**, *376*, 20200181. [CrossRef]
9. Thou, Z.Y. Achieving food security in China: Past three decades and beyond. *China Agr. Econ. Rev.* **2010**, *2*, 251–275.
10. Yang, C.Q.; Wang, X.J.; Li, J.N.; Zhang, G.W.; Shu, H.M.; Hu, W.; Han, H.Y.; Liu, R.X.; Guo, Z.C. Straw return increases crop production by improving soil organic carbon sequestration and soil aggregation in a long-term wheat-cotton cropping system. *J. Integr. Agr.* **2023**, *23*, 669–679. [CrossRef]
11. Pu, C.; Kan, Z.R.; Liu, P.; Ma, S.T.; Qi, J.Y.; Zhao, X.; Zhang, H.L. Residue management induced changes in soil organic carbon and total nitrogen under different tillage practices in the North China Plain. *J. Integr. Agr.* **2019**, *18*, 1337–1347. [CrossRef]
12. Stella, T.; Mouratiadou, I.; Gaiser, T.; Berg-Mohnicke, M.; Wallor, E.; Ewert, F.; Nendel, C. Estimating the contribution of crop residues to soil organic carbon conservation. *Environ. Res. Lett.* **2019**, *14*, 094008. [CrossRef]
13. Wang, X.D.; He, C.; Liu, B.Y.; Zhao, X.; Liu, Y.; Wang, Q.; Zhang, H.L. Effects of residue returning on soil organic carbon storage and sequestration rate in China's croplands: A meta-analysis. *Agronomy* **2020**, *10*, 691. [CrossRef]
14. Wang, X.J.; Jia, Z.K.; Liang, L.Y.; Yang, B.P.; Ding, R.X.; Nie, J.F.; Wang, J.P. Maize straw effects on soil aggregation and other properties in arid land. *Soil Till. Res.* **2015**, *153*, 131–136. [CrossRef]
15. Zhu, X.F.; Xie, H.T.; Masters, M.D.; Rui, Y.C.; Luo, Y.; He, H.B.; Zhang, X.D.; Liang, C. Microorganisms, their residues, and soil carbon storage under a continuous maize cropping system with eight years of variable residue retention. *Appl. Soil Ecol.* **2023**, *187*, 104846. [CrossRef]
16. Chen, J.H.; Gong, Y.Z.; Wang, S.Q.; Guan, B.Z.; Balkovic, J.; Kraxner, F. To burn or retain crop residues on croplands? An integrated analysis of crop residue management in China. *Sci. Total Environ.* **2019**, *662*, 141–150. [CrossRef] [PubMed]
17. Fu, B.; Chen, L.; Huang, H.Y.; Qu, P.; Wei, Z.G. Impacts of crop residues on soil health: A review. *Environ. Pollut. Bioavail.* **2021**, *33*, 164–173. [CrossRef]
18. Kumar, N.; Chaudhary, A.; Ahlawat, O.P.; Naorem, A.; Upadhyay, G.; Chhokar, R.S.; Gill, S.C.; Khippal, A.; Tripathi, S.C.; Singh, G.P. Crop residue management challenges, opportunities and way forward for sustainable food-energy security in India: A review. *Soil Till. Res.* **2023**, *228*, 105641. [CrossRef]
19. Wu, Y.P.; He, J.L.; Liu, W.; Cheng, W.G.; Shaaban, M.; Jiang, Y.B. The effects of continuous straw returning strategies on SOC balance upon fresh straw incorporation. *Environ. Res.* **2023**, *232*, 116225. [CrossRef]
20. Zhu, C.C.; Zhong, W.H.; Han, C.; Deng, H.; Jiang, Y.B. Driving factors of soil organic carbon sequestration under straw returning across China's uplands. *J. Environ. Manag.* **2023**, *335*, 117590. [CrossRef]
21. Chen, L.M.; Sun, S.L.; Yao, B.; Peng, Y.T.; Gao, C.F.; Qin, T.; Zhou, Y.Y.; Sun, C.R.; Quan, W. Effects of straw return and straw biochar on soil properties and crop growth: A review. *Front. Plant Sci.* **2022**, *13*, 986763. [CrossRef]

22. Liu, T.; He, G.J.; Lau, A.K.H. Statistical evidence on the impact of agricultural straw burning on urban air quality in China. *Sci. Total Environ.* **2020**, *711*, 134633. [CrossRef] [PubMed]
23. Seglah, P.A.; Wang, Y.J.; Wang, H.Y.; Bi, Y.Y.; Zhou, K.; Wang, Y.; Wang, H.; Feng, X.X. Crop straw utilization and field burning in Northern region of Ghana. *J. Clean. Prod.* **2020**, *261*, 121191. [CrossRef]
24. Yang, X.M.; Cheng, L.L.; Huang, X.L.; Zhang, Y.; Yin, C.B.; Lebailly, P. Incentive mechanism to promote corn stalk return sustainably in Henan, China. *Sci. Total Environ.* **2020**, *738*, 139775. [CrossRef] [PubMed]
25. Jaćimović, G.; Aćin, V.; Mirosavljević, M.; Brbaklić, L.; Vujić, S.; Dunđerski, D.; Šeremešić, S. Effects of combined long-term straw return and nitrogen fertilization on wheat productivity and soil properties in the wheat-maize-soybean rotation system in the Pannonian Plain. *Agronomy* **2023**, *13*, 1529. [CrossRef]
26. Yan, Y.; Ji, W.J.; Li, B.G.; Wang, G.M.; Chen, S.C.; Zhu, D.H.; Liu, Z. Quantification of the effects of long-term straw return on soil organicMatter spatiotemporal variation: A case study in typical black soil region. *Soil* **2023**, *9*, 351–364. [CrossRef]
27. He, H.; Peng, M.W.; Lu, W.D.; Ru, S.B.; Hou, Z.A.; Li, J.H. Organic fertilizer substitution promotes soil organic carbon sequestration by regulating permanganate oxidizable carbon fractions transformation in oasis wheat fields. *Catena* **2023**, *221*, 106784. [CrossRef]
28. Guo, Z.; Han, J.C.; Li, J.; Xu, Y.; Wang, X.L. Effects of long-term fertilization on soil organic carbon mineralization and microbial community structure. *PLoS ONE* **2019**, *14*, e0211163.
29. Salehi, A.; Fallah, S.; Sourki, A.A. Organic and inorganic fertilizer effect on soil CO_2 flux, microbial biomass, and growth of Nigella Sativa L. *Int. Agrophys.* **2017**, *31*, 103–116. [CrossRef]
30. Sarma, B.; Farooq, M.; Gogoi, N.; Borkotoki, B.; Kataki, R.; Garg, A. Soil organic carbon dynamics in wheat—Green gram crop rotation amended with vermicompost and biochar in combination with inorganic fertilizers: A comparative study. *J. Clean. Prod.* **2018**, *201*, 471–480. [CrossRef]
31. Su, G.C.; Ong, H.C.; Zulkifli, N.W.M.; Ibrahim, S.; Chen, W.H.; Chong, C.T.; Ok, Y.S. Valorization of animal manure via pyrolysis for bioenergy: A review. *J. Clean. Prod.* **2022**, *343*, 130965. [CrossRef]
32. Hayatu, N.G.; Liu, Y.R.; Han, T.F.; Daba, N.A.; Zhang, L.; Shen, Z.; Li, J.W.; Muazu, H.; Lamlom, S.F.; Zhang, H.M. Carbon sequestration rate, nitrogen use efficiency and rice yield responses to long-term substitution of chemical fertilizer by organic manure in a rice-rice cropping system. *J. Integr. Agr.* **2022**, *22*, 2848–2864. [CrossRef]
33. Hossain, M.E.; Mei, X.R.; Zhang, W.Y.; Dong, W.Y.; Yan, Z.X.; Liu, X.; Rachit, S.; Gopalakrishnan, S.; Liu, E.K. Substitution of chemical fertilizer with organic fertilizer affects soil total nitrogen and its fractions in Northern China. *Int. J. Environ. Res. Public Health* **2021**, *18*, 12848. [CrossRef]
34. Liao, B.; Cai, T.C.; Wu, X.; Luo, Y.; Liao, P.; Zhang, B.C.; Zhang, Y.T.; Wei, G.F.; Hu, R.G.; Luo, Y.F.; et al. A combination of organic fertilizers partially substitution with alternate wet and dry irrigation could further reduce greenhouse gases emission in rice field. *J. Environ. Manag.* **2023**, *344*, 118372. [CrossRef] [PubMed]
35. Ma, Y.J.; Shen, S.Z.; Wan, C.; Wang, S.Q.; Yang, F.X.; Zhang, K.Q.; Gao, W.X. Organic fertilizer substitution over six years improves the productivity of garlic, bacterial diversity, and microbial communities network complexity. *Appl. Soil Ecol.* **2023**, *182*, 104718. [CrossRef]
36. Zhai, L.C.; Wang, Z.B.; Zhai, Y.C.; Zhang, L.H.; Zheng, M.J.; Yao, H.P.; Lv, L.H.; Shen, H.P.; Zhang, J.T.; Yao, Y.R.; et al. Partial substitution of chemical fertilizer by organic fertilizer benefits grain yield, water use efficiency, and economic return of summer maize. *Soil Till. Res.* **2022**, *217*, 105287. [CrossRef]
37. Jia, W.; Qin, W.; Zhang, Q.; Wang, X.; Ma, Y.; Chen, Q. Evaluation of crop residues and manure production and their geographical distribution in China. *J. Clean. Prod.* **2018**, *188*, 954–965. [CrossRef]
38. Liu, D.T.; Song, C.C.; Xin, Z.H.; Fang, C.; Liu, Z.H.; Xu, Y.P. Agricultural management strategies for balancing yield increase, carbon sequestration, and emission reduction after straw return for three major grain crops in China: A meta-analysis. *J. Environ. Manag.* **2023**, *340*, 117965. [CrossRef] [PubMed]
39. Wang, Y.L.; Wu, P.N.; Mei, F.J.; Ling, Y.; Qiao, Y.B.; Liu, C.S.; Leghari, S.J.; Guan, X.K.; Wang, T.C. Does continuous straw returning keep China farmland soil organic carbon continued increase? A meta-analysis. *J. Environ. Manag.* **2021**, *288*, 112391. [CrossRef] [PubMed]
40. Luo, Y.Q.; Hui, D.F.; Zhang, D.Q. Elevated CO_2 stimulates net accumulations of carbon and nitrogen in land ecosystems: A meta-analysis. *Ecology* **2006**, *87*, 53–63. [CrossRef]
41. Wang, J.Z.; Wang, X.J.; Xu, M.G.; Feng, G.; Zhang, W.J.; Lu, C.A. Crop yield and soil organic matter after long-term straw return to soil in China. *Nutr. Cycl. Agroecosys.* **2015**, *102*, 371–381. [CrossRef]
42. Huang, T.T.; Yang, N.; Lu, C.; Qin, X.L.; Siddique, K.H.M. Soil organic carbon, total nitrogen, available nutrients, and yield under different straw returning methods. *Soil Till. Res.* **2021**, *214*, 105171. [CrossRef]
43. Pallmann, P. Applied meta-analysis with R. *J. Appl. Stat.* **2015**, *42*, 914–915. [CrossRef]
44. Wittig, V.E.; Ainsworth, E.A.; Naidu, S.L.; Karnosky, D.F.; Long, S.P. Quantifying the impact of current and future tropospheric ozone on tree biomass, growth, physiology and biochemistry: A quantitative meta-analysis. *Global Change Biol.* **2009**, *15*, 396–424. [CrossRef]
45. Fenoglio, M.S.; Rossetti, M.R.; Videla, M. Negative effects of urbanization on terrestrial arthropod communities: A meta-analysis. *Global Ecol. Biogeogr.* **2020**, *29*, 1412–1429. [CrossRef]
46. Rosenberg, M.S. The file-drawer problem revisited: A general weighted method for calculating fail-safe numbers in meta-analysis. *Evolution* **2005**, *59*, 464–468. [PubMed]

47. Che, W.K.; Piao, J.L.; Gao, Q.; Li, X.B.; Li, X.; Jin, F. Response of soil physicochemical properties, soil nutrients, enzyme activity and rice yield to rice straw returning in highly saline-alkali paddy soils. *J. Soil Sci. Plant Nut.* **2023**, *23*, 4396–4411. [CrossRef]
48. Yang, J.H.; Luo, Y.L.; Chen, J.; Jin, M.; Wang, Z.L.; Li, Y. Effects of main food yield under straw return in China: A meta-analysis. *Sci. Agric. Sin.* **2020**, *53*, 4415–4429. (In Chinese)
49. Cong, P.; Li, Y.Y.; Wang, J.; Gao, Z.J.; Pang, H.C.; Zhang, L.; Liva, N.; Dong, J. Increasing straw incorporation rates improves subsoil fertility and crop yield in the Huang-Huai-Hai Plain of China. *Arch. Agron. Soil Sci.* **2020**, *66*, 1976–1990. [CrossRef]
50. Tao, Z.Q.; Li, C.F.; Li, J.J.; Ding, Z.S.; Xu, J.; Sun, X.F.; Zhou, P.L.; Zhao, M. Tillage and straw mulching impacts on grain yield and water use efficiency of spring maize in Northern Huang–Huai–Hai Valley. *Crop J.* **2015**, *3*, 445–450. [CrossRef]
51. Wu, X.H.; Wang, W.; Xie, X.L.; Yin, C.M.; Hou, H.J. Effects of rice straw mulching on N_2O emissions and maize productivity in a rain-fed upland. *Environ. Sci. Pollut. R.* **2018**, *25*, 6407–6413. [CrossRef]
52. Berhane, M.; Xu, M.; Liang, Z.Y.; Shi, J.L.; Wei, G.H.; Tian, X.H. Effects of long-term straw return on soil organic carbon storage and sequestration rate in North China upland crops: A meta-analysis. *Global Change Biol.* **2020**, *26*, 2686–2701. [CrossRef]
53. Islam, M.U.; Guo, Z.C.; Jiang, F.H.; Peng, X.H. Does straw return increase crop yield in the wheat-maize cropping system in China? A meta-analysis. *Field Crop Res.* **2022**, *279*, 108447. [CrossRef]
54. Jiang, C.M.; Yu, W.T.; Ma, Q.; Xu, Y.G.; Zou, H. Alleviating global warming potential by soil carbon sequestration: A multi-level straw incorporation experiment from a maize cropping system in Northeast China. *Soil Till. Res.* **2017**, *170*, 77–84. [CrossRef]
55. Wang, X.J.; Jia, Z.K.; Liang, L.Y.; Zhao, Y.F.; Yang, B.P.; Ding, R.X.; Wang, J.P.; Nie, J.F. Changes in soil characteristics and maize yield under straw returning system in dryland farming. *Field Crop Res.* **2018**, *218*, 11–17. [CrossRef]
56. Lin, H.Y.; Zhou, M.H.; Zeng, F.R.; Xu, P.; Ma, S.L.; Zhang, B.W.; Li, Z.Y.; Wang, Y.Q.; Zhu, B. How do soil organic carbon pool, stock and their stability respond to crop residue incorporation in subtropical calcareous agricultural soils? *Agr. Ecosyst. Environ.* **2022**, *332*, 107927. [CrossRef]
57. Zhang, P.; Wei, T.; Li, Y.L.; Wang, K.; Jia, Z.K.; Han, Q.F.; Ren, X.L. Effects of straw incorporation on the stratification of the soil organic C, total N and C:N ratio in a semiarid region of China. *Soil Till. Res.* **2015**, *153*, 28–35. [CrossRef]
58. Metwally, M.S.; Shaddad, S.M.; Liu, M.Q.; Yao, R.J.; Abdo, A.I.; Li, P.; Jiao, J.G.; Chen, X.Y. Soil properties spatial variability and delineation of site-specific management zones based on soil fertility using fuzzy clustering in a hilly field in Jianyang, Sichuan, China. *Sustainability* **2019**, *11*, 7084. [CrossRef]
59. Han, X.; Xu, C.; Dungait, J.A.J.; Bol, R.; Wang, X.J.; Wu, W.L.; Meng, F.Q. Straw incorporation increases crop yield and soil organic carbon sequestration but varies under different natural conditions and farming practices in China: A system analysis. *Biogeosciences* **2018**, *15*, 1933–1946. [CrossRef]
60. Jin, Z.Q.; Shah, T.R.; Zhang, L.; Liu, H.Y.; Peng, S.B.; Nie, L.X. Effect of straw returning on soil organic carbon in rice-wheat rotation system: A review. *Food Energy Secur.* **2020**, *9*, e200. [CrossRef]
61. Yuan, G.Y.; Huan, W.W.; Song, H.; Lu, D.J.; Chen, X.Q.; Wang, H.Y.; Zhou, J.M. Effects of straw incorporation and potassium fertilizer on crop yields, soil organic carbon, and active carbon in the rice–wheat system. *Soil Till. Res.* **2021**, *209*, 104958. [CrossRef]
62. Zhao, X.L.; Yuan, G.Y.; Wang, H.Y.; Lu, D.J.; Chen, X.Q.; Zhou, J.M. Effects of full straw incorporation on soil fertility and crop yield in rice-wheat rotation for silty clay loamy cropland. *Agronomy* **2019**, *9*, 133. [CrossRef]
63. Fan, W.; Wu, J.G.; Ahmed, S.; Hu, J.; Chen, X.D.; Li, X.H.; Zhu, W.Y.; Opoku-Kwanowaa, Y. Short-term effects of different straw returning methods on the soil physicochemical properties and quality index in dryland farming in NE China. *Sustainability* **2020**, *12*, 2631. [CrossRef]
64. Zhang, B.Y.; Dou, S.; Guan, S.; Yang, C.; Wang, Z. Deep straw burial accelerates straw decomposition and improves soil water repellency. *Agronomy.* **2023**, *13*, 1927. [CrossRef]
65. Zhang, Y.; Zhao, J.; Wang, H.Y.; Pang, H.C. Pelletized straw incorporation in sandy soil increases soil aggregate stability, soil carbon, and nitrogen stocks. *Sustainability* **2023**, *15*, 7079. [CrossRef]
66. Wang, X.Q.; Lv, G.Y.; Zhang, Y.; Yu, Y.; Wang, X.B.; Peixoto, L.; Qian, C.R.; Pang, H.C. Annual burying of straw after pelletizing: A novel and feasible way to improve soil fertility and productivity in Northeast China. *Soil Till. Res.* **2023**, *230*, 105699. [CrossRef]
67. Wang, J.; Zhang, L.; Pang, H.C.; Zhang, J.T. Returning granulated straw for accelerating decomposition rate and improving soil fertility. *Trans. Chin. Soc. Agric. Eng.* **2017**, *33*, 177–183. (In Chinese)
68. Zhang, L.; Wang, J.; Pang, H.C.; Zhang, J.T.; Guo, J.J.; Dong, G.H.; Cong, P. Effects of pelletized straw on soil nutrient properties in relation to crop yield. *Soil Use Manag.* **2018**, *34*, 479–489. [CrossRef]
69. Cong, P.; Pang, H.C.; Wang, J.; Liu, N.; Li, Y.Y.; Zhang, L. Effect of returning chopped and pelletized straw at a high rate enhancing soil organic carbon in subsoil of farmlands of black soil. *Acta Pedol. Sin.* **2020**, *57*, 811–823. (In Chinese)
70. Xu, X.; Pang, D.W.; Chen, J.; Luo, Y.L.; Zheng, M.J.; Yin, Y.P.; Li, Y.X.; Li, Y.; Wang, Z.L. Straw return accompany with low nitrogen moderately promoted deep root. *Field Crop Res.* **2018**, *221*, 71–80. [CrossRef]
71. Zhou, G.X.; Zhang, J.B.; Zhang, C.Z.; Feng, Y.Z.; Chen, L.; Yu, Z.H.; Xin, X.L.; Zhao, B.Z. Effects of changes in straw chemical properties and alkaline soils on bacterial communities engaged in straw decomposition at different temperatures. *Sci. Rep.* **2016**, *6*, 22186. [CrossRef] [PubMed]
72. Cui, Y.F.; Meng, J.; Wang, Q.X.; Zhang, W.M.; Cheng, X.Y.; Chen, W.F. Effects of straw and biochar addition on soil nitrogen, carbon, and super rice yield in cold waterlogged paddy soils of North China. *J. Integr. Agr.* **2017**, *16*, 1064–1074. [CrossRef]
73. Zhang, R.X.; Qu, Z.Y.; Liu, L.; Yang, W.; Wang, L.P.; Li, J.J.; Zhang, D.L. Soil respiration and organic carbon response to biochar and their influencing factors. *Atmosphere* **2022**, *13*, 2038. [CrossRef]

74. Huang, C.X.; Cheng, C.P.; Wang, Z.Y.; Zhao, X.; Yang, Y.; Hu, L.L.; Li, Y.Z.; Ma, J.H.; Wang, L.L.; Chang, L.; et al. Straw strip mulching increased soil organic carbon components of a wheat field in dry farming regions of the Loess Plateau. *Water* **2022**, *14*, 2645. [CrossRef]
75. Zhou, Z.J.; Zeng, X.Z.; Chen, K.; Li, Z.; Guo, S.; Shangguan, Y.X.; Yu, H.; Tu, S.H.; Qin, Y.S. Long-term straw mulch effects on crop yields and soil organic carbon fractions at different depths under a no-till system on the Chengdu Plain, China. *J. Soil Sediment.* **2019**, *19*, 2143–2152. [CrossRef]
76. Zheng, Y.; Han, X.R.; Li, Y.Y.; Yang, J.F.; Li, N.; An, N. Effects of biochar and straw application on the physicochemical and biological properties of paddy soils in Northeast China. *Sci. Rep.* **2019**, *9*, 16531. [CrossRef] [PubMed]
77. Zhang, L.; Wang, J.; Pang, H.C.; Zhang, J.T.; Guo, J.J.; Dong, G.H. Effects of short-term granulated straw incorporation on grain yield and soil respiration in a winter wheat-summer maize cropping system. *Chin. J. Appl. Ecol* **2018**, *29*, 565–572. (In Chinese)
78. Lal, R. World crop residues production and implications of its use as a biofuel. *Environ. Int.* **2005**, *31*, 575–584. [CrossRef]
79. Wang, X.Y.; Yang, L.; Steinberger, Y.; Liu, Z.X.; Liao, S.H.; Xie, G.H. Field crop residue estimate and availability for biofuel production in China. *Renew. Sust. Energ. Rev.* **2013**, *27*, 864–875. [CrossRef]
80. Dong, L.L.; Wang, H.H.; Shen, Y.; Wang, L.Q.; Zhang, H.D.; Shi, L.L.; Lu, C.Y.; Shen, M.X. Straw type and returning amount affects SOC fractions and Fe/Al oxides in a rice-wheat rotation system. *Appl. Soil Ecol.* **2023**, *183*, 104736. [CrossRef]
81. Zhang, S.L.; Li, M.; Cui, X.Y.; Pan, Y.M. Effect of different straw retention techniques on soil microbial community structure in wheat-maize rotation system. *Front. Microbiol.* **2023**, *13*, 1069458. [CrossRef]
82. Shao, J.M.; Gao, C.Y.; Seglah, P.A.; Xie, J.; Zhao, L.; Bi, Y.Y.; Wang, Y.J. Analysis of the available straw nutrient resources and substitution of chemical fertilizers with straw returned directly to the field in China. *Agriculture* **2023**, *13*, 1187. [CrossRef]
83. Wang, X.Y.; Sun, B.; Mao, J.D.; Sui, Y.Y.; Cao, X.Y. Structural convergence of maize and wheat straw during two-year decomposition under different climate conditions. *Environ. Sci. Technol.* **2012**, *46*, 7159–7165. [CrossRef] [PubMed]
84. Mi, W.H.; Sun, Y.; Xia, S.Q.; Zhao, H.T.; Mi, W.T.; Brookes, P.C.; Liu, Y.L.; Wu, L.H. Effect of inorganic fertilizers with organic amendments on soil chemical properties and rice yield in a low-productivity paddy soil. *Geoderma* **2018**, *320*, 23–29. [CrossRef]
85. Wang, J.D.; Wang, K.H.; Wang, X.J.; Ai, Y.C.; Zhang, Y.C.; Yu, J.G. Carbon sequestration and yields with long-term use of inorganic fertilizers and organic manure in a six-crop rotation system. *Nutr. Cycl. Agroecosys.* **2018**, *111*, 87–98. [CrossRef]
86. Xu, Y.G.; Peng, Z.P.; Tu, Y.T.; Huang, J.C. Combining organic and inorganic fertilization increases rice yield and soil nitrogen and carbon: Dissolved organic matter chemodiversity and soil microbial communities. *Plant Soil* **2023**, *492*, 557–571. [CrossRef]
87. Hou, Q.; Ni, Y.M.; Huang, S.; Zuo, T.; Wang, J.; Ni, W.Z. Effects of substituting chemical fertilizers with manure on rice yield and soil labile nitrogen in paddy fields of China: A meta-analysis. *Pedosphere* **2023**, *33*, 172–184. [CrossRef]
88. Du, Y.D.; Cui, B.J.; Zhang, Q.; Wang, Z.; Sun, J.; Niu, W.Q. Effects of manure fertilizer on crop yield and soil properties in China: A meta-analysis. *Catena* **2020**, *193*, 104617. [CrossRef]
89. Qaswar, M.; Jing, H.; Ahmed, W.; Li, D.C.; Liu, S.J.; Lu, Z.; Cai, A.D.; Liu, L.S.; Xu, Y.M.; Gao, J.S.; et al. Yield sustainability, soil organic carbon sequestration and nutrients balance under long-term combined application of manure and inorganic fertilizers in acidic paddy soil. *Soil Till. Res.* **2020**, *198*, 104569. [CrossRef]
90. Hu, C.; Li, S.L.; Qiao, Y.; Liu, D.H.; Chen, Y.F. Effects of 30 years repeated fertilizer applications on soil properties, microbes and crop yields in rice-wheat cropping systems. *Exp. Agr.* **2015**, *51*, 355–369. [CrossRef]
91. Lu, X.L. Fertilizer types affect soil organic carbon content and crop production: A meta-analysis. *Agr. Res.* **2020**, *9*, 94–101. [CrossRef]
92. Cao, M.H.; Duan, Y.; Li, M.H.; Tang, C.G.; Kan, W.J.; Li, J.Y.; Zhang, H.L.; Zhong, W.L.; Wu, L.F. Manure substitution improves maize yield by promoting soil fertility and mediating the microbial community in lime concretion black soil. *J. Integr. Agr.* **2023**, *23*, 698–710. [CrossRef]
93. Cai, A.D.; Xu, M.G.; Wang, B.R.; Zhang, W.J.; Liang, G.P.; Hou, E.Q.; Luo, Y.Q. Manure acts as a better fertilizer for increasing crop yields than synthetic fertilizer does by improving soil fertility. *Soil Till. Res.* **2019**, *189*, 168–175. [CrossRef]
94. Qaswar, M.; Liu, Y.R.; Huang, J.; Liu, K.L.; Mudasir, M.; Lv, Z.Z.; Hou, H.Q.; Lan, X.J.; Ji, J.H.; Ahmed, W.; et al. Soil nutrients and heavy metal availability under long-term combined application of swine manure and synthetic fertilizers in acidic paddy soil. *J. Soil Sediment* **2020**, *20*, 2093–2106. [CrossRef]
95. Wei, Z.B.; Ying, H.; Guo, X.W.; Zhuang, M.H.; Cui, Z.L.; Zhang, F.S. Substitution of mineral fertilizer with organic fertilizer in maize systems: A meta-analysis of reduced nitrogen and carbon emissions. *Agronomy* **2020**, *10*, 1149. [CrossRef]
96. Ding, X.L.; Yuan, Y.R.; Liang, Y.; Li, L.J.; Han, X.Z. Impact of long-term application of manure, crop residue, and mineral fertilizer on organic carbon pools and crop yields in a Mollisol. *J. Soil Sediment* **2014**, *14*, 854–859. [CrossRef]
97. Chen, S.Y.; Zhang, X.Y.; Shao, L.W.; Sun, H.Y.; Niu, J.F.; Liu, X.W. Effects of straw and manure management on soil and crop performance in North China Plain. *Catena* **2020**, *187*, 104359. [CrossRef]
98. Dămătîrcă, C.; Moretti, B.; Bertora, C.; Ferrarini, A.; Lerda, C.; Mania, I.; Celi, L.; Gorra, R.; Zavattaro, L. Residue incorporation and organic fertilisation improve carbon and nitrogen turnover and stabilisation in maize monocropping. *Agr. Ecosyst. Environ.* **2023**, *342*, 108255. [CrossRef]
99. Sun, L.; Sun, Z.X.; Opoku-Kwanowaa, Y.; Hu, J.; Wu, J.G. Effects of returning organic waste on soil enzymes and microbial quantity in dryland farming. *Int. Agrophys.* **2021**, *35*, 279–287. [CrossRef]
100. Koga, N.; Tsuji, H. Effects of reduced tillage, crop residue management and manure application practices on crop yields and soil carbon sequestration on an Andisol in Northern Japan. *Soil Sci. Plant Nutr.* **2009**, *55*, 546–557. [CrossRef]

101. Pei, H.; Miao, Y.; Hou, R.X. Meta analysis of the effects of global organic material returning on soil organic carbon sequestration in Mollisols. *Trans. Chin. Soc. Agric. Eng.* **2023**, *39*, 79–88. (In Chinese)
102. Xia, L.L.; Lam, S.K.; Yan, X.Y.; Chen, D.L. How does recycling of livestock manure in agroecosystems affect crop productivity, reactive nitrogen losses, and soil carbon balance? *Environ. Sci. Technol.* **2017**, *51*, 7450–7457. [CrossRef] [PubMed]
103. Luo, G.W.; Li, L.; Friman, V.P.; Guo, J.J.; Guo, S.W.; Shen, Q.R.; Ling, N. Organic amendments increase crop yields by improving microbe-mediated soil functioning of agroecosystems: A meta-analysis. *Soil Biol. Biochem.* **2018**, *124*, 105–115. [CrossRef]
104. Xiao, G.M.; Liu, L.; Zhao, Y.; Ru, S.H.; Zhao, O.Y.; Hou, L.M.; Sun, S.Y.; Wang, L.; Wang, C. Effects of organic materials input on soil organic carbon in China based on meta-analysis. *Soil Fertil. Sci. China.* **2023**, *8*, 23–32. (In Chinese)
105. Liu, J.; Chen, R.L.; Chen, G.Y.; Zhao, D.C.; Zhang, X.J.; Li, F.Z.; Li, T.F. Effects of combined application of nitrogen and organic fertilizer and straw returning on wheat growth and yield. *J. Henan Agric. Sci.* **2015**, *48*, 48–51+64. (In Chinese)
106. Sun, X.Y. *Soil Science*, 1st ed.; China Forestry Publishing House: Beijing, China, 2005. (In Chinese)
107. Wu, H.B.; Guo, Z.T.; Peng, C.H. Distribution and storage of soil organic carbon in China. In *Global Biogeochemical Cycles*; Wiley Online Library: Hoboken, NJ, USA, 2003; Volume 17.
108. Liu, X.J.; Vitousek, P.; Chang, Y.H.; Zhang, W.F.; Matson, P.; Zhang, F.S. Evidence for a historic change occurring in China. *Environ. Sci. Technol.* **2016**, *50*, 505–506. [CrossRef]
109. He, Z.Q.; Pagliari, P.H.; Waldrip, H.M. Applied and Environmental Chemistry of Animal Manure: A Review. *Pedosphere* **2016**, *26*, 779–816. [CrossRef]

agriculture

MDPI

Article

Effect of Long-Term Different Land Uses on Improving Stable Humic Compounds in Arenosol

Kristina Amaleviciute-Volunge [1,*], Liudmila Tripolskaja [2], Asta Kazlauskaite-Jadzevice [2], Alvyra Slepetiene [1] and Eugenija Baksiene [2]

1 Chemical Research Laboratory, Institute of Agriculture, Lithuanian Research Centre for Agriculture and Forestry, Instituto al. 1, Akademija, LT-58344 Kedainiai, Lithuania; alvyra.slepetiene@lammc.lt
2 Voke Branch, Lithuanian Research Centre for Agriculture and Forestry, Zalioji 2, LT-02232 Vilnius, Lithuania; liudmila.tripolskaja@lammc.lt (L.T.); asta.kazlauskaite-jadzevice@lammc.lt (A.K.-J.); eugenija.baksiene@lammc.lt (E.B.)
* Correspondence: kristina.amaleviciute-volunge@lammc.lt; Tel.: +370-6458643

Abstract: There has not been enough research conducted on the effect of land use on the composition of humus in Arenosols. This long-term study (1995–2022) aimed to determine the differences in the formation of humic compounds in the natural and agricultural ecosystems of Arenosols. Soil samples were collected from six plots at two soil depths (0–15 and 15–25 cm), with four replicates. Conclusions were reached based on the results of the accumulation of humic substances (HSs) and their qualitative fractional composition, C/N ratio, humification degree (HD), and the optical properties of the humus substances. Afforestation, after 27 years, significantly increased (+6.7 g kg^{-1}) the soil organic carbon (SOC) and influenced the qualitative composition of HS: HA + FA 79.3% of the SOC. Grassland cultivation showed faster (+3.8 g kg^{-1}) SOC sequestration, a higher HA/FA ratio, and an increased HD. Arenosols may be used in crop rotation with approximately 40% leguminous plants to maintain a stable humus balance. Additionally, the effects of mineral fertilisers on the humification processes and humus quality of +2.59 g kg^{-1} SOC, +1.27 g kg^{-1} humin in crop rotation, +3.26 g kg^{-1} SOC, and 2.82 g kg^{-1} humin in a grass cultivation field were established. For SOC accumulation and a larger humus amount of a better quality, it is recommended that an Arenosol is used, as it is suitable for use in perennial cut grasslands, natural grasslands, and pine afforestation.

Keywords: arenosol; humic substances; SOC accumulation; agricultural ecosystem; natural ecosystem

Citation: Amaleviciute-Volunge, K.; Tripolskaja, L.; Kazlauskaite-Jadzevice, A.; Slepetiene, A.; Baksiene, E. Effect of Long-Term Different Land Uses on Improving Stable Humic Compounds in Arenosol. *Agriculture* **2024**, *14*, 250. https://doi.org/10.3390/agriculture14020250

Academic Editor: Riccardo Scotti

Received: 11 December 2023
Revised: 11 January 2024
Accepted: 1 February 2024
Published: 3 February 2024

1. Introduction

SOC sequestration has two important functions in the ecosystem: to increase soil fertility and to reduce the concentration of atmospheric carbon dioxide through its accumulation in the organic part of the soil. In agricultural arable soils, increasing the amount of SOC improves the physical, chemical, and biological properties of the soil and increases water absorption and erosion resistance [1–8]. This is particularly relevant for low-buffering soils with a light granulometric composition because agricultural activities often lead to the degradation of their properties.

Considerable attention is currently being paid to increasing the sustainability of ecosystems with low-fertility agricultural soils, and there is a need for optimal methods of using these soils to produce feed and food. The farmers of these soils often abandon arable agriculture and switch to other land uses that are suitable for SOC sequestration to prevent the fast mineralisation of soil organic matter (SOM).

Many studies have highlighted the effectiveness of grassy ecosystems in increasing SOC accumulation [6,9–14]. In forest ecosystems, compared with grasslands, the accumulation of SOC at the beginning is slower and becomes faster after the formation of abundant tree foliage, which provides a greater amount of organic fallout for the formation of humic substances (HSs) [15–20]. However, the ability of land use to accumulate SOC cannot be the main criterion for choosing the type of land use in a certain area, as it is equally important to ensure the necessary amount of food and feed production and to maintain the economic efficiency of these activities [1,12].

The type of land use determines the amount and quality of organic matter that returns to the soil after the harvest of plants or at the end of vegetation; this organic matter later humifies and replenishes the soil humus reserves. The hydrological conditions influencing the intensity of soil microorganism activity are also of great importance for this process, as positive temperatures and optimal soil moisture content are necessary for their optimal activity. In recent decades, climate warming has significantly changed the climatic norms for rainfall and air temperature in various European countries [21–24]. This not only increases the atmospheric CO_2 concentration, but also affects the microbial population and enzymatic activities in the soil [5].

To reduce the concentration of CO_2 in the atmosphere through the accumulation of SOC, it is important not only to increase the total amount of SOC, but also to promote the formation of stable humic substances (HSs), which are more resistant to mineralisation. This is because labile carbon compounds are released into the atmosphere during the further destruction of HSs. The stable HSs include humic acids (HAs), which are soluble in alkaline media and insoluble at pH 1.0, and humins (HNs), which are insoluble at all pH values [25,26].

In Lithuania, the fractional composition of humus was studied in [27–29]. Other researchers [27] investigated the humus status in clay soils (Endocalca-ri-Endohypogleyic Cambisol) under different tillage regimes. Mockeviciene et al. [28,29] hypothesised that acidic soils (natural acid Dystric Glossic Retisol) would promote organic carbon mineralisation and that it is likely that the concentration of carbon fractions of different stabilities would change during the carbon transformation processes. The changes in organic carbon in the mineral topsoil of formerly cultivated Arenosols under different land uses, the re-naturalisation of ex-arable Arenosols, nitrogen accumulation, and nitrogen leaching as affected by legume crop residues on sandy loam in the eastern Baltic region were analysed in Lithuania.

There is also a lack of knowledge on how humification takes place and on the distribution of humus fractions in differently used Arenosols (natural and agricultural) in Lithuania.

This study aimed to determine the differences in the formation of stable humic compounds in natural (pine afforestation and abandoned land) and agricultural (cropland and managed grassland) ecosystems in Arenosols.

2. Materials and Methods

2.1. Study Object and Experimental Site

This research was conducted as a long-term trial of the LAMMC Voke Branch. The experiments were located in eastern Lithuania (NW: 54°33′52.27″ N, 25°05′12.68″ E; NE: 54°33′52.01″ N, 25°05′14.60″ E; SE: 54° 33′48.23″ N, 25°05′12.97″ E, SW: 54°33′48.56″ N, 25°05′10.86″ E). The experimental site represents an agricultural landscape of a stretch of the morainic hummocky uplands of eastern Lithuania. This stretch was formed as a terminal moraine; therefore, it is characterised by an upper layer of sandy loam and loamy sand parent material, which form the sandy soils under study (Figure 1). The experiment was conducted on arable soil that has been used for more than 50 years (until 1995) to grow various grain crops and is fertilised with mineral fertilisers. Table 1 presents the treatments and experimental design.

The clay content of the upper part of the profile was low. In horizon "A", the content of sand (63 μm–2 mm) was 80.7–83.8%, that of the silt content (2 95–63 μm) was 11.8–14.3%, and the clay content was 4.5–5.4% (WRB 2022). According to the soil texture and structure of the profile (Ah-AB-B1-B2-2Cα1-2Cα2), the soil belonged to the Arenosol group (Eutric Endocalcaric Brunic Arenosol (Geoabruptic)) (WRB 2022) [30]. The upper layer of the soil was relatively low in organic carbon, and its amount was significantly reduced at the AB, B1, and B2 horizons. At a depth of approximately 40(60)–110 cm, the carbonate gravel horizon (2Cα1-2Cα2) began. The upper part of the profile consisted of non-carbonate sand, and the lower part consisted of carbonate pebbles and cobbles. The texture of the profile determined the presence of a geochemical barrier (calcium carbonate) at depths of 40–60 cm.

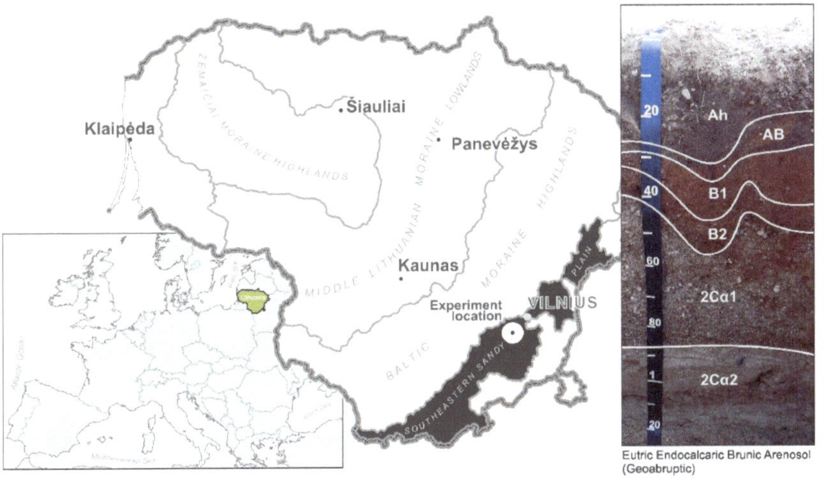

Figure 1. Study object and research site. Soil profile index according to WRB 2022: "A"—a mineral horizon in which decomposed organic material is accumulated; "B"—an illuvial horizon; "C"—an initial mineral horizon; "h"—a humic horizon with a significant amount of organic matter; "i"—organic material in the initial state of decomposition; "2"—a mineral horizon of another origin (in combination with horizon C); "α"—the primary carbonate.

Table 1. Description of the research site and land use.

Land Use Systems and Total Size of Each Land Use Site	Land Uses
Non-fertilised Cropland Cultivation (CCunfert) (200 m^2)	Various agricultural plants (*Secale cereale* L., *Hordeum vulgare* L., *Triticosecale wittmack*, *Triticum aestivum* L. *Fagopyrum esculentum* Moench, *Solanum tuberosum* L., *Brassica napus* L., *Lupinus angustifolius* L., *Trifolium pratense*, *Lupinus angustifolius* L., *Secale cereale* L., *Hordeum vulgare* L., and *Trifolium pratense*) were cultivated in the cropland site. In order to increase the accumulation of SOC, since 2016, in CC, the five-field crop rotation of a 40% legume plant family was performed. One part of this site was unfertilised. In another part, a fertiliser application was used: N 0–100 kg ha^{-1}, P—13–26 kg ha^{-1}, and K—25–100 kg ha^{-1} (ammonium nitrate, granulate superphosphate, and potassium chloride, respectively). In 1995 and 2000, the cropland soil was fertilised with 40 t ha^{-1} manure.
Fertilised Cropland Cultivation (CCfert) (200 m^2)	
Non-fertilised Cut Grassland Cultivation (GRunfert) (200 m^2)	A grass–legume mixture (*Medicago varia* L., *Festuca arundinacea* Schreb., *Bromus inermis* Leyss, *Festuca rubra* L., and *Dactylis glomerata* L.) was grown in a cut grassland site. It was reduced twice during the vegetation period and resown as needed for approximately three decades. One part of this site was unfertilised. In another part, a fertiliser application was used: N60+30P40K1020. N60P40K1020 was applied at the beginning of grass vegetation. The grass was fertilised for a second time (N30) after the first grass cutting.
Fertilised Cut Grassland Cultivation (GRfert) (200 m^2)	
Uncultivated Abandoned Land (UAL) (400 m^2)	Uncultivated uncontrolled wild grass. It was reduced to only shrubs and trees, and sporadic wood cutting was performed according to the need in order to avoid overgrowth of trees (self-afforestation process). Biomass was left in the experimental area.
Pine Afforestation Field (PA) (400 m^2)	Afforested by pine trees (*Pinus sylvestris* L.).

In the year of the beginning of the experiment (1995), the chemical properties of the soil were as follows: pH_{KCl} 6.0–6.8, available phosphorus 157–188 mg P_2O_5 kg^{-1}, available potassium 170–194 mg K_2O kg^{-1}, and SOC 9.5–9.9 g kg^{-1}. Depending on land use and fertilisation, the soil chemical properties changed over the last 25 years. Without using fertilisers in the agricultural land use (CCunfert and GRunfert), the amount of available phosphorus decreased by 70–106 mg P_2O_5 kg^{-1} and that of the available potassium decreased by 94–96 mg K_2O kg^{-1}. Fertilisation with mineral fertilisers increased the amount of nutritional elements in the soil. The amounts of mobile elements changed differently in the natural land use area. In the UAL soil, the concentration of phosphorus and potassium did not change significantly, while the PA decreased by 35 mg P_2O_5 kg^{-1} and 64 mg K_2O kg^{-1}, respectively. The amount of SOC at the beginning of the study was small (9.5–9.9 g kg^{-1}), and during the course of the experiment, it changed depending on the land use and applied agrotechnical measures. The carbon-to-nitrogen ratio (C/N) in 1995 ranged from 12.5 to 13.6, and this was favourable for the transformation of the organic matter in the soil. For the long-term experiment, the experimental land use changes and experimental design are described in more detail in [31].

2.2. *Soil Sampling and Chemical Analysis*

Soil samples were collected from each of the six plots at two soil depths (0–15 and 15–25 cm), with four replicates (n = 48). The air-dried soil samples were crushed and sieved through 2 mm and 0.25 mm sieves and homogeneously mixed before the visible roots and plant residues were removed manually. Total soil nitrogen (N) was determined using a spectrophotometric measurement procedure at a wavelength of 655 nm after mineralisation with sulfuric acid (H_2SO_4). The SOC content was determined according to

Nikitin's modified Tyurin method [32], which consisted of dichromate digestion at 160 °C for 30 min and a spectrophotometric measurement at a wavelength of 590 nm using glucose as a standard. The group and fractional compositions of the soil humus were determined using Ponomareva and Plotnikova's (1980) [33] version of the Tyurin method according to the scheme presented in Figure 2. The following humic acid fractions were identified: HA1—free and weakly bound with clay minerals, referred to as the mobile humic acids fraction; HA2—bound with calcium; HA3—strongly bound with soil clay minerals; the fulvic acid fractions: FA1a—the so-called aggressive fulvic acid fraction; FA1—mobile; FA2—bound with calcium; FA3—bound with soil clay mineral fulvic acid fractions. A more detailed description of the fractionation methodology can be found in the previous studies [27,28,32–34].

The humification properties were calculated using the following formula:

$$\text{Humification degree (HD, \%)} = \Sigma\text{HA}/\text{SOC} \times 100 \quad (1)$$

where HA is the humic acid C content (g kg^{-1}), and SOC is the soil organic carbon content (g kg^{-1}).

The "aggressiveness" of HS was calculated using the following formula [27]:

$$\text{Aggressiveness, \%} = \text{FA1a}/(\Sigma\text{HA} + \Sigma\text{FA}) \times 100 \quad (2)$$

where FA1a is the C content of the most aggressive FA1a fraction (g kg^{-1}), HA is the humic acid C content (g kg^{-1}), and FA is the fulvic acid C content (g kg^{-1}).

The C/N ratio was calculated as the relationship between the SOC (g kg^{-1}) and nitrogen content (N, g kg^{-1}) in the soils.

The optical density (specific extinction) of the HS solution was determined as described previously [35]. It is a classic parameter that is closely related to maturity and aromaticity. The absorbance of the HS was measured at 465 and 665 nm in 0.5 M NaOH extracts using a UV–VIS spectrophotometer, and the polydispersity or degree of polymerisation was calculated using the ratio E4/E6.

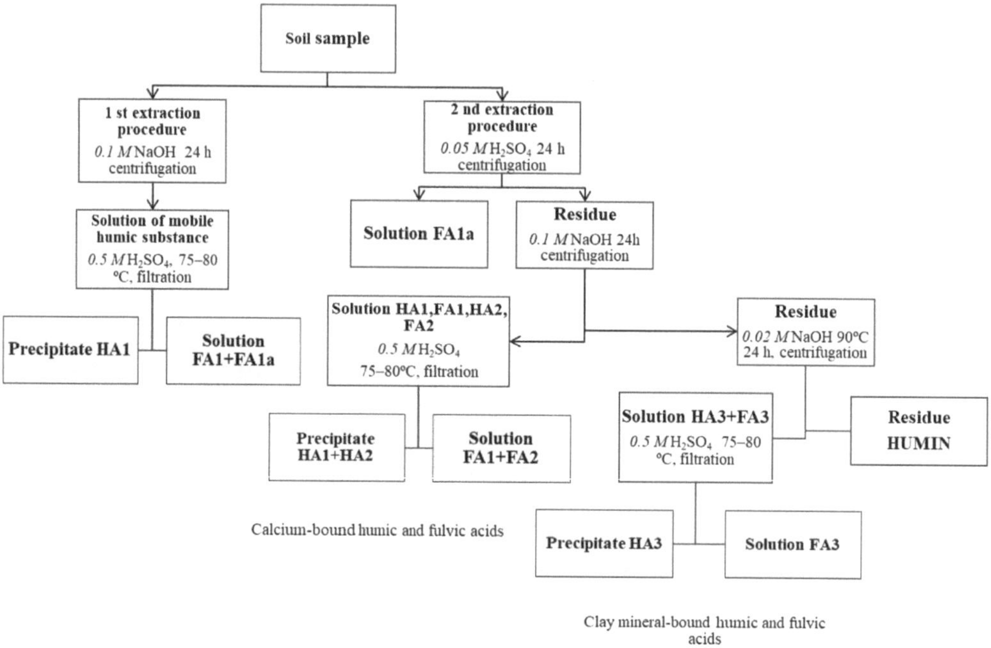

Figure 2. The scheme of humus fractional composition (according to Ponomareva and Plotnikova).

2.3. Statistical Analysis

All the data were analysed using SAS Enterprise software, version 7.1 (SAS Institute Inc., Cary, NC, USA). All the results for the SOC and humus fractional composition are the means of four field replicates. The differences between the experimental treatments were tested using one-way analysis of variance (ANOVA). The probability level was set at 0.05 and grouped according to letters by Duncan's test. Standard error (SE) values were used to estimate the deviations of the soil chemical parameters from the mean values.

2.4. Meteorological Conditions during the Experiment (1995–2022)

The study was conducted in east Lithuania, which is a part of Central Europe (the eastern Baltic region). These regions are characterised by a moderate climate, with a mean long-term (1991–2020) annual precipitation value of 678 mm, and an annual mean air temperature of 7.4 °C (standard climate norm—SCN).

During 1995–2022, annual precipitation ranged from 519 mm in 1996 and 1999 to 963 mm in 2010 (SKN 678 mm). The average annual precipitation during this period was 687 mm, which did not differ significantly from that of the 1991–2020 (678 mm) SKN. Based on the regression analysis of the precipitation change during the investigation period, the linear correlation function trend did not show any significant change in its quantity due to climate change ($r = 0.08$). The relatively rainy years (when the amount of precipitation was 20% higher than the SKN) were 2010, 2011, and 2017; the dry years (when the amount of precipitation was 20% lower than the SKN) were 1996, 1999, 2018, and 2019. The air temperature in Lithuania is rising, as is the case for the whole world. Larger changes began in the last decade of the 20th century and in the 21st century. Compared with the 1961–1990 air temperature SKN, the 1991–2020 SKN increased by +1.0 °C. The winter and spring seasons became warmer by +1.6 °C, summer became warmer by +1.4 °C, and autumn became warmer by +1.3 °C [36]. As the average annual air temperature rises, the duration of plant vegetation changes in Lithuania. Compared to 1961–1991, in the period of 1991–2020, the growing season was 13 days longer. This creates more favourable conditions for the growth of plant biomass and extends the period of the decomposition of organic residues in the soil, as the period of soil freezing is reduced when the microbiological processes in it are significantly slowed down due to negative air temperatures.

3. Results

3.1. Distribution of SOC Amounts for Different Land Uses and Depths

The variation in the SOC amount in 1995–2022, according to the data, is given in Figure 3. For the Arenosol, by applying arable farming practices, it is possible to maintain a stable amount if 40% of the crop rotation consists of soil-improving plants (red clover, mixtures of legumes, and cereal plants) and the recommended rates of mineral fertilisers are used.

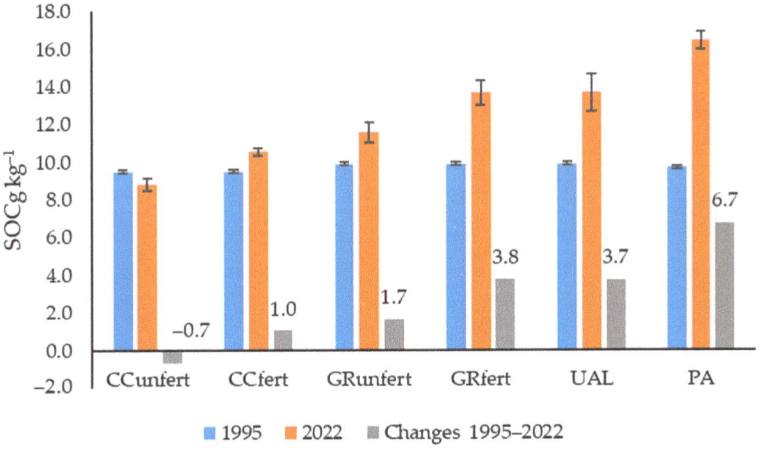

Figure 3. Changes in SOC amounts (1995–2022) in the topsoil of Arenosols with different uses. Notes: CCunfert—non-fertilised crop cultivation; CCfert—fertilised crop cultivation; GRunfert—non-fertilised cut grassland cultivation; GRfert—fertilised cut grassland cultivation; UAL—uncultivated abandoned land; PA—pine afforestation field. The standard errors are marked.

Without the use of fertilisers, plants produce less biomass, fewer post-harvest residues are returned to the soil, and the SOC content gradually decreases at a rate of −0.7 g kg^{-1}. The SOC amount increased the most, at a rate of +6.7 g kg^{-1}, according to the conversion of arable land use to pine afforestation. The cultivation of grasslands (UAL, GRfert, and GRunfert) on arable land was also effective; the amount of SOC increased substantially by an average of 1.0–3.8 g kg^{-1} (Figure 3).

The analysis of SOC in 2022 showed that its levels ranged from 8.50 to 17.32 g kg^{-1} (0–15 cm) and from 8.70 to 13.99 g kg^{-1} (15–25 cm) (Figure 4a,b; Table S1).

Cropland cultivation (CC) negatively affected the amount of SOC compared to the other land uses. The lowest concentration of SOC was established here in the 0–15 cm layer and in the 15–25 cm layer (8.50 and 8.70 g kg^{-1}, respectively). The accumulation of SOC in the upper layer of horizon A was established during the conversion of arable soil into PA or grass phytocenoses (UAL and GR). Pine cultivation, due to naturally occurring organic forest fallout, was determined to cause the highest SOC amount compared to those of the other treatments. In the upper layer, in all the treatments in this research, 0–15 cm had higher SOC amounts than the deeper 15–25 cm layer because of the accumulation of natural organic matter. The natural ecosystems (PA and UAL) and GRfert showed statistically significant differences in SOC accumulation in the 15–25 cm layers.

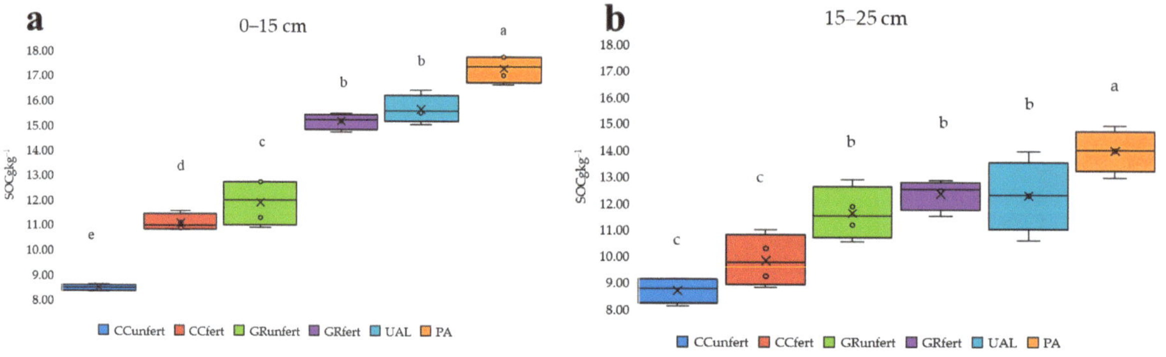

Figure 4. Distribution of SOC in differently used Arenosols at 0–15 cm (**a**) and 15–25 cm (**b**). The different letters (a–e) in the diagram indicate significant differences ($p < 0.05$) between the different Arenosol uses. CCunfert—non-fertilised crop cultivation; CCfert—fertilised crop cultivation; GRunfert—non-fertilised cut grassland cultivation; GRfert—fertilised cut grassland cultivation; UAL—uncultivated abandoned land; PA—pine afforestation field. The standard errors are marked.

3.2. Distribution of Labile and Stable Humus Fractions by Arenosol Use and Depth

One of the best methods used to assess soil quality is to establish the fractional composition of the humus. The HA1 and FA1 fractions are labile; they have the most soluble HA and FA, which are the most likely to form and are the quickest to mineralise. HA1 is free or faintly bound to non-silicate sesquioxides and is insoluble under acidic (pH < 2) conditions, but it is soluble in solutions with higher pH values [27–29]. During the mineralisation process, this labile fraction of humus makes the soil valuable in terms of nutrients. This is necessary for plant development and for the formation of aboveground biomass. In the investigated land types, labile humus fractions consist of 42–53% of the total SOC. The different distributions were due to the different uses of Arenosol. The data presented in Figure 5a,b show that the use of an Arenosol as an arable soil led to a decrease in the amounts of labile HA, FAa1, and FA, while in the uncultivated abandoned land and managed grassland, they increased: PA (HA1 6.19 C g kg^{-1}, FA1 2.29 C g kg^{-1} FAa1 0.79 C g kg^{-1}); UAL (HA1 4.88 C g kg^{-1}, FA1 1.98 C g kg^{-1} FAa1 0.59 C g kg^{-1}); and GRfert (HA1 3.99 C g kg^{-1}, FA1 1.97 C g kg^{-1}). In the deeper soil layer at 15–25 cm, the same trend as in the upper layer was determined. Their amounts varied as follows: HA1 2.28–4.93 C g kg^{-1}, FA1 1.04–1.72 C g kg^{-1}, and FA1 0.29–0.68 C g kg^{-1}. Significantly higher amounts of labile HS were found in the PA as well as in the GRfert and UAL in the deeper layers. The humus labile fractions (HA1, FAa1, and FA1) in the differently used Arenosols at 0–15 cm decreased in the following order: PA > UAL > GRfert > GRunfert < CCfert < CCunfert, respectively; 15–25 cm: PA > GRfert > UAL > GRunfert < CCfert < CCunfert (Table S2).

On closer examination, based on the data in Figure 5c,d, it could be stated that the HA2 quantity was higher in CCunfert at 0.89 C g kg^{-1} (0–15 cm) and UAL at 0.70 C g kg^{-1} (15–25 cm). GRunfert and GRfert contained the lowest amounts of HA2. Thus, the opposite regularity of the distribution of FA2 compared to that of HA2 by land use was established. A smaller amount of FA2 was found in the CCunfert land type compared to those of the other treatments. Larger amounts of FA2 were fo… in the upper layers of CC and PA than those in the lower layers, and the opposite wa… e in the grassland (Table S2). The fractions of HA2 and FA2 were associated with Ca an… ere the most agronomically valuable humus fractions. These humus fractions comprise t… mallest part of the humus (0–15 cm 4–12%; 15–25 cm 6–10%). Based on these results, it …n be stated that a higher amount of HA2 bound to Ca was found in the upper layer tha… that in the deeper layers in the CCunfert, CCfert, and GRunfert soils. The humus fractio… of HA2 in the Arenosols decreased in the following order: CCunfert > UAL > GRfert > CCfert and PA > GRfert. The depth also determined the differences between the soil uses, because at a depth of

15–25 cm, the following order was established: UAL > CCunfert > PA > GRfert > GRunfert (Figure 5c,d).

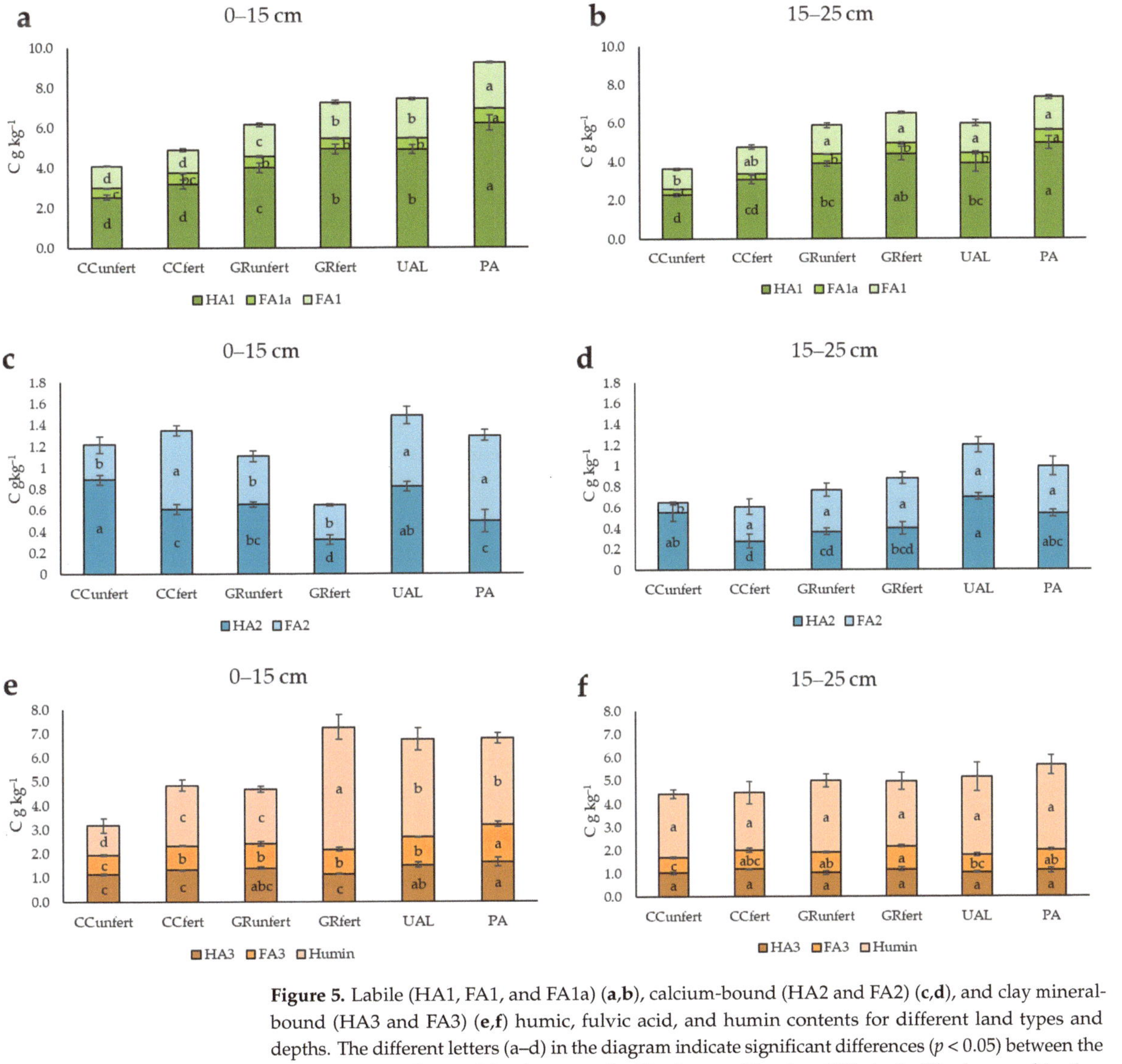

Figure 5. Labile (HA1, FA1, and FA1a) (**a**,**b**), calcium-bound (HA2 and FA2) (**c**,**d**), and clay mineral-bound (HA3 and FA3) (**e**,**f**) humic, fulvic acid, and humin contents for different land types and depths. The different letters (a–d) in the diagram indicate significant differences ($p < 0.05$) between the different Arenosol uses. CCunfert—non-fertilised crop cultivation; CCfert—fertilised crop cultivation; GRunfert—non-fertilised cut grassland cultivation; GRfert—fertilised cut grassland cultivation; UAL—uncultivated abandoned land; PA—pine afforestation field. The standard errors are marked.

In assessing these results, more attention should be paid to the dominant ratio rather than the amounts of HA2 and FA2. The Ca-bound humic-to-FA ratio (HA2/FA2) is characterised by the intensity of the second stage of humification and is used to assess the degree of polymerisation rate of HS and the formation of humates. Secondary humification occurred when HA2/FA2 > 1 in the CCunfert Arenosol: 3.19 (0–15 cm) and 5.44 (15–25 cm), GRunfert 1.51 (0–15 cm), UAL 1.52, and PA 1.44 (15–25 cm).

The HA3 and FA3 fractions accounted for 17–23% (0–15 cm) and 14–20% (15–25 cm) of the total SOC, respectively. The highest amounts of HA3 and FA3 were determined in the PA soil use groups at HA3 1.6 C g kg^{-1} and FA3 1.5 C g kg^{-1}, and conversely, the lowest were CCunfert HA3 1.14 C g kg^{-1} and FA3 0.799 C g kg^{-1}. In the other soil use groups, similar amounts of HA3 and FA3 prevailed.

Examining the differences between the layers in the HA3 and FA3 fractions, it was found that CCunfert, CCfert, and GRfert negatively affected the amounts of HA3 and FA3 in the upper layers. The HA3 amounts were higher in the upper layers of CCfert, GRunfert, and UAL. The FA3 amount was lower in the deeper layers of CCfert, UAL, and PA (Figure 5e,f; Table S2).

Humin is a component of SOM that is insoluble in aqueous bases at any pH and is a heterogeneous mixture of old and new macro-organic substances [37]. The humin fraction represents the most resistant pool of SOM; it plays a crucial role in soil carbon sequestration and, due to its high functional group content, contributes to the maintenance of soil and its ecosystem services [38]. The knowledge regarding the properties of humin and its importance in soil quality is lacking. As reported previously [27,39,40], humin represents the highest proportion of soil humus composition. The results of our experiment on Arenosols showed that the humin fraction ranged from 15% to 33% in the 0–15 cm layer and from 23% to 31% in the 15–25 cm layer of SOC in all the soils used (Figure 5e,f). The highest amount of humin was found in GRfert and UAL (0–15 cm) and in PA in the lower layer (15–25 cm). In the non-fertilised treatments, a higher amount of humin formed in the deeper layer than that in the upper layer (Table S2).

3.3. *Qualitative Characteristics of Humus According to Different Land Uses of Arenosol*

The main indicators that indicate the qualitative composition of humus are the amounts of ΣHA and ΣFA and their ratios. Table 2 presents the summarised data for all the humic ΣHA (HA1 + HA2 + HA3) and fulvic ΣFA (FA1a + FA1 + FA2 + FA3) acids and their distributions according to land use.

Table 2. Quality and properties of humic substances according to land use and depth in Arenosols.

Land Use	ΣHA		ΣFA		ΣHA + ΣFA		ΣHA/ΣFA		"Aggressiveness"	
	g kg^{-1}								%	
					0–15 cm					
CCunfert	4.57	dA	2.68	dA	7.25	eA	1.71	aA	6.31	aA
CCfert	5.12	dA	3.47	cA	8.59	dA	1.48	aA	6.2	aA
GRunfert	6.06	cA	3.61	cA	9.67	cA	1.69	aA	5.76	abA
GRfert	6.39	bcA	3.71	cA	10.1	cA	1.73	aA	5.28	bcA
UAL	7.22	bA	4.40	bA	11.6	bA	1.64	aA	5.03	cB
PA	8.33	aA	5.41	aA	13.7	aA	1.54	aA	5.49	bcB
					15–25 cm					
CCunfert	3.86	dB	2.1	dB	5.96	dB	1.85	aA	4.89	cB
CCfert	4.52	cdA	2.82	cB	7.35	cB	1.61	bA	4.12	dB
GRunfert	5.27	bcA	3.27	bA	8.54	bcA	1.61	bA	5.78	abA
GRfert	5.97	abA	3.6	abA	9.57	abA	1.65	abA	5.52	bcA
UAL	5.65	abB	3.30	bB	8.95	bB	1.70	abA	6.03	abA
PA	6.59	aB	3.74	aB	10.3	aB	1.76	abA	6.60	aA

Notes. CCunfert—non-fertilised crop cultivation; CCfert—fertilised crop cultivation; GRunfert—non-fertilised cut grassland cultivation; GRfert—fertilised cut grassland cultivation; UAL—uncultivated abandoned land; PA—pine afforestation field. ΣHA—sum of humic acid; ΣFA—sum of fulvic acids; ΣHA + ΣFA—sum of humic and fulvic acids; ΣHA/ΣFA—ratio of sum of humic and sum of fulvic acids. Lowercase letters indicate differences ($p < 0.05$) between land use in the 0–15 and 15–25 cm layers, and capital letters indicate differences ($p < 0.05$) between soil layers within the same land type.

The amounts of ΣHA and ΣFA were the highest in PA (ΣHA3 8.33 and 6.59; ΣFA3 5.41 and 3.47), and, in contrast, they were the lowest in CCunfert and CCfert. Fertilisation affected the amounts of ΣHA and ΣFA, as the fertilised treatments contained more ΣHA and ΣFA than the unfertilised treatments. Owing to the roots and fallows, more ΣHA and ΣFA accumulated in the upper layer than those in the lower layer in the natural ecosystem soil. This was set up for the cultivated leguminous plants, and in CCunfert, the soil used contained more ΣHA and ΣFA than in the lower layer. Significantly more ΣFAs were found in the upper layer of the CCfert than in the deeper layers.

By summarising all the fractions of humus, PA, and UAL maintenance, the largest amounts of HS (ΣHA + ΣFA) and their accumulation were determined. The CCunfert soil was the most inappropriate land use type in terms of HSs. An increased ($p < 0.05$) accumulation of HSs was found in the upper layers than that in the deeper layers (Table 2).

The ΣHA/ΣFA ratio is the main indicator of humus quality. The higher the ratio is, the better the humus quality is, and the greater the possibility of preserving the humus is. If the HA/FA ratio is <0.5, FA predominates in the soil humus; when the ratio varies from 0.5 to 1.0, the humus is called humic fulvic type. When HA/FA > 1, it predominates the HA [6,11,27,28]. All the variants in this research are dominated by HA acids, as the ΣHA/ΣFA ratio ranges from 1.46 to 1.73 (0–15 cm soil layer) and from 1.61 to 1.85 (15–25 cm soil layer) (Table 2). Therefore, they were evaluated as having high-quality HSs. Significant differences between the soils used in the upper layer were not observed; however, in the 15–25 cm soil layer, CCunfert was different. It had a larger HA/FA ratio than that of the other treatments. No differences were found between the depths, except in CCfert, and there was a set ΣHA/ΣFA ratio in the deeper layer that was larger than that in the upper layer. Therefore, the ΣHA/ΣFA ratio should not be unambiguously evaluated. This is because the cultivation of leguminous plants positively influences humification in crop cultivation. In the arable soil in the upper layer (0–15 cm), the HS was characterised by a greater degree of aggressiveness (6.20–6.31), while in the grass cenoses (UAL and GRfert) and PA, the aggressiveness of HS was significantly reduced (5.03–5.49) (Table 2). In the deeper layer (15–25 cm), the highest amounts of aggressive compounds were found in the PA soil, but here they should be considered as naturally occurring fulvic compounds due to pine cultivation.

3.4. Cultivation and Abandonment Effects on Humification and Optical Properties of HS

The proportion of SOC and N content (C/N) is a simple and popular indicator of SOM quality, the potential humification of organic residues, and N mineralisation [27,40–42]. The C/N ratio indicates the SOM composition, stability, and mineralisation. Therefore, this ratio helps predict the impact of land use change on SOC sequestration and greenhouse gas emissions [41,43].

The C/N ratio ranged from 14.7 to 19.8. The UAL and PA soils had higher C/N ratios than the cropland (CC) and GRunfert soils, which may have been due to N supplementation and its complexation with the SOM (Table 3). The most stable SOCs, according to the C/N ratio, were formed when the Arenosol was used in PA (19.8 and 18.6) and UAL (17.0 and 16.5). A less-stable SOC was formed in the Arenosol when it was used for crop cultivation. The lowest C/N ratio (14.7–15.8) was determined in the upper layers of the Arenosol in the CC. According to [44], the organic matter in the studied soils was characterised by a high C/N ratio (approximately 20), which indicates the low availability of nitrogen for microorganisms and plants.

The relative proportion of HA to SOC, which is comprehensible as the humification degree (HD), indicates the stability of the SOM [6,28]. An increase in the HD is both agronomically and ecologically valuable. The data show that the highest degree of humification (HD) was in the PA (0–15 cm and 15–25 cm). In other land use systems, no significant differences were established, although a tendency towards a greater accumulation of HA in the herbaceous phytocenoses compared to that of CC can be noted. The most commonly used indicator, E4/E6, provides information on the molecular size and weight of the HSs [11].

Lower values of this ratio indicate the advanced transformation of the SOM and the prevalence of humus compounds in the mature phase of humification [11,29]. Other authors have suggested that a low E4/E6 ratio indicates the accumulation of stable forms of SOM because of the potentially higher proportion of HA [20,26,34,44]. According to the previous studies [20,44–46], the E4/E6 ratio is inversely related to the degree of condensation and aromaticity of HS and the degree of humification. HSs with high molecular weights (HA) have low ratios (E4/E6 < 5), whereas those with low molecular weights (FA) have high ratios (5 < E4/E6 < 10) [20,47].

Table 3. Indicators of humification and optical properties of HS of differently used Arenosols.

Land Use	C/N		HD, %		E465		E665		E4/E6	
	0–15 cm	15–25 cm	0–15 cm	15–25 cm	0–15 cm	15–25 cm	0–15 cm	15–25 cm	0–15 cm	15–25 cm
CCunfert	14.7 cA	16.0 bcA	30 bA	26 bA	0.117dA	0.108 cA	0.018 cA	0.015 cA	6.71 abA	7.05 aA
CCfert	15.5 bcA	14.8 bcA	29 bA	31 abA	0.170 dA	0.163 cA	0.025 cA	0.025 cA	6.87 aA	6.53 abcA
GRunfert	14.7 cA	15.5 bcA	33 abA	33 aA	0.262 cA	0.244 bA	0.041 bA	0.038 bA	6.39 abA	6.44 bcA
GRfert	15.8 bcA	13.6 cB	32 abA	35 aA	0.320bcA	0.276 bA	0.048 bA	0.041 bA	6.73 abA	6.90 abA
UAL	17.0 bA	16.5 abA	31 abA	32 abA	0.334 bA	0.263 bA	0.050 bA	0.040 bA	6.77 abA	6.67 abcA
PA	19.8 aA	18.6 aA	36 aA	26 bA	0.465 aA	0.375 aB	0.074 aA	0.060 aB	6.26 bA	6.31 cA

Notes. C/N—ratio of soil organic carbon and nitrogen; HD—humification degree; E465—humic solution absorbance at 465 nm; E655—humic solution absorbance at 665 nm; E4/E6—ratio of humic solution absorbance from 465 nm to 665 nm. CCunfert—non-fertilised crop cultivation; CCfert—fertilised crop cultivation; GRunfert—non-fertilised cut grassland cultivation; GRfert—fertilised cut grassland cultivation; UAL—uncultivated abandoned land; PA—pine afforestation field. Lowercase letters indicate differences ($p < 0.05$) between land use in the 0–15 and 15–25 cm layers, and capital letters indicate differences ($p < 0.05$) between soil layers within the same land type.

The measurement results of the absorbance of HSs were 0.117–0.465 and 0.108–0.375 at E465 and 0.018–0.074 and 0.015–0.060 at E655 at 0–15 cm and 15–25 cm, respectively. The highest values of E4 and E6 were determined in the PA soil, and, conversely, the lowest values were found in the fertilised and non-fertilised CC. UAL- and MGunfert and -fert were similar. Statistical differences between the depths were determined only for PA soil use. During secondary humification, some of the organic compounds are transformed into FA, and later, into HA; lower values of the E4/E6 ratio are typical for the more transformed soils. The E4/E6 ratio of the studied soils was significantly lower in PA (0–15 and 15–25 cm), suggesting secondary humification. The lowest E4/E6 ratios were characteristic of large molecules and indicated increased levels of molecular association and humification (Table 3). These results show variations in the low molecular weights of the particles, which had a high E4/E6 ratio of 6.26–7.05. No significant differences were observed in the optical densities of E465 and E665 of the HSs at the 0–15 cm soil depth, except for the PA treatment at the soil depth of 15–25 cm.

4. Discussion

The results of a study of the humus status of soil 27 years after the conversion of arable soil into various types of land use made it possible to assess the influence of phytocenoses and agrotechnical practices on the accumulation of humic substances and their fractional composition in Arenosols. Our study confirmed the previously formulated conclusion that the more active accumulation of humus in sandy loam soils occurs in forests. In herbaceous phytocenoses, the accumulation of humus occurs more slowly than it does in PA. Similar data on faster SOC accumulation in forest and grass phytocenoses have been reported by many researchers [4,13,48].

Research on the quality of HSs has revealed that in Arenosols, the most effective way to increase the amount of HSs was to perform pine afforestation because, in 27 years, the SOC amount increased by 6.3 g kg^{-1}, and the highest amounts of the following contents were also determined: the SOC, agronomically valuable labile HA1, FA1, HA3, and FA3 bound to clay, ΣHA + ΣFA, C/N ratio, and HD. At the same time, the low-level aromaticity of HSs was discovered.

4.1. Humic Substance Quality in Agricultural Ecosystems of Arenosols

Relatively more (c 85.3%) SOC was formed in the CCunfert soil. The lowest amount was found in the GRfert and UAL lands; they accumulated more humin 25.9–33.4% SOC. The most pronounced changes in humus fractions were observed in the GRfert soil. There were more labile HS and fewer aggressive compounds and stable compounds with calcium and clay, but, at the same time, almost twice as much humin 33.4% from SOC than that in CC was formed in GRfert. This confirms and contributes to the research conducted in [49,50] in sandy soils, and it was concluded that the use of mineral fertilisers has a beneficial effect on the amounts of stable SOM fractions; however, the application of mineral fertilisers reduced the humification processes of SOM, supported the mineralisation processes, and led to a decrease in the quality of the humus.

4.2. Humic Substance Quality in Natural Ecosystems of the Arenosol

A similar composition of HS was found in the UAL soil, although, unlike in the GRfert, the aboveground biomass of the plant was not removed from the field, and after the end of vegetation, it remained mineralised on the soil surface. However, the application of mineral fertilisers increases [51] the biomass of grasses by 2–3 times in GRfert, which leads to the formation of a larger root mass and a higher amount of post-harvest residues in this land use. Therefore, the mineralisation processes of plant residues in both land types had a similar effect on the relative amount of HS fractions: they formed similar amounts of HA, FA, and HD. The compositions of the HA and FA fractions also showed no significant differences, except for those of the labile HA1, FA1, and FA fractions. These fractions, which were more labile in the soil, were less abundant in the UAL soil. According to [11] and [52,53], abandoning land usually causes an increase in the SOC content, but SOM transformation and accumulation depend on many factors, such as temperature, microbial activity, soil texture, soil moisture, and the composition of the plant and microbial residues. The quality indicators of HSs include not only the amounts of HA and FA and their ratio, but also the HD and E6/E6 ratio. This research confirmed and supplemented the results of the other studies [12,27] that demonstrated that ploughing negatively affects humification. The results showed that the use of Arenosol led to a low C/N ratio and HD and, at the same time, a higher E4/E6 ratio, and humification was accelerated by pine cultivation (high C/N ratio of 19.6 and HD of 36% at 0–15 cm, and C/N ratio of 18.6 and HD of 35% at 15–25 cm and, at the same time, lower E4/E6 values) and had the same indicator trend as grassland cultivation.

4.3. Afforestation Effect on Humic Fractional Composition and Quality

In the PA soil, the qualitative composition of HSs was not significantly different from that of the GRfert soil. However, PA had relatively more HA and FA 79.3% SOC, higher HDs (36% at 0–15 cm; 35% at 15–25 cm), C/N ratios of 17.0 and 16.5, and less humin 20.7% SOC. This indicates a relatively lower formation of difficult-to-decompose HS (low E4/E6 ratio) in the PA soil. Kukuļs and their team [4] studied the influence of afforestation on the SOC content and the properties of SOM in the mineral topsoil in Latvia. They found that a gradual decrease in the HA-to-FA ratio after land use change showed that the SOM was transported to the mineral topsoil, mainly through the leaching of the soluble FA fraction. The FTIR spectra revealed that the SOM in the mineral topsoil was degraded to small compounds. However, the HA extracted from forest soils had a higher abundance of –OH groups, more aromatic C–C structures, and a relatively lower abundance of N–H groups in the amides [54]. The soil properties may also affect the chemical properties of the SOM during afforestation [55]. The molecular weight of the HA fraction gradually increased with the age of the forest land in the sandy soils. These results suggest that sandy soils have the most rapid changes in SOM properties after afforestation. According to a study conducted in Lithuania [19], the findings suggest the encouragement of the afforestation of former agricultural land according to the climate and soil characteristics of the territory, but the conversion of perennial grasslands to forests should be performed with care. The SOC values in the Arenosols 30 years after afforestation did not differ significantly from those in the croplands or grasslands [56]. In the Arenosols, there was higher SOC accumulation in the forest topsoil with an increasing stand age, whereas the proportion of SOC stocks in the mineral topsoil layers was similar to that in the grasslands.

In the Arenosol soil, according to the effect on SOC sequestration and the qualitative composition of HSs, the formation of the resistance to decay and the sequence of land use was as follows: GRfert > PA > UAL > GRunfert > CCunfert > CCfert.

5. Conclusions

Long-term research (1995–2022) on the transformation of arable lands into natural ecosystems (uncultivated abandoned land and pine afforested field) or into land to be used for agriculture (fertilised and unfertilised cut grassland) has confirmed the differences in humus accumulation and revealed trends in the change in the HS fractional composition.

The different uses of Arenosols changes not only the accumulation of SOC in Arenosols, but also affect the qualitative composition of the humus compounds.

Arenosols used for PA have an increased SOC (0–15 cm +6.23 g kg^{-1}), and the use of herbaceous phytocenoses positively impacts the SOC by +3.8 g kg^{-1} in GRfert and +3.7 g kg^{-1} in UAL compared to the arable land (CCfert).

When the arable soil is converted into PP, the concentrations of humin, ΣHA + ΣFA, and aggressive humus fraction FA1 also increase significantly. These processes were established in the 0–25 cm soil layer.

Humin formed more intensively in herbaceous phytocenoses (GR and UAL) compared to that in CCfert, but only in GRfert and UAL (5.08 g kg^{-1} and 4.06 g kg^{-1}, respectively). The relative amount of ΣHA + ΣFA reactions changed slightly. The plant residues remaining in the UAL after the end of the plant growing season influenced the formation of humic substances; the SOC amount in the soil was higher by +3.74 g kg^{-1}, with more humin (+1.80 g kg^{-1}) and ΣHA + ΣFA (+1.95 g kg^{-1}) compared with those of GRunfert.

The use of fertilisers in CC and GR contributes to the more intensive accumulation of SOC (+2.59 g kg^{-1} in CC and +3.26 g kg^{-1} in GR, respectively), including humin (+1.27 g kg^{-1} in CC and 2.82 g kg^{-1} in GR, respectively), but, at the same time, the relative amount of the ΣHA + ΣFA fraction decreases.

Supplementary Materials: The following supporting information can be downloaded at: https://www.mdpi.com/article/10.3390/agriculture14020250/s1, Table S1. Significant ($p < 0.05$) differences in SOC were observed between the soil layers within the same land type. The standard errors are marked; Table S2. Significant ($p < 0.05$) differences of HS between soil layers within the same land type.

Author Contributions: Conceptualization, K.A.-V., L.T. and A.K.-J.; methodology, K.A.-V., L.T. and A.K.-J.; software, K.A.-V., L.T. and A.K.-J.; validation, L.T., A.K.-J., K.A.-V., A.S. and E.B.; formal analysis, K.A.-V. and L.T.; investigation, L.T., K.A.-V., A.K.-J. and E.B.; resources, L.T., A.K.-J., K.A.-V., A.S. and E.B.; data curation, K.A.-V., L.T. and A.K.-J.; writing—original draft preparation, L.T., K.A.-V. and A.K.-J.; writing—review and editing, L.T., K.A.-V., A.K.-J., A.S. and E.B.; visualization, K.A.-V. and L.T.; supervision, L.T., K.A.-V. and A.K.-J. All authors have read and agreed to the published version of the manuscript.

Funding: This research received no external funding.

Institutional Review Board Statement: Not applicable.

Data Availability Statement: The original contributions presented in the study are included in the article/Supplementary Material, further inquiries can be directed to the corresponding author.

Acknowledgments: The experimental findings were obtained through the research programme "Biopotential of plants for multifunctional use and sustainability of agroecosystems" implemented by the Lithuanian Research Centre for Agriculture and Forestry. The authors greatly thank M. Petrovas for the experimental technique development (1994) and for conducting the experiment up to 2001 and S. Marcinkonis for experimental execution during the period 2002–2012; we also thank J. Volungevicius for the soil profile characterisation in 2015.

Conflicts of Interest: The authors declare no conflicts of interest.

References

1. Stockmann, U.; Adams, M.A.; Crawford, J.W.; Field, D.J.; Henakaarchchi, N.; Jenkins, M.; Minasny, B.; McBratney, A.B.; de Courcelle, S.V.R.; Singh, K.; et al. The knowns, known unknowns and unknowns of sequestration of soil organic carbon. *Agric. Ecosyst. Environ.* **2013**, *164*, 80–99. [CrossRef]
2. Lal, R.; Wakene Negassa, W.; Klaus Lorenz, K. Carbon sequestration in soil. *Curr. Opin. Environ. Sustain.* **2015**, *15*, 79–86. [CrossRef]
3. Lal, R.; Smith, P.; Jungkunst, H.F.; Mitsch, W.J.; Lehmann, J.; Nair, P.K.R.; McBratney, B.; de Moraes Sá, J.C.; Schneider, J.; Zinn, Y.L.; et al. The carbon sequestration potential of terrestrial ecosystems. *J. Soil. Water Conserv.* **2018**, *73*, 145A–152A. [CrossRef]
4. Kukuļs, I.; Kļaviņš, M.; Nikodemus, O.; Kasparinskis, R.; Brūmelis, G. Changes in soil organic matter and soil humic substances following the afforestation of former agricultural lands in the boreal-nemoral ecotone (Latvia). *Geoderma Reg.* **2019**, *15*, e00213. [CrossRef]
5. Malhi, G.S.; Kaur, M.; Kaushik, P. Impact of Climate Change on Agriculture and Its Mitigation Strategies: A Review. *Sustainability* **2021**, *13*, 1318. [CrossRef]
6. Raiesi, F. The quantity and quality of soil organic matter and humic substances following dry-farming and subsequent restoration in an upland pasture. *Catena* **2021**, *202*, 105249. [CrossRef]
7. Ukalska-Jaruga, A.; Bejger, R.; Debaene, G.; Smreczak, B. Characterization of Soil Organic Matter Individual Fractions (Fulvic Acids, Humic Acids, and Humins) by Spectroscopic and Electrochemical Techniques in Agricultural Soils. *Agronomy* **2021**, *11*, 1067. [CrossRef]
8. Malou, O.P.; Chevallier, T.; Moulin, P.; Sebag, D.; Rakotondrazafy, M.N.; Badiane-Ndour, N.Y.; Thiam, A.; Chapuis-Lardy, L. Measuring the stability of soil organic carbon in Arenosols in the Senegalese Groundnut Basin. *J. Arid. Environ.* **2023**, *213*, 104978. [CrossRef]
9. Dilly, O.; Gnaß, A.; Pfeiffer, E.-M. Humus accumulation and microbial activities in Calcari-Epigleyic Fluvisols under grassland and forest diked in for 30 years. *Soil Biol. Biochem.* **2005**, *37*, 2163–2166. [CrossRef]
10. Don, A.; Scholten, T.; Schulze, E.-D. Conversion of cropland into grassland: Implications for soil organic-carbon stocks in two soils with different texture. *J. Plant. Nutr. Soil Sci.* **2009**, *172*, 53–62. [CrossRef]
11. Mielnik, L.; Hewelke, E.; Weber, J.; Oktaba, L.; Jonczak, J.; Podlasiński, M. Changes in the soil hydrophobicity and structure of humic substances in sandy soil taken out of cultivation. *Agric. Ecosyst. Environ.* **2021**, *319*, 107554. [CrossRef]
12. Guillaume, T.; Makowski, D.; Libohova, Z.; Elfouki, S.; Fontana, M.; Leifeld, J.; Bragazza, L.; Sinaj, S. Carbon storage in agricultural topsoils and subsoils is promoted by including temporary grasslands into the crop rotation. *Geoderma* **2022**, *422*, 115937. [CrossRef]
13. Rambaut, L.-A.E.; Vayssières, J.; Versini, A.; Salgado, P.; Lecomte, P.H.; Tillard, E. 15-year fertilization increased soil organic carbon stock even in systems reputed to be saturated like permanent grassland on Andosols. *Geoderma* **2022**, *425*, 116025. [CrossRef]
14. Papierowska, E.; Szatyłowicz, J.; Ruta, M.; Łachacz, A.; Gnatowski, T.; Stańczyk, T. Water repellency of soils on unpaved roads in coniferous forests. *Catena* **2020**, *195*, 104784. [CrossRef]
15. Vasconez Navas, L.K.; Becker, J.N.; Heger, A.; Grongroft, A.; Eschenbach, A. Are active and former floodplain soils of the lower middle Elbe similar? A study of soil characteristics and possible implications for forest restoration. *Catena* **2023**, *222*, 106814. [CrossRef]
16. An, S.; Mentler, A.; Mayer, H.; Blum, W.E.H. Soil aggregation, aggregate stability, organic carbon and nitrogen in different soil aggregate fractions under forest and shrub vegetation on the Loess Plateau. *Catena* **2010**, *81*, 226–233. [CrossRef]
17. Barcena, T.G.; Kier, L.P.; Vesterdal, L.; Stefansdottir, H.M.; Gundersen, P.; Sigurdsson, B.D. Soil carbon stock change following afforestation in Northern Europe: A meta-analysis. *Glob. Chang. Biol.* **2014**, *18*, 2393–2405. [CrossRef]
18. Lafleur, B.; Labrecque, M.; Arnold, A.A.; Bélanger, N. Organic Carbon Accumulation in Topsoil Following Afforestation with Willow: Emphasis on Leaf Litter Decomposition and Soil Organic Matter Quality. *Forests* **2015**, *6*, 769–793. [CrossRef]
19. Varnagiryte-Kabašinskiene, I.; Žemaitis, P.; Armolaitis, K.; Stakenas, V.; Urbaitis, G. Soil Organic Carbon Stocks in Afforested Agricultural Land in Lithuanian Hemiboreal Forest Zone. *Forests* **2021**, *12*, 1562. [CrossRef]
20. Adiyah, F.; Mich'eli, E.; Csorba, A.; Weldmichael, T.G.; Gyuricza, C.; Ocansey, C.M.; Dawoe, E.; Owusu, S.; Fuchs, M. Effects of land-use change and topography on the quantity and distribution of soil organic carbon stocks on Acrisol catena's in tropical small-scale shade cocoa systems of the Ashanti region of Ghana. *Catena* **2022**, *216*, 106366. [CrossRef]
21. Knist, S.; Goergen, K.; Simmer, C. Evaluation and projected changes of precipitation statistics in convection permitting WRF climate simulations over Central Europe. *Clim. Dyn.* **2020**, *55*, 325–341. [CrossRef]
22. Riedel, T.; Weber, T.K.D. Review: The influence of global change on Europe's water cycle and groundwater recharge. *Hydrogeol. J.* **2020**, *28*, 1939–1959. [CrossRef]
23. Bevacqua, E.; Zappa, G.; Lehner, F.; Zscheischler, J. Precipitation trends determine future occurrences of compound hot–dry events. *Nat. Clim. Chang.* **2022**, *12*, 350–355. [CrossRef]
24. Plunge, S.; Gudas, M.; Povilaitis, A. Expected Climate Change Impacts on Surface Water Bodies in Lithuania. *Ecohydrol. Hydrobiol.* **2021**, *22*, 246–268. [CrossRef]
25. Tao, Z.-Y.; Zhang, J.; Zhai, J.-J. Characterization and differentiation of humic acids and fulvic acids in soils from various regions of China by nuclear magnetic resonance spectroscopy. *Anal. Chim. Acta* **1999**, *395*, 199–203. [CrossRef]

26. Zalba, P.; Amiotti, N.M.; Juan, A.; Galantini, J.A.; Pistolaore, S. Soil Humic and Fulvic Acids from Different Land-Use Systems Evaluated By E4/E6 Ratios. *Commun. Soil Sci. Plant Anal.* **2016**, *47*, 1675–1679. [CrossRef]
27. Liaudanskiene, I.; Zukaitis, T.; Velykis, A.; Satkus, A.; Parasotas, I. The impact of tillage practices on the distribution of humified organic carbon in a clay loam. *Zemdirbyste* **2021**, *108*, 11–18. [CrossRef]
28. Mockeviciene, I.; Karcauskiene, D.; Slepetiene, A.; Vilkiene, M.; Repsiene, R.; Braziene, Z.; Anne, O. Influence of Liming Intensity on Fractions of Humified Organic Carbon in Acid Soil: A Case Study. *Sustainability* **2022**, *14*, 5297. [CrossRef]
29. Mockeviciene, I.; Repsiene, R.; Amaleviciute-Volunge, K.; Karcauskiene, D.; Slepetiene, A.; Lepane, V. Effect of long-term application of organic fertilizers on improving organic matter quality in acid soil. *Arch. Agron. Soil Sci.* **2022**, *68*, 1192–1204. [CrossRef]
30. IUSS Working Group WRB. World Reference Base for Soil Resources. In *International Soil Classification System for Naming Soils and Creating Legends for Soil Maps*, 4th ed.; International Union of Soil Sciences (IUSS): Vienna, Austria, 2022.
31. Tripolskaja, L.; Kazlauskaite-Jadzevice, A.; Baksiene, E.; Razukas, A. Changes in organic carbon in mineral topsoil of a formerly cultivated Arenosol under different land uses in Lithuania. *Agriculture* **2022**, *12*, 488. [CrossRef]
32. Nikitin, B.A. Methods for soil humus determination. *AgroChemistry* **1999**, *3*, 156–158. (In Russian)
33. Ponomareva, V.V.; Plotnikova, T.A. *Humus and Soil Formation*; Nauka: Leningrad, Russia, 1980; p. 220.
34. Slepetiene, A.; Slepetys, J. Status of humus in soil under various long-term tillage systems. *Geoderma* **2005**, *127*, 207–215. [CrossRef]
35. Kalisz, B.; Urbanowicz, P.; Smólczyński, S.; Orzechowski, M. Impact of siltation on the stability of organic matter in drained peatlands. *Ecol. Indic.* **2021**, *130*, 108149. [CrossRef]
36. Bukantis, A.; Kažys, J. 330 Years of Vilnius' Climate: History and Future. *Vilnius Univ. Proc.* **2020**, *10*, 10. [CrossRef]
37. Bagherifam, S.; Brown, T.C.; Bagherifam, S.; Baglieri, A. Sequential extraction of labile and recalcitrant fractions of soil organic matter: A case study focusing on antimony (Sb) in humic acids, fulvic acids and humin fractions of long-term aged, contaminated soils. *Environ. Pollut.* **2023**, *327*, 121610. [CrossRef] [PubMed]
38. Weber, J.; Jerzykiewicz, M.; Ukalska-Jaruga, A.; Ćwieląg-Piasecka, I.; Jamroz, E.; Kocowicz, A.; Debicka, M.; Bekier, J.; Mielnik, L.; Bejger, R.; et al. Molecular characteristics of humin fraction isolated from soils of temperate climate: A study on Chernozems and Phaeozems in Poland. *Authorea* **2023**. [CrossRef]
39. Litvinovich, A.; Pavlova, O.; Lavrishchev, A.; Bure, V.; Saljnikov, E. Empirical models of transformations of humic acids and humin in Umbric Albeluvisol Abruptic as influenced by liming. *Zemdirbyste* **2017**, *104*, 115–122. [CrossRef]
40. Han, B.; Kitamura, K.; Hirota, M.; Shen, H.; Tang, Y.; Suzuki, T.; Fujitake, N. Humus composition and humification degree of humic acids of alpine meadow soils in the northeasten part of the Qinghai-Tibet Plateau. *Soil Sci. Plant Nutr.* **2019**, *65*, 11–19. [CrossRef]
41. Radočaj, D.; Jurišić, M.; Antonić, O. Determination of soil C: N suitability zones for organic farming using an unsupervised classification in eastern Croatia. *Ecol. Indic.* **2021**, *123*, 107382. [CrossRef]
42. Amorim, H.C.S.; Araujo, M.A.; Lal, R.; Zinn, Y.L. What C: N ratios in soil particle-size fractions really say: N is preferentially sorbed by clays over organic C. *Catena* **2023**, *230*, 107230. [CrossRef]
43. Amorim, H.C.S.; Hurtarte, L.C.C.; Souza, I.F.; Zinn, Y.L. C: N ratios of bulk soils and particle-size fractions: Global trends and major drivers. *Geoderma* **2022**, *425*, 116026. [CrossRef]
44. Oktaba, L.; Odrobińska, D.; Uzarowicz, Ł. The impact of different land uses in urban area on humus quality. *J. Soils Sediments* **2018**, *18*, 2823. [CrossRef]
45. Kunlanit, B.; Butnan, S.; Vityakon, P. Land-use changes influencing C sequestration and quality in topsoil and subsoil. *J. Agron.* **2019**, *9*, 520. [CrossRef]
46. Chen, Y.; Senesi, N.; Schnitzer, M. Information provided on humic substances by E4/E6 ratios. *Soil Sci. Soc. Am. J.* **1977**, *41*, 352. [CrossRef]
47. Stevenson, F.J. *Humus Chemistry: Genesis, Composition and Reactions*; John Wiley and Sons Publications: New York, NY, USA, 1994; p. 496.
48. Murty, D.; Kirschbaum, M.U.F.; McMurtrie, R.E. Does conversion of forest to agricultural land change soil carbon and nitrogen? A review of the literature. *Glob. Chang. Biol.* **2002**, *8*, 105–123. [CrossRef]
49. Datta, A.; Choudhury, M.; Sharma, P.C.; Priyanka, H.S.; Jat, M.L.; Kar, S. Stability of humic acid carbon under conservation agriculture practices. *Soil Tillage Res.* **2022**, *216*, 105240. [CrossRef]
50. Ukalska-Jaruga, A.; Klimkowicz-Pawlas, A.; SMreczak, B. Characterization of organic matter fractions in the top layer of soils under different land uses in Central-Eastern Europe. *Soil Use Manag.* **2019**, *35*, 595–606. [CrossRef]
51. Šimanský, V.; Wójcik-Gront, E.; Horváthová, J.; Pikuła, D.; Lošák, T.; Parzych, A.; Lukac, M.; Aydın, E. Changes in Relationships between Humic Substances and Soil Structure following Different Mineral Fertilization of *Vitis vinifera* L. in Slovakia. *Agronomy* **2022**, *12*, 1460. [CrossRef]
52. Šimanský, V.; Juriga, M.; Jonczak, J.; Uzarowicz, Ł.; Stępień, W. How relationships between soil organic matter parameters and soil structure characteristics are affected by the long-term fertilization of a sandy soil. *Geoderma* **2019**, *342*, 75–84. [CrossRef]
53. Di Donato, N.; Chen, H.; Waggoner, D.; Patrick, G.; Hatcher, P.G. Potential origin and formation for molecular components of humic acids in soils. *Geochim. Cosmochim. Acta.* **2016**, *178*, 210–222. [CrossRef]
54. Martin, D.; Srivastava, P.C.; Ghosh, D.; Zech, W. Characteristics of humic substances in cultivated and natural forest soils of Sikkim. *Geoderma* **1998**, *84*, 345–362. [CrossRef]

55. Nikodemus, O.; Kasparinskis, R.; Kukuls, I. Influence of afforestation on soil genesis, morphology and properties in glacial till deposits. *Arch. Agron. Soil Sci.* **2013**, *59*, 449–465. [CrossRef]
56. Poeplau, C.; Don, A. Sensitivity of soil organic carbon stocks and fractions to different land-use changes across Europe. *Geoderma* **2013**, *192*, 189–201. [CrossRef]

agriculture

MDPI

Article

Effects of Soil Amendments on Soil Properties, Soil-Borne Pathogens, and Strawberry Growth after Dazomet Fumigation

Qingjie Li [1], Okbagaber Andom [1], Wensheng Fang [2], Dongdong Yan [2], Yuan Li [2], Qiuxia Wang [2], Xi Jin [3,*] and Aocheng Cao [2,*]

1 State Key Laboratory of Efficient Utilization of Arid and Semi-Arid Arable Land in Northern China, The Institute of Agricultural Resources and Regional Planning, Chinese Academy of Agricultural Sciences, Beijing 100081, China; 82101221185@caas.cn (Q.L.); 2020y90100055@caas.cn (O.A.)
2 Institute of Plant Protection, Chinese Academy of Agricultural Sciences, Beijing 100193, China; fangwensheng@caas.cn (W.F.); yandongdong@caas.cn (D.Y.); yli@ippcaas.cn (Y.L.); qxwang@ippcaas.cn (Q.W.)
3 Hebei Technology Innovation Center for Green Management of Soil-Borne Diseases, Baoding University, Baoding 071000, China
* Correspondence: jinxi@bdu.edu.cn (X.J.); aochengcao@ippcaas.cn (A.C.)

Abstract: Soil fumigation can kill soil pathogens and solve the problem of crop continuous cropping. However, soil fumigation also has negative effects on the soil environment. One way to mitigate the negative effects is to apply soil amendments, but there is limited evidence of the effects of combining soil fumigation and amendments. This study was a controlled environmental pot trial. We measured the effects of dazomet fumigation combined with soil amendments on soil-borne pathogens, soil nutrients, enzyme activities, and strawberry growth. The results showed that dazomet fumigation combined with soil amendments significantly increased the content of ammonium nitrogen, available phosphorus and organic matter and increased soil activities by varying degrees. We also found that the control effect of soil-borne pathogens *Fusarium* spp. and *Phytophthora* spp. was further enhanced, reaching 88.97–96.88%. Correlation analysis showed that the growth indices of strawberries such as plant height, stem diameter, chlorophyll content, and fresh weight were negatively correlated with *Fusarium* spp. ($R = -0.75$, $R= -0.62$, $R = -0.71$, $R = -0.88$; $p < 0.01$) and *Phytophthora* spp. ($R = -0.72$, $R= -0.72$, $R = -0.78$, $R = -0.91$; $p \leq 0.001$), respectively. The effect of fumigation combined with soil amendments was better than that of fumigation alone, and silicon fertilizer had the best effect. Our study suggests that dazomet fumigation combined with soil amendments can improve soil nutrient supply, activate soil enzyme activities, enhance the control effect of soil-borne pathogens, and thus promote strawberry growth.

Keywords: soil fumigation; silicon fertilizer; potassium humate; biofertilizer; soil enzyme activity; strawberry

Citation: Li, Q.; Andom, O.; Fang, W.; Yan, D.; Li, Y.; Wang, Q.; Jin, X.; Cao, A. Effects of Soil Amendments on Soil Properties, Soil-Borne Pathogens, and Strawberry Growth after Dazomet Fumigation. *Agriculture* **2024**, *14*, 9. https://doi.org/10.3390/agriculture14010009

Academic Editors: Anetta Siwik-Ziomek and Anna Figas

Received: 25 October 2023
Revised: 7 December 2023
Accepted: 16 December 2023
Published: 21 December 2023

1. Introduction

Strawberry (*Fragaria* × *ananassa* Duchesne) is an important high-value crop that is widely cultivated worldwide [1]. Strawberry fruit has high nutritional value, rich in sugar, vitamins, anthocyanins, and dietary fiber, which are known to have a protective role in anti-cancer, antioxidant, antibacterial, and cardiovascular disease prevention [2,3]. At present, facility cultivation has become the main cultivation pattern of strawberries as it is not limited by the external environment [4]. However, strawberries are usually planted continuously in the same field of the facility, resulting in the spread of soil pathogens, soil quality degradation, and nutrient imbalance [5,6]. If there is a lack of timely and effective control, soil-borne diseases can easily develop, resulting in reduced strawberry yield and quality [7].

Soil fumigation can kill soil pathogens and solve the problem of crop continuous cropping [8]. It involves applying soil fumigants to the soil with professional application equipment and covering it with a plastic film to produce volatile gases in a closed environment to prevent soil-borne diseases, insects, grass, and other hazards [9]. Dazomet (DZ) is a fumigant that hydrolyzes methyl isothiocyanate in soil to control soil-borne diseases [10]. It has been shown to be highly effective, low-toxicity, and residue-free on many crops such as ginger, tomato, and cucumber [11]. However, DZ fumigation not only eliminates soil-borne pathogens but also disrupts the soil environment, often leading to a "near vacuum" state after fumigation [12].

Soil amendments can activate soil beneficial microorganisms, improve the soil environment, and control soil pests and diseases [13,14]. For example, Cheng et al. [15] found that the biofertilizers *Trichoderma harzianum* and *Bacillus subtilis* could increase soil nutrients and reduce the abundance of soil-borne pathogens. Gao et al. [16] reported that biochar could reduce the negative impact of fumigants on the soil environment without reducing nematode control. Wu et al. [17] found that *Trichoderma* combined with dimethyl disulfide and chloropicrin could help control soil-borne pathogens and root-knot nematodes, improve the soil's ecological function, and thus increase the cucumber yield. While the combined use of certain soil amendments and fumigants has significantly enhanced the control effect of soil pests and diseases, additional soil amendments need to be identified for the restoration of the soil micro-ecological environment after fumigation.

Silicon, as the second most abundant element in the Earth's crust, can increase the chlorophyll content of plant leaves and reduce the uptake and transport of harmful substances in the soil by plants [18,19]. Furthermore, silicon fertilizer can enhance the ability of crops to survive adversity [20]. Potassium humate is a soil amendment used to prevent plant diseases and reduce the uptake and utilization of harmful substances by plants [21]. Jin et al. [22] reported that potassium humate can reduce potassium loss and potassium fixation, improve soil nutrient cycling function, and thus promote crop growth. It has been reported that as a beneficial microorganism of the soil, the biofertilizers produced by *Bacillus* can activate the micro-ecological function of the soil and inhibit the propagation of soil-borne pathogens [23]. Although silicon fertilizer, potassium humate, and *Bacillus* biofertilizer can be used to improve the soil environment, the effect of their application on fumigated soil is unclear.

In this study, we investigated the effects of DZ fumigation combined with silicon fertilizer, potassium humate, and *Bacillus* biofertilizer on the soil environment and strawberry growth. A pot experiment with different amendments applied after fumigation was conducted and the growth indexes, soil physicochemical properties, enzyme activities, and control effect of soil-borne pathogens of strawberry were analyzed. The objective of this study was to select the best amendments for soil remediation after fumigation.

2. Materials and Methods

2.1. Soil Collection

In early August 2020, soil was collected from five randomly selected points 5–20 cm deep in a greenhouse used for strawberry cultivation in Hebei, China. The mean annual temperature is 12.9 °C, with a mean annual precipitation of 1431.4 mm. Strawberries have been grown continuously in the greenhouse for more than 15 years. The soil is composed of 60.8% silt, 30.8% sand, and 8.4% clay, which is silty loam. The physicochemical properties of the soil (Table 1) were determined according to our previous research [11]. The soil was mixed through a 2 mm sieve before the experiment, which was cleared of stones and weeds.

Table 1. The basic properties of greenhouse soil.

Ammonium Nitrogen (AN, mg/kg)	Nitrate Nitrogen (NN, mg/kg)	Available Phosphorus (AP, mg/kg)	Available Potassium (AK, mg/kg)	Organic Matter (OM, g/kg)	pH (1:2.5)	Electrical Conductivity (EC, μS/cm)
0.52	44.02	56.14	864.17	16.85	7.37	303.50

2.2. Experimental Design

2.2.1. Soil Fumigation

About 60 kg of soil was divided evenly into five parts and placed in desiccators (30 cm × 30 cm), and DZ (98% purity, obtained from Nantong Shizhuang Chemical Co Ltd., Nantong, Jiangsu, China) was added to the soil to reach a final concentration of 82 mg/kg and then mixed manually. The control soil was without DZ fumigation (12 kg soil), and there were a total of six desiccators. The moisture content of all soils was adjusted to 13%, which was equivalent to the normal field water capacity. All desiccators were sealed and incubated at 25 °C for 10 days, followed by fume hood aeration for 8 days.

2.2.2. Soil Amendments Application

The aerated soil was distributed into plastic pots, and 0.1 g urea and 0.25 g phosphorus pentoxide were applied in each pot as a base fertilizer to ensure the normal growth of the strawberries. After the fertilized soil was stable at 25 °C for 7 days, strawberry seedlings (cv hongyan) at the same growth stage were transplanted into plastic pots, and one strawberry seedling was planted in each pot. All strawberries were incubated in the greenhouse for 60 days and watered regularly with deionized water. The strawberry seedlings used for the experiment were obtained from Wanjian strawberry farmers' professional cooperative in Shunping, Hebei, China.

We weighed 1 g of silicon fertilizer, 1 g of *Bacillus* biofertilizer, and 1 mL of potassium humate into 50 mL volumetric flasks, respectively, and then filled them up to volume with deionized water. In this way, we obtained soil amendments diluted 50 times. Silicon fertilizer was a black particle, potassium humate was a black viscous liquid, and *Bacillus* biofertilizer was a white powder. The composition and source of the soil amendments are shown in Table 2. A randomized complete block design was used with six treatments in this pot experiment: (1) Si: 25 mL of silicon fertilizer diluted 50 times; (2) KH: 25 mL of potassium humate diluted 50 times; (3) T: 25 mL of *Bacillus* biofertilizer diluted 50 times; (4) KHT: 25 mL of a mixture of potassium humate and *Bacillus* biofertilizer diluted 50 times (m: m = 1:1); (5) DZ: 25 mL of deionized water in fumigated soil; and (6) CK: 25 mL of deionized water in control soil. Five treated soils were fumigated with DZ. Each treatment had 12 replicates, for 72 pots in total. Different soil amendments were applied to the root of the strawberry seedlings after transplanting the strawberry seedlings for a week. Soil amendments were uniformly applied to the roots of strawberry seedlings with a syringe on the 7th, 14th, and 21st day after transplanting, respectively. Soil amendments were prepared and used as needed and stored indoors in a cool and dry place.

Table 2. The composition and source of soil amendments.

Soil Amendments	Composition	Source
Silicon fertilizer	SiO_2 (17%), MgO (10%), Ca (8%), humic acid (20%), and compound amino acid (5%)	Shanxi Jifei Industry Co., Ltd., Taiyuan, Shanxi, China
Potassium humate	Humic acid ≥ 60% and potassium ≥ 11%	Inner Mongolia Shengtian Agricultural Technology Co., Ltd., Baotou, Inner Mongolia, China
Bacillus biofertilizer	*Bacillus amyloliquefaciens* (0.9×10^8 CFU/g), *Bacillus subtilis* (0.9×10^8 CFU/g), and *Paenibacillus kribbensis* (0.2×10^8 CFU/g)	Inner Mongolia Shengtian Agricultural Technology Co., Ltd., Baotou, Inner Mongolia, China

The study was conducted in a greenhouse located in Haidian District, Beijing, China (116°18′33.0″ N, 40°4′26.4″ E). The experiment was conducted from September to November 2020. Throughout the experiment, a seedling bed made of steel wire with a height of 120 cm was used to place all potted plants, and the position of the potted plants was changed regularly to reduce errors caused by light. The greenhouse was maintained at a constant temperature of 25 °C by air conditioning with a light intensity of 60 klx/m^2 and L:D 12:12.

2.3. Soil and Strawberry Plant Sampling

Strawberry growth indices (plant height, stem diameter, chlorophyll content, and fresh weight) were measured 60 days after planting. At the same time, the rhizosphere soil of strawberries was collected by a gentle shaking method [24] to determine the soil properties and soil pathogens.

2.4. Soil Properties Analysis

Soil physicochemical properties were determined by Bao's method [25]. Soil pH and EC were determined in a 1:2.5 soil/water suspension. AN and NN were determined with a flow analyzer (Alliance Instruments, Eragny Sur-Oise, France). AP and AK were determined by a UV2012-PC spectrophotometer (UNICO, Fairfield, NJ, USA) and an FP640 Flame Photometer (Shanghai Instruments Group Co., Ltd., Shanghai, China), respectively. OM was determined according to Liu et al. [26].

According to the instruction manual provided by Beijing Soleibao Technology Co., Ltd., Beijing, China., soil catalase (S-CAT), sucrase (S-SC), and urease (S-UE) were extracted with corresponding kits, and their absorbance was measured at 630 nm, 540 nm, and 240 nm.

2.5. Soil-Borne Pathogens

Soil-borne pathogens were separated according to the method of Yun et al. [27]: 5 g of a soil sample was added to 95 mL of 0.7‰ sterilized agar water, and then the soil suspension was obtained by shaking in a shaker for 30 min. On the aseptic operating table, 2.5 mL of medium B component was mixed with 47.5 mL of medium A component, then 1 mL of the soil suspension was added, shaken well, and poured evenly into the petri dish. The number of colonies of *Fusarium* spp. and *Phytophthora* spp. was recorded after an incubation period of 3 days at 28 °C. The A component (taking 2L as an example) of *Fusarium* spp. consisted of 1 g of KCl, 1 g of $MgSO_4$, 2 g of KH_2PO_4, 4 g of L-asparagine, 30 g of agar, and 40 g of D-galactose; *Phytophthora* spp. consisted of 34 g of agar and 40 g of glucose. The B component (100 mL) of *Fusarium* spp. consisted of 2 g of $Na_2B_4O_7{\cdot}10H_2O$, 0.02 g of Fe-Na EDTA, 1 g of Oxgall, 1 g of Streptomycin sulfate, and 1.5 g of $C_6Cl_5NO_2$; the *Phytophthora* spp. consisted of 0.15 g of $C_6Cl_5NO_2$, 0.02 g of Rifampicin, and 0.03 g of Ampicillin.

2.6. Strawberry Growth Indices

The plant height, stem diameter, and fresh weight of strawberry plants were determined using a measuring tape, digital electronic vernier caliper, and electronic balance, respectively. The strawberry plant height was determined by the height of the unearthed part of the plant to the longest leaf when the plant was held upright by hand. Three healthy leaves were selected from each strawberry plant, and the chlorophyll content (SPAD) of the strawberry leaves was determined using a non-destructive handheld meter SPAD-502 chlorophyll meter.

2.7. Data Analysis

Soil properties, soil-borne pathogens, and strawberry growth indices were analyzed by one-way analysis of variance (ANOVA) and the Tukey test included in the SPSS statistical software package (V 20.0, IBM Corp., Armonk, NY, USA). Origin 2017 was used to draw graphs. The formula for calculating the control effect of soil-borne pathogens is:

$$Y = \frac{A_2 - A_1}{A_2} \times 100\%$$

where Y is the control effect (%), A_1 is the number of colonies of soil-borne pathogens in the treatment group, and A_2 is the number of colonies of soil-borne pathogens in the control group.

3. Results

3.1. Changes in Soil N, P, and K Contents

The content of ammonium nitrogen in fumigated and amended soil was significantly higher than that in the control, but the content of nitrate nitrogen was significantly lower (Figure 1A,B). The contents of ammonium nitrogen and nitrate nitrogen in the Si treatment were significantly higher than that in the DZ treatment by 40.21% and 24.14%, respectively. The content of ammonium nitrogen and nitrate nitrogen was significantly reduced by 14.86% and 28.35% in the KHT treatment, respectively, and the content of nitrate nitrogen in the T treatment was significantly decreased by 7.75%. There was no significant difference in the soil's available phosphorus content between the CK and DZ treatments (Figure 1C), but DZ treatment significantly decreased the soil's available potassium content (Figure 1D). Compared to DZ treatment, soil amendments applied after fumigation significantly increased the content of available phosphorus and available potassium except for T treatment. The content of available phosphorus and available potassium in the Si treatment was the highest, at 263.35 mg/kg and 503.43 mg/kg, respectively.

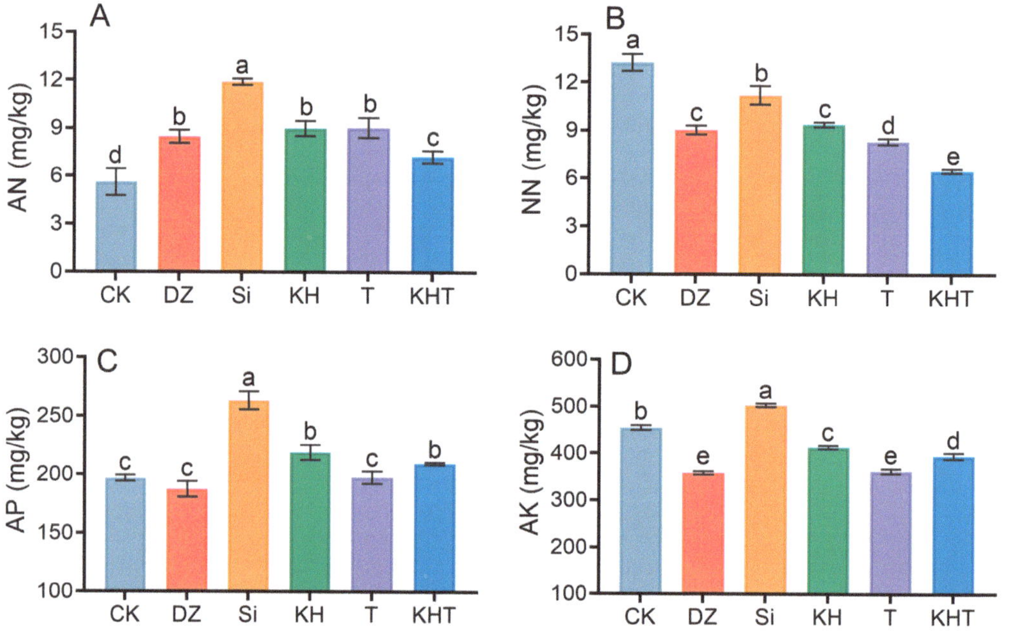

Figure 1. Changes in soil ammonium nitrogen (AN, (**A**)), NN (nitrate nitrogen, (**B**)), available phosphorus (AP, (**C**)), and available potassium (AK, (**D**)). Values are mean ± standard error. Different letters indicate significant differences in results, and the same letters indicate non-significant differences ($p < 0.05$). CK = control; DZ = dazomet fumigation; Si = silicon fertilizer applied after fumigation; KH = potassium humate applied after fumigation; T = *Bacillus* biofertilizer applied after fumigation; KHT = potassium humate mixed with *Bacillus* biofertilizer applied after fumigation.

3.2. Changes in Soil Organic Matter, pH, and Electrical Conductivity

Soil organic matter was significantly increased by fumigation and soil amendments applied after fumigation compared to the control (Figure 2A). Compared to DZ treatment, Si treatment and KHT treatment significantly increased organic matter content by 15.14% and 4.37%, respectively. There was no significant difference in soil pH (Figure 2B). Compared to the control, the Si treatment significantly increased the soil electrical conductivity by 8.56%, while the other treatments significantly decreased the soil electrical conductivity

(Figure 2C). The electrical conductivity of the soil was significantly lower in the T and KHT treatments than in the DZ treatment.

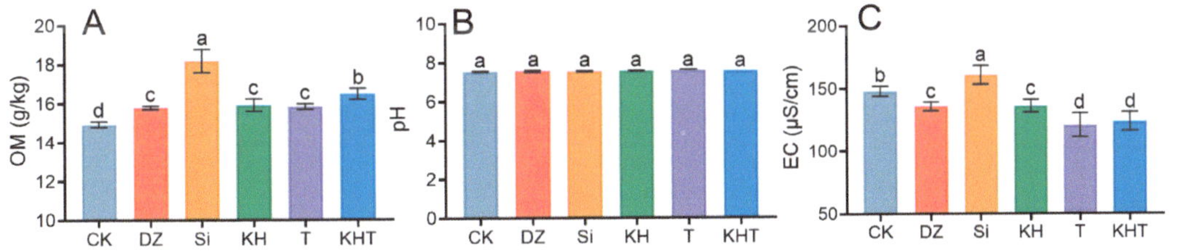

Figure 2. Changes in soil organic matter (OM, (**A**)), pH (**B**), and electrical conductivity (EC, (**C**)). Values are mean ± standard error. Different letters indicate significant differences in results, and the same letters indicate non-significant differences ($p < 0.05$). CK = control; DZ = dazomet fumigation; Si = silicon fertilizer applied after fumigation; KH = potassium humate applied after fumigation; T = *Bacillus* biofertilizer applied after fumigation; KHT = potassium humate mixed with *Bacillus* biofertilizer applied after fumigation.

3.3. Soil Enzyme Activities

The soil catalase activity of the Si, KH, and T treatments was significantly increased by 2.49%, 3.85%, and 7.85% respectively, compared to the DZ treatment (Figure 3A). Soil sucrase activity was significantly decreased by 13.75% after DZ fumigation. However, the application of amendments significantly increased the sucrase activity, among which the activity of the T treatment was the highest, significantly increased by 92.95% (Figure 3B). Soil urease activity in the Si treatment was significantly higher than that of the control, while the other treatments significantly decreased the soil urease activity (Figure 3C). Soil urease activity was significantly increased by 8.02–27.87% after DZ fumigation, except for KHT treatment.

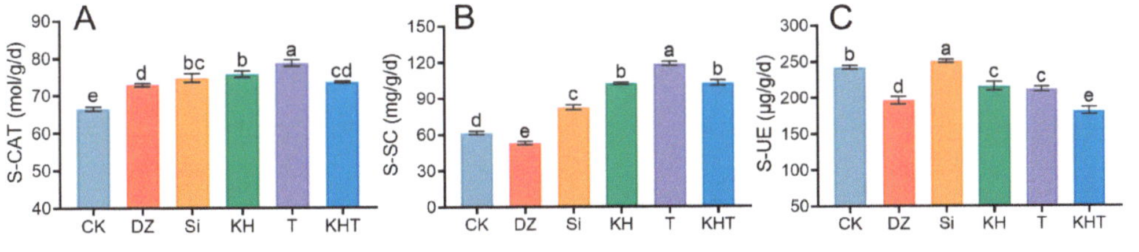

Figure 3. Changes in soil catalase (S-CAT, (**A**)), sucrase (S-SC, (**B**)), and urease (S-UE, (**C**)). Values are mean ± standard error. Different letters indicate significant differences in results, and the same letters indicate non-significant differences ($p < 0.05$). CK = control; DZ = dazomet fumigation; Si = silicon fertilizer applied after fumigation; KH = potassium humate applied after fumigation; T = *Bacillus* biofertilizer applied after fumigation; KHT = potassium humate mixed with *Bacillus* biofertilizer applied after fumigation.

3.4. Analysis of *Fusarium* spp. and *Phytophthora* spp. in Soil

Compared to the control, all treatments significantly reduced *Fusarium* spp. and *Phytophthora* spp. (Table 3). The control effect of the DZ treatment on *Fusarium* spp. and *Phytophthora* spp. was 87.32% and 76.97%, respectively. Compared to DZ treatment, DZ fumigation combined with soil amendments significantly reduced *Fusarium* spp., except for the KHT treatment, and the control effect reached 95.73–96.71%. *Phytophthora* spp. was further reduced by the application of amendments after fumigation. T treatment had the best control effect on *Phytophthora* spp. (96.88%). It can be seen that *Fusarium* spp. and *Phytophthora* spp. can be further reduced by the application of amendments after fumigation.

Table 3. Changes in soil-borne pathogens.

Treatments	*Fusarium* spp.		*Phytophthora* spp.	
	CFU/g Soil	Control Effect (%)	CFU/g Soil	Control Effect (%)
CK	1420.00 ± 11.55 a	/	3734.00 ± 76.74 a	/
DZ	180.00 ± 11.55 b	87.32	860.00 ± 23.09 b	76.97
Si	49.33 ± 5.21 c	96.53	333.34 ± 24.04 c	91.07
KH	46.67 ± 6.67 c	96.71	300.00 ± 11.55 cd	91.97
T	60.67 ± 6.36 c	95.73	116.56 ± 8.82 e	96.88
KHT	156.66 ± 12.02 b	88.97	220.00 ± 11.55 de	94.11

Note: Different letters indicate significant differences in results, and the same letters indicate non-significant differences ($p < 0.05$).

3.5. Changes in Strawberry Growth Indices

Strawberry plant height, stem diameter, chlorophyll content, and fresh weight were significantly higher in the DZ, Si, KH, T, and KHT treatments. Compared to the DZ treatment, the Si, KH, and T treatments significantly increased strawberry plant height by 17.24%, 15.67%, and 6.49%, respectively (Figure 4A). The stem diameter (22.09–53.68%), chlorophyll content (13.28–18.54%), and fresh weight (13.68–20.58%) of strawberry plants were significantly increased by the application of soil amendments after fumigation compared to DZ treatment (Figure 4B–D).

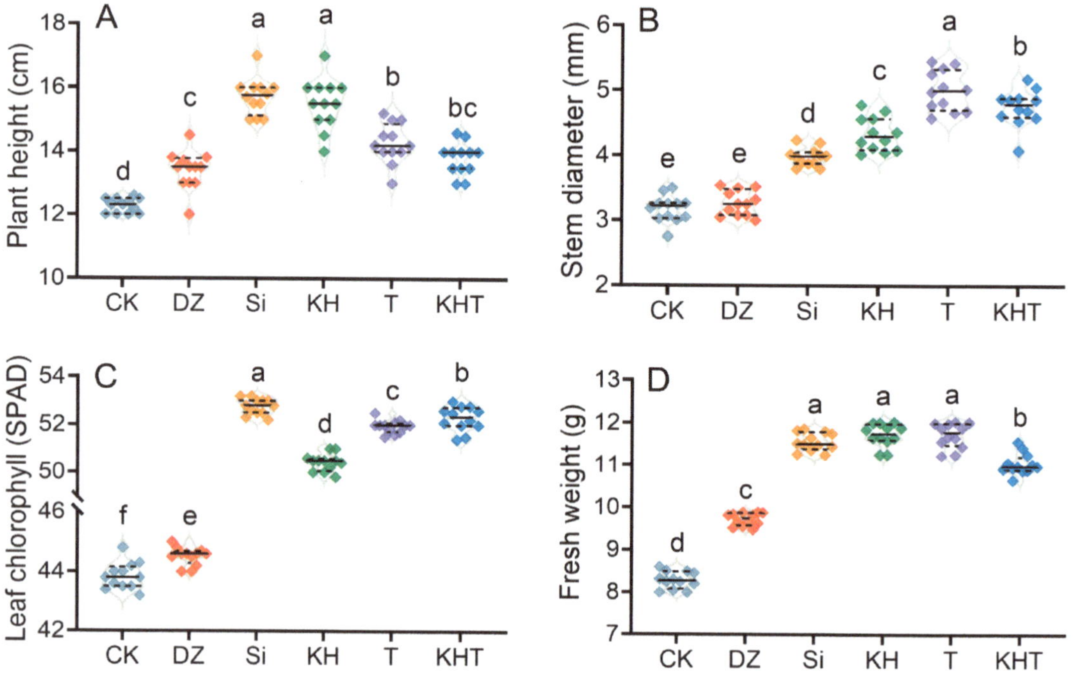

Figure 4. Changes in strawberry growth indices ((**A**): strawberry plant height; (**B**): stem diameter; (**C**): leaf chlorophyll; (**D**): fresh weight). Values are mean ± standard error. Different letters indicate significant differences in results, and the same letters indicate non-significant differences ($p < 0.05$). CK = control; DZ = dazomet fumigation; Si = silicon fertilizer applied after fumigation; KH = potassium humate applied after fumigation; T = *Bacillus* biofertilizer applied after fumigation; KHT = potassium humate mixed with *Bacillus* biofertilizer applied after fumigation.

3.6. Correlation Analysis of Strawberry Growth Indexes with Soil Environmental Factors and Soil-Borne Pathogens

Plant height was significantly positively correlated with AN, AP, OM, S-CAT, and S-SC (R = 0.84, R = 0.70, R = 0.66, R = 0.71, R = 0.49; $p < 0.05$) (Figure 5, Table S1). Strawberry stem diameter was significantly positively correlated with S-CAT and S-SC (R = 0.77, R = 0.97; $p < 0.001$), and was significantly negatively correlated with NN and EC (R = −0.70, R = −0.58; $p < 0.05$). The leaf chlorophyll content of strawberry was significantly positively correlated with AP, OM, S-CAT, and S-SC (R = 0.61, R = 0.67, R = 0.74, R = 0.83; $p < 0.01$) and was significantly negatively correlated with NN (R = −0.51, $p < 0.05$). Fresh weight showed a similar correlation with leaf chlorophyll content. Strawberry growth indices were significantly negatively correlated with soil-borne pathogens ($p < 0.01$). In addition, the soil-borne pathogens were significantly negatively correlated with AN, OM, S-CAT, and S-SC ($p < 0.05$) and significantly positively correlated with NN ($p < 0.001$).

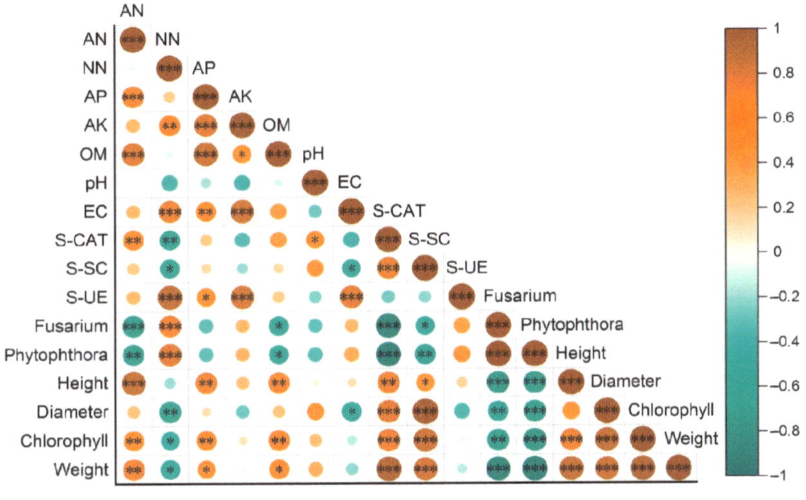

Figure 5. Pearson correlation analysis of strawberry growth indices with soil environmental factors and soil-borne pathogens. Green indicates a negative correlation and yellow indicates a positive correlation. Larger circles indicate a stronger correlation and smaller circles indicate a weaker correlation. Significance levels: * $p \leq 0.05$, ** $p \leq 0.01$, *** $p \leq 0.001$. Abbreviations: AN, ammonium nitrogen; NN, ammonium nitrogen; AP, available phosphorus; AK, available potassium; OM, organic matter; EC, electrical conductivity; S-CAT, soil catalase; S-SC, soil sucrase; S-UE, soil urease.

4. Discussion

Urea and phosphorus pentoxide can be converted into ammonium ions and phosphate in soil, increasing the content of ammonium nitrogen and available phosphorus in soil. The nitrate nitrogen in the soil was lower than that in the initial soil, probably due to the activity of some microorganisms inhibiting nitrification [28]. The abundance of soil microorganisms and genes related to the nitrogen cycle decreased after soil fumigation, which inhibited soil nitrification and resulted in an increase in soil ammonium nitrogen content [29]. In this study, DZ fumigation increased the content of ammonium nitrogen and decreased the content of nitrate nitrogen, which could be the reason why nitrification was inhibited. The contents of nitrogen, phosphorus, potassium, and organic matter in the Si treatment were significantly higher than that of DZ fumigation alone, which is consistent with the results of Liang et al. [30]; that is, silicon fertilizer can optimize the soil environment and increase soil nutrient supply. Kuang et al. [31] reported that potassium humate could improve the soil environment and increase soil organic matter and available potassium content. KH treatment significantly increased the contents of ammonium nitrogen, available phosphorus, and available potassium in the soil, possibly because humic acid had the effect of improving the soil structure and providing essential nutrients for plants [32]. *Bacillus* biofertilizers contained three species of *Bacillus* that could directly increase the absorption of nutrients in different ways. Our results showed that T treatment significantly increased the content of ammonium nitrogen and organic matter in soil compared with the control. *Bacillus amyloliquefaciens* could improve soil nutrient availability, including improving nitrogen supply and solubilizing phosphate and potassium [33]. *Bacillus subtilis* could improve the soil carbon sequestration process and promote soil nutrient cycling [34]. *Paenibacillus kribbensis* generally increased soil nutrients by increasing the number of beneficial microorganisms [35]. Our results showed that silicon fertilizer had the best effect on increasing soil nutrients.

Soil enzyme activity is mainly involved in the life activities of soil microorganisms and nutrient cycling, which can indicate the health degree of the soil [36]. Soil catalase can prevent the toxic effect of hydrogen peroxide on the soil and crop [37]. Soil sucrase can hydrolyze sucrose into corresponding monosaccharides to provide nutrients for soil microorganisms and affect the mineralization process of organic carbon [38]. We found that soil catalase and sucrase activities were highest in the T treatment compared to DZ fumigation alone, suggesting that biofertilizer can increase microbial activity. Huang et al. [39] found that biofertilizers prepared by the co-fermentation of seaweed polysaccharides, bacterial bran, and biochar activated the activity of soil beneficial microorganisms and reduced the disturbance effect of fumigation on soil. Cheng et al. [15] showed that *Bacillus subtilis* could significantly increase soil enzyme activity and thus promote soil nutrient conversion and microbial activity. Soil urease can hydrolyze urea in soil and increase the content of inorganic nitrogen content in soil [40]. In this study, DZ fumigation decreased soil sucrase and urease activities, possibly because the fumigation killed the soil microorganisms that could hydrolyze urea and sucrose, resulting in a decrease in enzyme activities. Among the amendments, silicon fertilizer had the strongest promotion effect on soil urease, which could be due to the improvement in soil fertility status. Silicon fertilizer contains humic acid and compound amino acids that could promote the formation of soil aggregates and activate the activity of beneficial microorganisms in the soil, which could also lead to an increase in soil enzyme activity [41,42].

Soil fumigation can control and kill root-knot nematodes, soil-borne pathogens, and weeds [43]. Wang et al. [44] reported that the inhibition rate of *Phytophthora* spp. was still 97.8% and 75.0% in the second and third years after chloropicrin fumigation, respectively. We found that the fumigant significantly reduced *Fusarium* spp. and *Phytophthora* spp., and the inhibitory effect was still in effect two months after strawberry planting. The application of amendments after fumigation further reduced *Fusarium* spp. and *Phytophthora* spp. and was significantly less than those in the DZ fumigation alone, possibly due to the active ingredients in the amendments that inhibited the growth of these pathogens [45–47].

Soil fumigation and soil amendments can both be used to promote crop growth, but they have different mechanisms of action. Soil fumigation mainly promotes crop growth by killing soil-borne pathogens, pests, and weeds [48], while soil amendments are substances that alter soil properties in order to enhance soil fertility and crop yield [49]. DZ fumigation alone and in combination with soil amendments promoted strawberry growth, but the strawberries grew better in the soil treated with the amendments. There was a significant negative correlation between the strawberry growth indices and the number of *Fusarium* spp. and *Phytophthora* spp. The soil-borne pathogens were killed by fumigation and the application of amendments after fumigation so that the plants could better absorb the nutrients in the soil to promote their growth [50]. At the same time, soil amendments contain amino acids, organic matter, and beneficial microorganisms, which can increase the activity of soil microorganisms and improve the soil micro-ecological environment [51,52]. According to soil nutrients, enzyme activities, and strawberry growth indices, silicon fertilizer had the best effect.

5. Conclusions

In this study, silicon fertilizer, potassium humate, *Bacillus* biofertilizer, and a combination of the latter two were added to soil after DZ fumigation. The results showed that DZ fumigation combined with soil amendments significantly increased the activities of catalase, sucrase, and urease in the soil by varying degrees. Among them, silicon fertilizer significantly increased soil nutrients, and the contents of nitrogen, phosphorus, potassium, and organic matter in the Si treatment increased significantly. At the same time, the soil amendments further improved the control effect of soil-borne pathogens and significantly promoted the growth of strawberries, and the effect was better than that of the DZ fumigation alone. The improvement in enzyme activities in the soil indicated the increase in soil biological activity, which directly or indirectly promoted strawberry growth. Correlation analysis showed that there was a significant negative correlation between the strawberry growth indices and soil-borne pathogens, so the control of soil-borne pathogens was an important measure to promote strawberry growth. Our results showed that DZ fumigation combined with silicon fertilizer had the best effect, which could not only improve the soil environment but also reduce the number of soil-borne pathogens and promote the growth of strawberry plants. Furthermore, soil amendments can also replace some fertilizers and promote green, efficient, and sustainable development. However, our experiments were conducted in a controlled environment. When these amendments are applied to the field, a small-scale test should be carried out first to avoid economic losses due to the complexity of the field environment.

Supplementary Materials: The following supporting information can be downloaded at: https://www.mdpi.com/article/10.3390/agriculture14010009/s1, Table S1: Correlation analysis of strawberry growth indices with soil environmental factors and soil-borne pathogens.

Author Contributions: Conceptualization, Q.L. and W.F.; methodology, D.Y.; formal analysis, O.A.; data curation, Y.L.; writing—original draft preparation, Q.L.; validation, Q.W.; writing—review and editing, and funding acquisition, X.J. and A.C. All authors have read and agreed to the published version of the manuscript.

Funding: This research was funded by the National Natural Science Foundation of China (31972313) and the Hebei Technology Innovation Center for Green Management of Soil-borne Diseases, Baoding University.

Institutional Review Board Statement: Not applicable.

Data Availability Statement: All data are contained within the manuscript.

Acknowledgments: We thank Wenbiao Liu for his help with this research.

Conflicts of Interest: The authors declare no conflict of interest.

References

1. Liu, Z.C.; Liang, T.; Kang, C.Y. Molecular bases of strawberry fruit quality traits: Advances, challenges, and opportunities. *Plant Physiol.* **2023**, *193*, 900–914. [CrossRef]
2. Basu, A.; Nguyen, A.; Betts, N.M.; Lyons, T.J. Strawberry as a functional food: An evidence-based review. *Crit. Rev. Food Sci. Nutr.* **2014**, *54*, 790–806. [CrossRef]
3. Newerli-Guz, J.; Śmiechowska, M.; Drzewiecka, A.; Tylingo, R. Bioactive ingredients with health-promoting properties of strawberry fruit (*Fragaria × ananassa* Duchesne). *Molecules* **2023**, *28*, 2711. [CrossRef]
4. Kang, J.X.; Wang, Z.; Zhong, Y.Y. Arthropod population dynamics and community structure characteristics in different strawberry variety orchards. *Shaanxi J. Agric. Sci.* **2023**, *69*, 68–72.
5. Chen, P.; Li, H.Q.; Li, X.Y.; Zhou, X.H.; Zhang, X.X.; Zhang, A.S.; Liu, Q.Z. Transcriptomic analysis provides insight into defensive strategies in response to continuous cropping in strawberry (*Fragaria × ananassa* Duch.) plants. *BMC Plant Biol.* **2022**, *22*, 476. [CrossRef] [PubMed]
6. Huang, Y.; Jing, J.Q.; Yan, M.L.; Hazard, C.; Chen, Y.H.; Guo, C.B.; Xiao, X.; Lin, J.J. Contribution of pathogenic fungi to N_2O emissions increases temporally in intensively managed strawberry cropping soil. *Appl. Microbiol. Biotechnol.* **2021**, *105*, 2043–2056. [CrossRef] [PubMed]

7. Pastrana, A.M.; Borrero, C.; Pérez, A.G.; Avilés, M. Soilborne pathogens affect strawberry fruit flavor and quality. *Plant Sci.* **2023**, *326*, 111533. [CrossRef] [PubMed]
8. Pu, R.F.; Wang, P.P.; Guo, L.P.; Li, M.H.; Cui, X.M.; Wang, C.X.; Liu, Y.; Yang, Y. The remediation effects of microbial organic fertilizer on soil microorganisms after chloropicrin fumigation. *Ecotoxicol. Environ. Saf.* **2022**, *231*, 113188. [CrossRef]
9. Zhang, D.L.; Ji, X.X.; Meng, Z.; Qi, W.Z.; Qiao, K. Effects of fumigation with 1,3-dichloropropene on soil enzyme activities and microbial communities in continuous-cropping soil. *Ecotoxicol. Environ. Saf.* **2019**, *169*, 730–736. [CrossRef]
10. Fu, C.H.; Hu, B.Y.; Chang, T.T.; Hsueh, K.L.; Hsu, W.T. Evaluation of dazomet as fumigant for the control of brown root rot disease. *Pest. Manag. Sci.* **2012**, *68*, 959–962. [CrossRef]
11. Li, Q.J.; Zhang, D.Q.; Cheng, H.Y.; Song, Z.X.; Ren, L.R.; Hao, B.Q.; Zhu, J.H.; Fang, W.S.; Yan, D.D.; Li, Y.; et al. Chloropicrin alternated with dazomet improved the soil's physicochemical properties, changed microbial communities and increased strawberry yield. *Ecotoxicol. Environ. Saf.* **2021**, *220*, 112362. [CrossRef]
12. Chen, H.J.; Zhao, S.; Zhao, J.M.; Zhang, K.K.; Jiang, J.; Guan, Z.Y.; Chen, S.M.; Chen, F.D.; Fang, W.M. Deep tillage combined with biofertilizer following soil fumigation improved chrysanthemum growth by regulating the soil microbiome. *MicrobiologyOpen* **2020**, *9*, e1045. [CrossRef] [PubMed]
13. Sun, Z.C.; Zhang, C.L.; Li, G.T.; Lin, Q.M.; Zhao, X.R. Does soil amendment alter reactive soil N dynamics following chloropicrin fumigation? *Chemosphere* **2018**, *212*, 563–571. [CrossRef] [PubMed]
14. Du, S.; Ge, A.H.; Liang, Z.H.; Xiang, J.F.; Xiao, J.L.; Zhang, Y.; Liu, Y.R.; Zhang, L.M.; Shen, J.P. Fumigation practice combined with organic fertilizer increase antibiotic resistance in watermelon rhizosphere soil. *Sci. Total Environ.* **2022**, *805*, 150426. [CrossRef] [PubMed]
15. Cheng, H.Y.; Zhang, D.Q.; Ren, L.R.; Song, Z.X.; Li, Q.J.; Wu, J.J.; Fang, W.S.; Huang, B.; Yan, D.D.; Li, Y.; et al. Bio-activation of soil with beneficial microbes after soil fumigation reduces soil-borne pathogens and increases tomato yield. *Environ. Pollut.* **2021**, *283*, 117160. [CrossRef] [PubMed]
16. Gao, S.D.; Doll, D.A.; Stanghellini, M.S.; Westerdahl, B.B.; Wang, D.; Hanson, B.D. Deep injection and the potential of biochar to reduce fumigant emissions and effects on nematode control. *J. Environ. Manag.* **2018**, *223*, 469–477. [CrossRef] [PubMed]
17. Wu, J.J.; Zhu, J.H.; Zhang, D.Q.; Cheng, H.Y.; Hao, B.Q.; Cao, A.C.; Yan, D.D.; Wang, Q.X.; Li, Y. Beneficial effect on the soil microenvironment of *Trichoderma* applied after fumigation for cucumber production. *PLoS ONE* **2022**, *17*, e0266347. [CrossRef]
18. Sun, X.; Liu, Q.; Tang, T.T.; Chen, X.; Luo, X. Silicon fertilizer application promotes phytolith accumulation in rice plants. *Front. Plant Sci.* **2019**, *10*, 425. [CrossRef]
19. Zhao, K.Q.; Yang, Y.; Peng, H.; Zhang, L.H.; Zhou, Y.Y.; Zhang, J.C.; Du, C.Y.; Liu, J.W.; Lin, X.; Wang, N.Y.; et al. Silicon fertilizers, humic acid and their impact on physicochemical properties, availability and distribution of heavy metals in soil and soil aggregates. *Sci. Total Environ.* **2022**, *822*, 153483. [CrossRef]
20. Yuan, Z.N.; Pang, Z.Q.; Fallah, N.; Zhou, Y.M.; Dong, F.; Lin, W.X.; Hu, C.H. Silicon fertilizer mediated structural variation and niche differentiation in the rhizosphere and endosphere bacterial microbiome and metabolites of sugarcane. *Front. Microbiol.* **2022**, *13*, 1009505. [CrossRef]
21. Hegab, R.H.; Eissa, D.; Abou-Shady, A. Effects of foliar application of selenium and potassium-humate on oat growth in Baloza, North Sinai, Egypt. *Sci. Rep.* **2022**, *12*, 15119. [CrossRef] [PubMed]
22. Jin, Q.; Zhang, Y.Y.; Wang, Q.X.; Li, M.J.; Sun, H.; Liu, N.; Zhang, L.L.; Zhang, Y.; Liu, B.Z. Effects of potassium fulvic acid and potassium humate on microbial biodiversity in bulk soil and rhizosphere soil of Panax ginseng. *Microbiol. Res.* **2022**, *254*, 126914. [CrossRef] [PubMed]
23. Zhu, Y.Q.; Lv, X.; Song, J.H.; Li, W.D.; Wang, H.J. Application of cotton straw biochar and compound *Bacillus* biofertilizer decrease the bioavailability of soil cd through impacting soil bacteria. *BMC Microbiol.* **2022**, *22*, 35. [CrossRef] [PubMed]
24. Yu, F.M.; Fu, M.Y.; Tang, C.J.; Mo, C.J.; Li, S.Y.; Luo, S.Y.; Qin, P.Q.; Zhao, Y.J.; Li, Y. Potential impact of polyethylene microplastics on the growth of water spinach (*Ipomoea aquatica* F.): Endophyte and rhizosphere effects. *Chemosphere* **2023**, *330*, 138737. [CrossRef] [PubMed]
25. Bao, S.D. *Soil Agrochemical Analysis*, 3rd ed.; China Agricultural Press: Beijing, China, 2010.
26. Liu, Y.W.; Feng, Y.; Cheng, D.M.; Xue, J.M.; Wakelin, S.A.; Hu, H.Y.; Li, Z.J. Gentamicin degradation and changes in fungal diversity and physicochemical properties during composting of gentamicin production residue. *Bioresour. Technol.* **2017**, *244*, 905–912. [CrossRef]
27. Yun, C.X.; Yan, C.R.; Xu, M.Z.; Liu, E.K.; Mormil, P.; Dong, W.Y.; Liu, Q. Effects of different soil disinfection methods on soil enzyme activities and soil-borne diseases. *J. Chin. Agric. Univ.* **2020**, *25*, 86–96.
28. Subbarao, G.V.; Yoshihashi, T.; Worthington, M.; Nakahara, K.; Ando, Y.; Sahrawat, K.L.; Rao, I.M.; Lata, J.C.; Kishii, M.; Braun, H.J. Suppression of soil nitrification by plants. *Plant Sci.* **2015**, *233*, 155–164. [CrossRef]
29. De Neve, S.; Csitári, G.; Salomez, J.; Hofman, G. Quantification of the effect of fumigation on short- and long-term nitrogen mineralization and nitrification in different soils. *J. Environ. Qual.* **2004**, *33*, 1647–1652. [CrossRef]
30. Liang, Y.Q.; Liao, M.; Fang, Z.P.; Guo, J.W.; Xie, X.M.; Xu, C.X. How silicon fertilizer improves nitrogen and phosphorus nutrient availability in paddy soil? *J. Zhejiang Univ. Sci. B* **2021**, *22*, 521–532. [CrossRef]
31. Kuang, G.; Wang, H.H.; Liu, D.Y.; Li, S.F.; Wang, B.; Yuan, Y.; Liu, G.S.; Ren, T.B. Effects of potassium humate on flue-cured tobacco quality and soil environment in western henan. *Chin. Agric. Sci. Bull.* **2017**, *33*, 60–66.

32. Li, Y.; Fang, F.; Wei, J.L.; Wu, X.B.; Cui, R.Z.; Li, G.S.; Zheng, F.L.; Tan, D.S. Humic Acid Fertilizer Improved Soil Properties and Soil Microbial Diversity of Continuous Cropping Peanut: A Three-Year Experiment. *Sci. Rep.* **2019**, *9*, 12014. [CrossRef] [PubMed]
33. Luo, L.; Zhao, C.Z.; Wang, E.T.; Raza, A.; Yin, C.Y. *Bacillus amyloliquefaciens* as an excellent agent for biofertilizer and biocontrol in agriculture: An overview for its mechanisms. *Microbiol. Res.* **2022**, *259*, 127016. [CrossRef] [PubMed]
34. Mahapatra, S.; Yadav, R.; Ramakrishna, W. *Bacillus subtilis* impact on plant growth, soil health and environment: Dr. Jekyll and Mr. Hyde. *J. Appl. Microbiol.* **2022**, *132*, 3543–3562. [CrossRef] [PubMed]
35. Ali, M.A.; Lou, Y.; Hafeez, R.; Li, X.; Hossain, A.; Xie, T.; Lin, L.; Li, B.; Yin, Y.; Yan, J.; et al. Functional Analysis and Genome Mining Reveal High Potential of Biocontrol and Plant Growth Promotion in Nodule-Inhabiting Bacteria Within *Paenibacillus polymyxa* Complex. *Front. Microbiol.* **2021**, *11*, 618601. [CrossRef] [PubMed]
36. Du, J.J.; Hou, F.; Zhou, Q.X. Response of soil enzyme activity and soil bacterial community to PCB dissipation across different soils. *Chemosphere* **2021**, *283*, 131229. [CrossRef]
37. Wu, L.X.; Wang, Y.; Lyu, H.; Chen, X.D. Effects of a compound *Trichoderma* agent on *Coptis chinensis* growth, nutrients, enzyme activity, and microbial community of rhizosphere soil. *PeerJ* **2023**, *11*, e15652. [CrossRef]
38. Yang, R.; Wang, J.H.; Zhu, L.S.; Wang, J.; Yang, L.L.; Mao, S.S.; Conkle, J.L.; Chen, Y.Y.; Kim, Y.M. Effects of interaction between enrofloxacin and copper on soil enzyme activity and evaluation of comprehensive toxicity. *Chemosphere* **2021**, *268*, 129208. [CrossRef]
39. Huang, B.; Chen, Y.X.; Pei, Z.Y.; Jiang, L.Q.; Zhang, Y.; Wang, J.; Wang, J. Application of microbial organic fertilizers promotes the utilization of nutrients and restoration of microbial community structure and function in rhizosphere soils after dazomet fumigation. *Front. Microbiol.* **2023**, *13*, 1122611. [CrossRef]
40. Guo, H.; Yao, J.; Cai, M.M.; Qian, Y.G.; Guo, Y.; Richnow, H.H.; Blake, R.E.; Doni, S.; Ceccanti, B. Effects of petroleum contamination on soil microbial numbers, metabolic activity and urease activity. *Chemosphere* **2012**, *87*, 1273–1280. [CrossRef]
41. Huang, C.P.; Wang, L.; Gong, X.Q.; Huang, Z.T.; Zhou, M.R.; Li, J.; Wu, J.S.; Chang, S.X.; Jiang, P.K. Silicon fertilizer and biochar effects on plant and soil PhytOC concentration and soil PhytOC stability and fractionation in subtropical bamboo plantations. *Sci. Total Environ.* **2020**, *715*, 136846. [CrossRef]
42. Fallah, N.; Pang, Z.Q.; Dong, F.; Zhou, Y.M.; Lin, W.X.; Fabrice, K.M.A.; Hu, C.H.; Yuan, Z.N. Niche differentiation modulates metabolites abundance and composition in silicon fertilizer amended soil during sugarcane growth. *BMC Plant Biol.* **2022**, *22*, 497. [CrossRef] [PubMed]
43. Li, X.P.; Skillman, V.; Dung, J.; Frost, K. Legacy effects of fumigation on soil bacterial and fungal communities and their response to metam sodium application. *Environ. Microbiome* **2022**, *17*, 59. [CrossRef] [PubMed]
44. Wang, X.; Wang, Q.; Li, W.J.; Zhang, D.Q.; Fang, W.S.; Li, Y.; Wang, Q.X.; Cao, A.C.; Yan, D.D. Long-term effects of chloropicrin fumigation on soil microbe recovery and growth promotion of Panax notoginseng. *Front. Microbiol.* **2023**, *14*, 1225944. [CrossRef] [PubMed]
45. Wang, M.; Gao, L.M.; Dong, S.Y.; Sun, Y.M.; Shen, Q.R.; Guo, S.W. Role of silicon on plant-pathogen interactions. *Front. Plant Sci.* **2017**, *8*, 701. [CrossRef] [PubMed]
46. Wang, A.D.; Zhang, L.P.; Xu, L.J.; Wang, Z.M. Experiment on yield-increasing effect of humic acid patassium and humic acid magnesium on rice. *North. Rice.* **2008**, *2*, 40–41.
47. Ongena, M.; Jacques, P. *Bacillus* lipopeptides: Versatile weapons for plant disease biocontrol. *Trends Microbiol.* **2008**, *16*, 115–125. [CrossRef] [PubMed]
48. Lembright, H.W. Soil fumigation: Principles and application technology. *J. Nematol.* **1990**, *22*, 632–644. [PubMed]
49. Palansooriya, K.N.; Shaheen, S.M.; Chen, S.S.; Tsang, D.C.W.; Hashimoto, Y.; Hou, D.; Bolan, N.S.; Rinklebe, J.; Ok, Y.S. Soil amendments for immobilization of potentially toxic elements in contaminated soils: A critical review. *Environ. Int.* **2020**, *134*, 105046. [CrossRef]
50. Pu, Y.Y.; Lv, X.M.; Wu, M.C.; Wang, D.S.; Li, W.M.; Wang, S.Q.; Li, H.X.; Hu, F.; Jiao, J.G. Effects of partial substitution for chemical fertilizer by organic manure on the growth and nutrient use of watermelon under fumigation condition. *J. Soil. Water Conserv.* **2017**, *31*, 306–311.
51. Liu, Y.X.; Shi, J.X.; Feng, Y.G.; Yang, X.M.; Li, X.; Shen, Q.R. Tobacco bacterial wilt can be biologically controlled by the application of antagonistic strains in combination with organic fertilizer. *Biol. Fertil. Soils.* **2013**, *49*, 447–464. [CrossRef]
52. Urra, J.; Alkorta, I.; Mijangos, I.; Garbisu, C. Commercial and farm fermented liquid organic amendments to improve soil quality and lettuce yield. *J. Environ. Manag.* **2020**, *264*, 110422. [CrossRef] [PubMed]

agriculture

MDPI

Article

Evaluation of Microbiological and Chemical Properties of Soils as a Result of Anthropogenic Denudation

Piotr Kanarek [1,*], Barbara Breza-Boruta [1], Mateusz Pawłowski [2] and Mirosław Kobierski [2]

[1] Department of Microbiology and Food Technology, Faculty of Agriculture and Biotechnology, Bydgoszcz University of Science and Technology, 6 Bernardyńska Street, 85-029 Bydgoszcz, Poland; breza@pbs.edu.pl
[2] Department of Biogeochemistry and Soil Science, Faculty of Agriculture and Biotechnology, Bydgoszcz University of Science and Technology, 6 Bernardyńska Street, 85-029 Bydgoszcz, Poland; matpaw007@pbs.edu.pl (M.P.); kobierski@pbs.edu.pl (M.K.)
* Correspondence: piokan004@pbs.edu.pl; Tel.: +48-52-3734-9548

Abstract: Excessive agricultural intensification adversely affects soil quality, particularly in hilly terrain, leading to increased erosion. Anthropogenic denudation, intensified by tillage erosion, results in the displacement of soil material from hilltops and shoulders to their bases. The research hypothesis posits that tillage erosion adversely affects the microbiological and chemical properties of soils, especially at the hilltops of intensively cultivated areas. The study aimed to assess the microbiological and chemical properties of Luvisols cultivated under conventional plowing in the moraine region of the Southern Krajna Lakeland, Poland. The evaluation focused on the results of soil sample analyses taken from the hilltops and foothills of eroded mounds. Microbiological investigations included determining the abundance of actinomycetes, filamentous fungi, heterotrophic bacteria, cellulolytic microorganisms, copiotrophs, and oligotrophs. Additionally, pH values and the contents of phosphorus, potassium, magnesium, total organic carbon, and nitrogen were determined. A higher abundance of bacteria, actinomycetes, and copiotrophs was observed at the foothills. Statistically significant differences due to slope effects were noted for all chemical parameters, with higher concentrations of organic carbon, nitrogen, potassium, and phosphorus found in the foothill areas. Understanding denudation processes can contribute to sustainable soil resource use and agrocenosis conservation.

Keywords: denudation; soil microbiota; soil degradation; chemical properties of soils; Luvisols; soil biodiversity

Citation: Kanarek, P.; Breza-Boruta, B.; Pawłowski, M.; Kobierski, M. Evaluation of Microbiological and Chemical Properties of Soils as a Result of Anthropogenic Denudation. *Agriculture* **2023**, *13*, 2247. https://doi.org/10.3390/agriculture13122247

Academic Editor: Eric Blanchart

Received: 12 November 2023
Revised: 30 November 2023
Accepted: 4 December 2023
Published: 6 December 2023

1. Introduction

The increasing pressure on natural resources, driven by population growth and the consequent need to enhance food production, is one of the contemporary challenges facing the maintenance, protection, and restoration of agrobiocenosis [1,2]. Agricultural production itself is critically tied to the soil; initially, the growth of production relied on expanding the cultivated land area (currently approximately 1.6 billion hectares). Over time, however, investments in agricultural intensification, understood as an increase in crop production per unit area, have increased [3]. Such a condition has the potential to disrupt the sustainable development of agriculture, characterized by the system's capacity to consistently uphold food production without exerting degradative impacts on the natural environment [4]. Sustaining global efforts to monitor changes occurring directly in the soil and indirectly in agriculture allows for early identification of threats related to environmental protection and the global food production system. There is a scientific consensus confirming that systematic monitoring of soil erosion is a key element in the effective implementation of sustainable agricultural practices. Understanding erosion processes and their impact on soil structure and chemical and microbial composition is fundamental to achieving a sustainable ecological balance [5]. Detailed microbiological and chemical analyses of soil enable an understanding of the causes, prevention, minimization, and combatting of the adverse effects of soil degradation [6–8]. To this end, tools based on indicators of potential risk of soil productivity loss have been implemented to enable the predictive control of agroecosystems [9].

The intensification of agricultural production, particularly in developed countries, coupled with the occurrence of extreme weather events, has the potential to disrupt water relations and induce alterations in the soil environment, including phenomena such as cracking, crusting, and degradation of soil structure and aggregation [8,10]. The erosion rate is not only dependent on the topography and climatic conditions. Numerous factors influence the intensity of soil degradation processes, including soil type, texture, aggregate structure, cultivation system, cultivation practices, and the type of cultivated plant [9,11]. An additional factor influencing the acceleration of soil erosion processes is the socio-economic factors that impact the selection of environmentally unfavorable practices within the farming system [12]. The problem of soil erosion is so significant that it has been classified as the greatest global threat to soil functionality and health, thus affecting humanity's food security [13,14].

One of the key indicators of soil quality is soil microbiota. A detailed understanding of the dynamics of changes within specific groups of microorganisms in the soil environment is crucial, given that soil microbiota constitutes a significant component of the ecosystem's response to erosive processes [15]. The rapid response of soil microorganisms is attributed to their reaction to the degradative impact of erosive processes on abiotic factors (pH, particle size composition, carbon, and nitrogen content), which directly shape the populations of fungi and bacteria [16]. Subsequently, alterations and disturbances in soil microbiota result in changes related to the availability of elements, nitrogen and phosphorus binding, organic matter accumulation, and mass and energy cycling. These processes ultimately exert a negative impact on the physiological stress of plants, leading to a decline in crop yield [17,18].

Luvisols, the prevailing soil group in the lowland and upland areas of Poland, significantly impact agricultural production due to their classification as relatively fertile soils [4]. Currently, soil degradation in Europe is a recognized and growing issue. This phenomenon is particularly noticeable in moraine areas, where slope processes lead to the displacement of soil material from the summit of the slope towards the foothills [6,19]. Inappropriate cultivation practices intensify the denudation effect on the soil, recognized as the surface lowering of the slope in a terrain with relief. The process of slope erosion is comprehensive and can involve the entire soil profile, ranging from the topsoil layer to the bedrock [6,20]. Anthropogenic denudation (accelerated erosion) is recognized as one of the more serious threats to the geoenvironment, given its impact on the productivity of plant production [2]. It is worth noting that areas subjected to significant anthropogenic pressure, particularly since the beginning of human settlement, are especially vulnerable. Currently, these processes intensify due to deforestation, extensive drainage networks, and, above all, as a result of agricultural intensification and the use of advanced agricultural machinery. This accelerates the truncation of pedons and leads to the exposure of deeper genetic horizons in soil profiles on elevated terrain, potentially resulting in physical, chemical, and biological changes in soil properties. This phenomenon applies not only to the summit but also to the foothills of the hillslope [3,6,21].

Changes in soil management practices can affect soil morphology, water retention, chemical properties, and biological activity. In particular, tillage practices lead to the intensification of slope processes [3,22,23]. Soil erosion leads to transformations of the soil cover of young morainic terrains of mesoregions in northern Poland. Main alterations relate to soil truncation on summits of slopes, whereas at footslopes and toeslopes and within local depressions, colluvial material is accumulated [24,25].

Given the challenges posed by climate change, a growing population, and soil degradation, it is essential to focus more on protecting and monitoring agricultural soils. These soils, which are particularly susceptible to erosion processes, require increased attention to ensure sustainable crop production. Monitoring fields in moraine terrain is among the top priorities for modern agriculture in Poland. Sustainable agricultural practices enable more efficient use of production resources and better environmental protection. Therefore, it is essential to have knowledge about the actual quality of soils and biological activity, especially in soils prone to erosion.

The research hypothesis states that agricultural erosion has a detrimental effect on the microbiological and chemical properties of soils subjected to intensive cultivation under a conventional agricultural system. This phenomenon is particularly pronounced in regions characterized by undulating topography, such as hilly or moraine landscapes. Under such topographic conditions, the erosive forces associated with agricultural activities are amplified because the post-glacial terrain promotes the formation of runoff paths for soil particles, nutrients, and organic matter. The undulating nature of the terrain contributes to accelerating erosion, intensifying the movement of essential components and affecting the composition of the soil microbiota.

The aim of the study was to assess the microbiological and chemical properties of the topsoil and subsoil layers of Luvisols subjected to anthropogenic denudation in the moraine region of the Southern Krajna Lakeland (northern Poland). The assessment included the quantification of actinobacteria, the total count of fungi and bacteria, as well as the count of copiotrophic, cellulolytic, and oligotrophic microorganisms in the soil at the summit and foothills of the hills. The study compared the content of plant-available forms of phosphorus, potassium, and magnesium, as well as the content of total organic carbon and nitrogen, along with pH values. In addition, the grain size composition was analyzed and the electrical conductivity (EC) of soil was determined. Comparisons of microbiological and physicochemical parameters were conducted for two genetic horizons: topsoil (0–30 cm) and subsoil (30–50 cm). Monitoring potential microbiological and chemical changes may contribute to a better understanding of the erosion dynamics in Luvisols subjected to intense anthropogenic pressure in areas with diverse topography.

2. Materials and Methods

2.1. Study Sites and Sampling

The research was conducted in the mesoregion of the South Baltic Lakeland (northern Poland)—specifically, the Southern Krajna Lakeland, with an average elevation of 115.6 m above sea level [26,27]. The climate of the studied region is temperate, with an average annual temperature of 7.8 °C (average annual minimum temperature of 3.8 °C and average annual maximum temperature of 12.8 °C) and a long-term average annual rainfall of 504 mm. Farmers do not implement proper crop rotation on the examined fields, which would enrich the soil with organic matter serving as a water reservoir. Winter wheat and rapeseed dominate the crop rotation on the surveyed fields. Unfortunately, farmers choose this option to ensure the financial stability of their farms. While this positively affects the profitability of production and enables farm development, it deteriorates soil fertility and the potential for biodiversity.

Three arable fields were selected (designated as R, S, T) that exhibited the typical mosaic topography characteristic of this area (Figure 1).

A total of 6 soil pits were excavated, representative of the soil cover in the studied region. All soil samples were collected after the winter wheat harvest, which was fertilized with a similar dose of NPK (average NPK fertilizer consumption was 140 kg: 80 kg N, 25 kg P, 35 kg K). A conventional cultivation system and similar agrotechnical treatments were applied uniformly across all locations. For each location (arable field), two soil pits were excavated: one at the summit of the hill and another at its foothills. Soil samples for testing were taken from the hills on the northern side of slopes ranging from 82 to 150 m above sea level. The length of the slope ranged from 90 to 150 m (Figure 2).

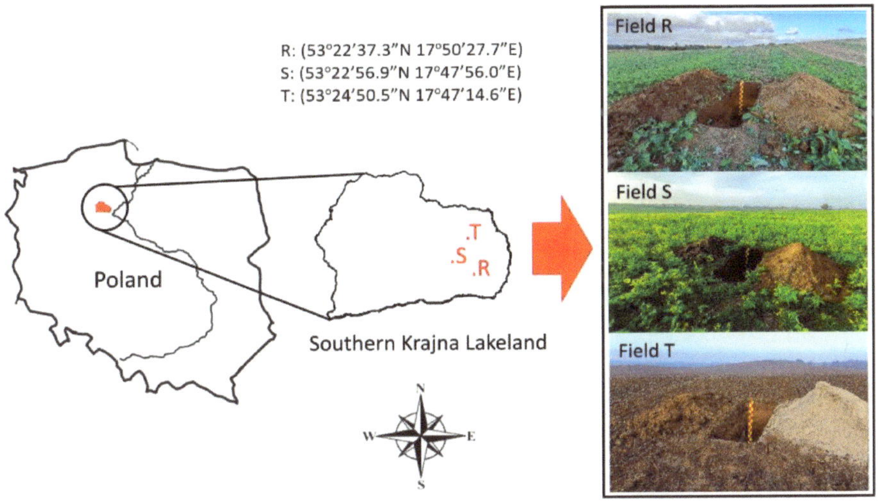

Figure 1. Location of tested fields (photos of profile pits at the foot of the hills).

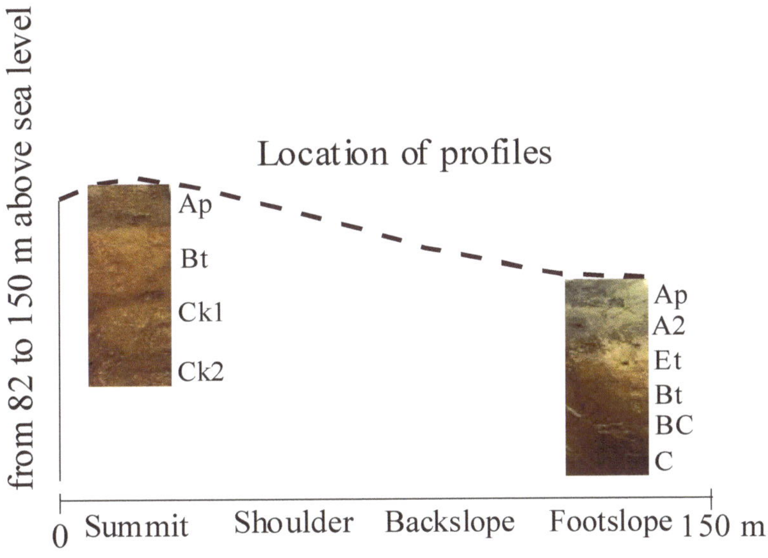

Figure 2. Locations of soil profiles (the example given here is site R). Abbreviations: Ap—plow horizon, A2—hummus horizon, Et—luvic horizon, Bt—argic horizon, BC—transitional horizon between argic and parent material, C—parent material, Ck1—parent material with calcium carbonate, Ck2—second parent material with calcium carbonate.

The locations of the soil pits were distant from buildings and roads. From each soil pit, soil samples were collected from two genetic horizons (Ap-surface and subsoil), corresponding to depths of 0–30 cm and 30–50 cm, respectively, in October 2022 (Table 1).

Table 1. The compilation of media used for the isolation and cultivation of microorganisms from the investigated soil samples.

Site	Profile	Soil Classification WRB 2022 [28]	Sampling
R	Ap-Bt-Ck1-Ck2 Ap-A2-Et-BC-C	Epicutanic Luvisol Haplic Luvisol	A1—Summit: B1—Ap horizon; B2—Bt horizon A2—Footslope: B1—Ap horizon: B2—A2 horizon
S	Ap-Ck1-Ck2-Ck3 Ap-A2-Et-EB-Bt-C	Calcaric, Eutric Regosol Haplic Luvisol	A1—Summit: B1—Ap horizon; B2—Ck1 horizon A2—Footslope: B1—Ap horizon: B2—A2 horizon
T	Ap-Ck1-Ck2-2Ck Ap-A2-Et-EB-Bt1-Bt2-C	Calcaric, Eutric Regosol Haplic Luvisol	A1—Summit: B1—Ap horizon; B2—Ck1 horizon A2—Footslope: B1—Ap horizon: B2—A2 horizon

These samples were stored in plastic containers, allowing gas exchange, then cooled to 4 °C and transported to the laboratory. Soil samples for chemical analysis were placed in plastic bags, air-dried, and sieved through a 2-mm mesh screen.

2.2. *Microbiological Analyses*

Microbiological analyses of soil samples included quantitative determinations for the following groups of microorganisms:

1. Actinobacteria (Ac),
2. Total bacterial count (B),
3. Total fungal count (Ff),
4. Cellulolytic bacteria (Ce),
5. Oligotrophs (Ol),
6. Copiotrophs (Co).

For all determinations, 10 g of soil were weighed and introduced into 90 mL of Ringer's solution. Subsequently, the samples were shaken for 25 min at room temperature. Following the homogenization of the suspension, a series of 10-fold dilutions (10^{-1} to 10^{-7}) were prepared. Subsequently, 1000 μL of the soil solution was transferred onto dedicated selective media (Table 2).

Table 2. The compilation of media used for the isolation and cultivation of microorganisms from the investigated soil samples.

Tested Groups of Microorganisms	Applied Medium	Reference
Ac	Modified yeast extract-glucose medium (YGA)	[29]
B	Standard nutrient agar	[30]
Ff	Rose–Bengal agar with 30 μg mL^{-1} streptomycin	[30]
Ce	Agar containing 0.1% sodium carboxymethylcellulose	[31]
Ol	Modified medium with limited nutrients	[32]
Co	Modified medium with additional nutrients	[32]

The incubation of microorganisms was carried out at 25–28 °C for 4 days for B, Ff, Co, and Ce; 10 days for actinobacteria, and 14 days for oligotrophic microorganisms. All analyses were performed in three replicates. The count of colony-forming units (cfu) obtained was determined per 1 g of soil dry matter. (cfu × g^{-1} d.m. of soil).

2.3. *Soil Sample Chemical Analyses*

The texture of soil samples was determined using a laser diffraction particle size analyzer (Mastersizer 2000, Malvern Instruments, Malvern, UK). Potentiometric measurements of pH values were conducted in a 1 M KCl (1:2.5 *w*/*v*) solution [33] using a pH Meter (CP-551—Elmetron, Zabrze, Poland). The total organic carbon (TOC) and total nitrogen (TN) content were determined using a CN dry combustion analyzer (Vario Max CN, Elementar

Analysensysteme GmbH, Hanau, Germany). The content of available magnesium (Mg) was assessed using the Schachtschabel method [34]. The contents of available forms of phosphorus (P) [35] and potassium (K) were determined by the Egner–Riehm method [36]. The content of forms available to plants was determined by atomic absorption spectroscopy and atomic emission spectroscopy using a Thermo Solaar S4 Atomic Absorption Spectrometer (Thermo Electron Corporation, Waltham, MA, USA). Measuring ECs with a conductivity meter in a soil-water extract was based on a fixed soil:solution ratio (1:5). The water-soluble forms of cations were based on the extraction with distilled water. Soluble salts were measured by the 1:5 (*v:v*) soil:de-ionized water extraction method. All the measurements were performed in three replications. Finally, Ca, Mg, K, and Na contents were measured using a Thermo Solaar S4 Atomic Absorption Spectrometer (Thermo Electron Corporation, Waltham, MA, USA).

2.4. Statistical Analyses

The obtained results of microbiological and chemical parameters underwent statistical analysis through a two-way analysis of variance (ANOVA). The first factor (A) was associated with the location of the soil pit (A1—summit and A2—footslope), while the second factor was related to the depth (B1—surface level 0–30 cm and B2—subsurface level 30–50 cm). To assess the significance of these factors and their mutual interactions, Tukey's post hoc test with a 95% confidence interval was conducted, enabling a comparison of the mean values of the analyzed parameters.

In order to better explain the soil variation in terms of selected parameters (biological and chemical) and the relationship between them, the multivariate statistical method Principal Component Analysis (PCA) was used. The use of PCA also made it possible to present correlations between the studied parameters. Differences between objects were analyzed using PCA based on the mean values of all investigated soil features. Two principal components (PC1 and PC2) were used to rank the cases. A cluster analysis (CA) was also carried out. CA allowed the separation of groups of sites based on the diversity of variables, which was graphically presented in the form of a dendrogram. The Ward method [37] was employed to calculate the distances between individual clusters. All statistical analyses were conducted using Statistica 12 PL software from StatSoft [38].

3. Results

3.1. Microbiological Analysis

Analyses of microbiological data revealed statistically significant differences in the abundance of selected microbial groups both between surface and subsurface levels and between locations at the summit and footslope (Table 3). Concerning the overall bacterial count, test field T exhibited the most substantial decrease between surface levels, with a notable increase in colony-forming units (cfu) in soil samples collected at the footslope by 674.20% ($LSD_{0.05}$ = 10.868).

A similar phenomenon was also observed for actinobacteria in field S, where a statistically significant decrease in cfu was noted with an increase in horizontal position in the surface level ($LSD_{0.05}$ = 0.431) and in the case of fungi abundance (field T). Tendencies related to changes in bacterial abundance between the examined levels at the summit and foothill were also observed; in all cases, a decrease in cfu was noted with the depth of sample collection.

Table 3. The number of bacteria, fungi, and actinomycetes in the soil, depending on the soil profile and location on the slope.

Level of Factor B	Test Site R			Test Site S			Test Site T		
	Level of Factor A *								
	A1	A2	Mean	A1	A2	Mean	A1	A2	Mean
	Heterotrophic bacteria (B) (10^6 cfu g^{-1})								
B1	27.3 ^{a}	23.0 ^{a}	25.2 ^{a}	18.7 ^{a}	18.7	18.7 ^{a}	3.1 ^{a}	24.0 ^{a}	13.6 ^{a}
B2	6.1 ^{a}	5.2 ^{a}	5.7 ^{b}	2.3 ^{b}	2.4	2.3 ^{b}	2.5 ^{a}	1.3 ^{b}	1.9 ^{b}
Mean	16.7	14.1		10.5	10.5		2.8	12.6	
$LSD_{0.05}$	A = n.s.; B = 5.499; A/B = n.s.; B/A = n.s.			A = n.s.; B = 3.991; A/B = n.s.; B/A = n.s.			A = 7.685; B = 7.685; A/B = 10.868; B/A = 10.868		
	Actinobacteria (Ac) (10^5 cfu g^{-1})								
B1	28.7 ^{a}	30.7 ^{a}	29.7 ^{a}	18.0 ^{a}	28.7 ^{a}	23.3 ^{a}	27.0 ^{a}	52.0 ^{a}	3.9 ^{b}
B2	18.7 ^{a}	26.3 ^{a}	22.5 ^{a}	8.6 ^{b}	6.8 ^{b}	7.6 ^{b}	2.0 ^{a}	10.8 ^{a}	6.4 ^{a}
Mean	23.7	28.5		18.6	12.3		14.0	31.0	
$LSD_{0.05}$	A = n.s.; B = n.s.; A/B = n.s.; B/A = n.s.			A = 0.305; B = 0.305; A/B = 0.431; B/A = 0.431			A = 1.252; B = 1.252; A/B = n.s.; B/A = n.s.		
	Filamentous fungi (Ff) (10^5 cfu g^{-1})								
B1	5.0 ^{a}	3.0 ^{a}	4.0 ^{a}	3.0 ^{a}	4.0 ^{a}	3.2 ^{a}	2.0 ^{a}	13.0 ^{a}	7.5 ^{a}
B2	1.0 ^{a}	1.0 ^{a}	1.4 ^{b}	1.0 ^{a}	2.0 ^{a}	1.2 ^{b}	1.0 ^{b}	1.0 ^{b}	1.0 ^{b}
Mean	3.1	2.3		1.6	2.8		1.30	0.7	
$LSD_{0.05}$	A = n.s.; B = 0.133; A/B = n.s.; B/A = n.s.			A = 0.108; B = 0.108; A/B = n.s.; B/A = n.s.			A = 0.436; B = 0.436; A/B = 0.616; B/A = 0.616		

* Levels of factor A: A1 (summit), A2 (footslope); Levels of factor B: B1 (surface level), B2 (subsurface level); LSD—Least Significant Difference, significance of differences tested with Tukey's test at $p \leq 0.05$; n.s.—not significant; ^{a,b}—letters in columns indicate significant differences at $p \leq 0.05$.

In the case of trophic groups of microorganisms (Ce, Co, and Ol), the research results demonstrated greater diversity (Table 4). The abundance of cellulose-degrading microorganisms was higher at the summit of the slope than at its footslope. This trend reversed in soil samples from subsurface levels, where a higher abundance of cellulolytic microbes was observed at the foot of the slope. On the other hand, no statistically significant differences were found in any of the tested fields for this trophic group. Copiotrophs, however, predominated at the foot of the slope, regardless of the surveyed location. The soil samples from field R revealed statistically significant differences ($LSD_{0.05}$ = 13.543) between the summit and the footslope in the presence of oligotrophic microorganisms. These microbes, inhabiting an environment with a reduced supply of energy substrates, clearly dominated on the slope of the field.

Table 4. The number of cellulolytic, copiotrophic, and oligotrophic microorganisms in the soil, depending on the soil profile and location on the slope.

Level of Factor B	Test Site R			Test Site S			Test Site T		
	Level of Factor A								
	A1	A2	Mean	A1	A2	Mean	A1	A2	Mean
	Cellulolytic microorganisms (Ce) (10^6 cfu g^{-1})								
B1	8.0 ^{a}	5.3 ^{a}	6.7 ^{a}	6.3 ^{a}	3.3 ^{a}	4.8 ^{a}	5.1 ^{a}	4.8 ^{a}	5.0 ^{a}
B2	1.0 ^{a}	1.2 ^{a}	1.1 ^{b}	1.4 ^{b}	3.3 ^{a}	1.6 ^{b}	0.5 ^{a}	0.8 ^{a}	0.6 ^{b}
Mean	4.5	3.3		3.9	1.7		2.8	2.8	
$LSD_{0.05}$	A = n.s.; B = 2.255; A/B = n.s.; B/A = n.s.			A = 1.360; B = 1.360; A/B = 1.923; B/A = 1.923			A = n.s; B = 4.218; A/B = n.s; B/A = n.s.		

Table 4. *Cont.*

Level of Factor B	Test Site R			Test Site S			Test Site T		
	Level of Factor A								
	A1	A2	Mean	A1	A2	Mean	A1	A2	Mean
	Copiotrophic microorganisms (Co) (10^6 cfu g^{-1})								
B1	14.3 ^{a}	24.7 ^{a}	19.5 ^{a}	15.0 ^{a}	19.3 ^{a}	17.2 ^{a}	11.3 ^{a}	24.0 ^{a}	15.5 ^{a}
B2	2.9 ^{a}	6.0 ^{a}	4.4 ^{b}	2.6 ^{a}	3.2 ^{a}	3.0 ^{b}	1.1 ^{a}	1.3 ^{a}	1.2 ^{b}
Mean	8.6	15.3		8.8	11.3		4.0	12.7	
$LSD_{0.05}$	A = n.s.; B = 9.367; A/B = n.s.; B/A = n.s.			A = n.s.; B = 6.255; A/B = n.s.; B/A = n.s.			A = n.s.; B = 12.619; A/B= n.s.; B/A = n.s.		
	Oligotrophic microorganisms (Ol) (10^6 cfu g^{-1})								
B1	65.0 ^{a}	40.0 ^{a}	52.6 ^{a}	51.0 ^{a}	34.0 ^{a}	42.5 ^{a}	37.3 ^{a}	64.0 ^{a}	50.7 ^{a}
B2	13.9 ^{a}	6.9 ^{a}	10.4 ^{b}	6.0 ^{b}	14.2 ^{b}	10.3 ^{b}	2.2 ^{a}	0.6 ^{a}	1.4 ^{b}
Mean	39.5	23.4		28.6	24.1		19.8	32.3	
$LSD_{0.05}$	A = 13.543; B = 13.543 A/B = n.s.; B/A = n.s.			A = n.s.; B = 9.224; A/B = 13.045; B/A = 13.045			A = n.s.; B = 14.400; A/B = n.s.; B/A = n.s.		

Abbreviations: see Table 3. [a,b]—letters in columns indicate significant differences at $p \leq 0.05$.

3.2. Chemical Analysis

Statistically significant differences were demonstrated for selected chemical parameters of soils between the summit and footslope of the hills, as well as between the genetic levels (Table 5). In terms of pH values, statistically lower pH values were observed at the footslope in two fields (S and T). Both TOC and TN contents indicated higher accumulation at the footslope, regardless of the genetic level. No statistically significant differences were noted in the content of available Mg; however, on fields R and S, Mg content was higher in the surface level at the summit, while for field T, the same trend was observed for TOC and TN. The content of available forms of K and P differed significantly between slope locations, irrespective of depth.

Table 5. pH level, TOC, TN, available Mg, P, and K content of the soil, depending on the soil profile and location on the slope.

Level of Factor B	Test Site R			Test Site S			Test Site T		
	Level of Factor A								
	A1	A2	Mean	A1	A2	Mean	A1	A2	Mean
	pH value								
B1	6.33 ^{a}	7.14 ^{a}	6.73 ^{a}	8.18 ^{b}	6.48 ^{a}	7.33 ^{a}	8.36 ^{b}	7.01 ^{a}	7.66 ^{a}
B2	6.65 ^{a}	6.94 ^{a}	6.79 ^{a}	8.52 ^{a}	6.17 ^{b}	7.34 ^{a}	8.72 ^{a}	6.67 ^{b}	7.69 ^{a}
Mean	6.49	7.04		8.35	6.33		8.54	6.82	
$LSD_{0.05}$	A = 0.313; B = n.s. A/B = n.s.; B/A = n.s.			A = 0.233; B = n.s. A/B = 0.330; B/A = 0.330			A = 0.055; B = n.s. A/B = 0.077; B/A = 0.077		
	Total Organic Carbon (TOC) (g·kg^{-1})								
B1	5.94 ^{a}	7.30 ^{a}	6.62 ^{a}	6.92 ^{a}	9.79 ^{a}	8.33 ^{a}	5.49 ^{a}	9.21 ^{a}	7.35 ^{a}
B2	2.85 ^{b}	6.38 ^{b}	4.62 ^{b}	2.71 ^{b}	8.25 ^{b}	5.48 ^{b}	3.94 ^{b}	5.90 ^{b}	4.92 ^{b}
Mean	4.40	6.84		4.82	8.99		4.72	7.56	
$LSD_{0.05}$	A = 0.117; B = 0.117; A/B = 0.165; B/A = 0.165			A = 1.055; B = 1.055; A/B = 1.492; B/A = 1.492			A = 0.233; B = 0.233; A/B = 0.329.; B/A = 0.329		

Table 5. *Cont.*

Level of Factor B	Test Site R			Test Site S			Test Site T		
	Level of Factor A								
	A1	A2	Mean	A1	A2	Mean	A1	A2	Mean
	Total Nitrogen (TN) (g·kg^{-1})								
B1	0.69 [a]	0.71 [a]	0.70 [a]	1.05 [a]	1.10 [a]	1.08 [a]	0.81 [a]	0.90 [a]	0.86 [a]
B2	0.50 [a]	0.70 [a]	0.60 [a]	0.29 [b]	0.92 [b]	0.61 [b]	0.40 [b]	0.72 [b]	0.56 [b]
Mean	0.60	0.71		0.67	1.01		0.61	0.81	
$LSD_{0.05}$	A = n.s.; B = n.s.; A/B = n.s.; B/A = n.s.			A = 0.129; B = 0.129; A/B = 0.182; B/A = 0.182			A = 0.091; B = 0.091; A/B = 0.129; B/A = 0.129		
	Magnesium (Mg) (mg·kg^{-1})								
B1	30.7 [a]	22.6 [a]	26.65 [a]	30.2 [a]	27.6 [a]	28.88 [a]	17.5 [a]	24.8 [a]	21.13 [a]
B2	41.1 [a]	19.8 [a]	30.45 [a]	21.00 [a]	29.4 [a]	25.18 [a]	29.4 [a]	30.2 [a]	29.78 [a]
Mean	35.90	21.20		25.55	29.40		23.45	27.45	
$LSD_{0.05}$	A = 11.851; B = n.s.; A/B = n.s.; B/A = n.s.			A = n.s.; B = n.s.; A/B = n.s.; B/A = n.s.			A = n.s; B = n.s.; A/B = n.s.; B/A = n.s.		
	Phosphorus (P) (mg·kg^{-1})								
B1	97.5 [a]	119.1 [a]	108.28 [a]	32.5 [a]	125.5 [a]	79.00 [a]	60.5 [a]	122.7 [a]	91.60 [a]
B2	51.7 [b]	94.9 [b]	73.28 [b]	21.5 [a]	98.9 [a]	54.68 [b]	31.1 [b]	111.3 [b]	71.20 [b]
Mean	74.58	106.98		21.48	112.20		45.80	117.00	
$LSD_{0.05}$	A = 9.908; B = 9.908; A/B = 14.012; B/A = 14.012			A = 10.175; B = 10.175; A/B = n.s.; B/A = n.s.			A = 6.276; B 6.276; A/B = 8.875; B/A = 8.875		
	Potassium (K) (mg·kg^{-1})								
B1	128.4 [a]	185.4 [a]	156.9 [a]	212.7 [a]	251.0 [a]	231.85 [a]	125.8 [a]	223.3 [a]	174.55 [a]
B2	117.5 [a]	101.2 [b]	109.4 [b]	116.3 [a]	142.8 [a]	129.55 [b]	85.7 [b]	221.1 [a]	153.40 [b]
Mean	122.93	143.30		164.50	196.90		105.75	222.20	
$LSD_{0.05}$	A = 13.381; B = 13.381; A/B = 18.923; B/A = 18.923			A = 20.524; B = 20.524; A/B = n.s.; B/A = n.s.			A = 11.377; B = 11.377; A/B = 16.090; B/A = 16.090		

Abbreviations: see Table 3. [a,b]—letters in columns indicate significant differences at $p \leq 0.05$.

3.3. TOC:TN Ratio and Granulometric Composition

The highest values of the TOC:TN ratio for locations R and T were observed in the surface level at the footslope, while for location S, they were found in the subsurface level at the summit. Similarly high values were noted in the subsurface profile at the footslope. The lowest values were recorded for the surface level (fields S and T) at the summit of the hill (Figure 3).

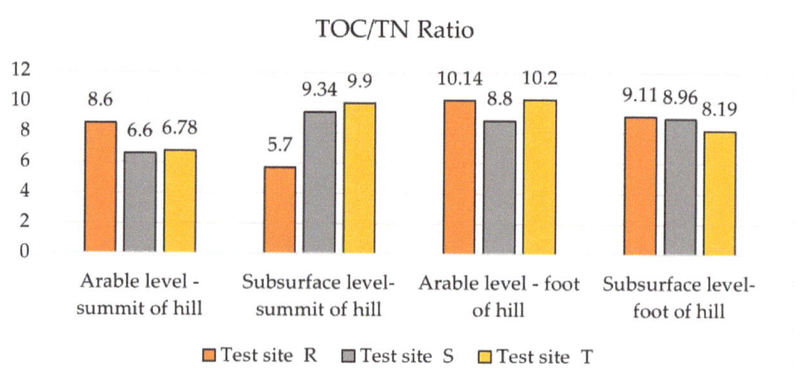

Figure 3. Carbon to nitrogen ratio values.

The particle size analysis revealed a higher proportion of sand in the surface level at the footslope, contrasting with the surface layer at the summit of the slope (Figure 4). The silt fraction, on the other hand, was higher in the surface and subsurface levels at the summit of the hill. Variations in the clay fraction were minor and did not exceed 6% in the composition of all examined samples. Nevertheless, higher values were also observed in the upper parts of the surveyed elevations.

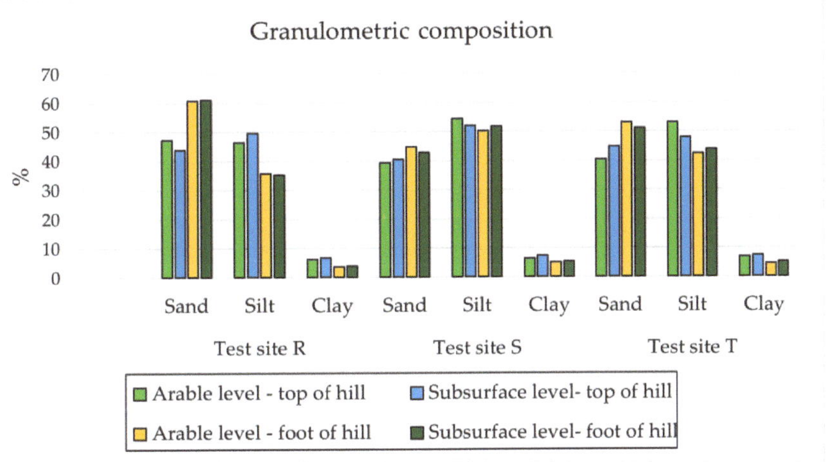

Figure 4. Particle size composition. Percentage content of sand (2.0–0.05 mm), silt (0.05–0.002 mm), clay (<0.002 mm).

3.4. Soil Electrical Conductivity

The surface layer of footslopes soil exhibited significantly lower average content of water-soluble forms of magnesium and calcium compared to summits (Table 6). However, in the soil material of the 0–30 cm layer of footslopes, a higher average content of water-soluble sodium was observed (Table 7). The average content of water-soluble potassium was lower in the surface layer of footslopes compared to summits. In the subsurface layer of the investigated footslope soils, a significantly higher average content of Na and Mg, and a significantly lower average content of Ca were recorded compared to summits. The highest content of water-soluble calcium cations was observed in the S and T subsurface horizons containing calcium carbonate.

Table 6. Sodium, potassium, magnesium, and calcium in the water-soluble fraction of surface soil horizon and values of soil electrical conductivity.

Location	Na	K	Mg	Ca	ECs
			$g{\cdot}kg^{-1}$		$\mu S{\cdot}cm^{-1}$
R A1B1	6.6 ± 3.39	44.3 ± 1.27	15.9 ± 5.80	79.0 ± 1.84	93.4 ± 8.20
S A1B1	12.7 ± 4.17	77.9 ± 3.81	17.3 ± 1.27	201.0 ± 9.90	118.0 ± 3.68
T A1B1	8.7 ± 2.05	36.7 ± 0.92	16.4 ± 4.24	210.5 ± 2.12	166.9 ± 3.46
Mean	9.33	53.0	16.3	163.5	126.1
R A2B1	16.4 ± 4.38	38.3 ± 4.38	15.7 ± 1.63	40.8 ± 3.53	101.7 ± 1.34
S A2B1	11.1 ± 3.32	49.0 ± 1.63	13.7 ± 3.53	47.7 ± 3.61	148.6 ± 1.27
T A2B1	10.0 ± 0.35	21.2 ± 1.55	15.3 ± 0.56	33.4 ± 0.28	105.9 ± 0.78
Mean	12.5	36.2	14.0	40.6	118.7
	$p = 0.30$	$p = 0.32$	$p = 0.022$ *	$p = 0.045$ *	$p = 0.79$

Abbreviations: see Table 3; ECs—soil electrical conductivity, p—significance factor, *—statistically significant.

Table 7. Sodium, potassium, magnesium, and calcium in the water-soluble fraction of subsurface soil horizon and values of soil electrical conductivity.

Location	Na	K	Mg	Ca	ECs
		$g \cdot kg^{-1}$			$\mu S \cdot cm^{-1}$
R A1B2	9.5 ± 1.83	32.5 ± 1.55	11.6 ± 1.48	78.7 ± 4.10	70.2 ± 6.22
S A1B2	10.7 ± 2.19	40.5 ± 3.60	17.3 ± 2.33	240.5 ± 3.53	68.6 ± 7.71
T A1B2	10.6 ± 0,21	23.2 ± 0.99	13.6 ± 4.31	233.0 ± 2.12	159.3 ± 7.78
Mean	10.3	34.1	14.2	184.1	99.4
R A2B2	12.5 ± 0.85	42.7 ± 0.28	19.7 ± 1.77	32.8 ± 1.91	117.8 ± 6.64
S A2B2	12.3 ± 0.85	95.6 ± 0.56	22.5 ± 3.68	44.7 ± 1.77	148.5 ± 2.90
T A2B2	12.8 ± 0.71	41.5 ± 0.14	18.6 ± 3.82	32.6 ± 0.21	105.7 ± 4.38
Mean	12.5	59.9	20.3	36.7	124.0
	$p = 0.005$ *	$p = 0.21$	$p = 0.04$ *	$p = 0.49$	$p = 0.49$

Abbreviations: see Table 3; Table 6. *—statistically significant.

The soil material of the investigated areas did not exhibit salinity characteristics. In soil samples taken from the plow layer in summits, higher average values of ECs were recorded compared to footslopes. However, in the subsurface layer (B2), higher average values of EC were found in footslope soil. Tillage erosion processes did not significantly determine the values of soil electrical conductivity in the soil material on cultivated hillslopes.

3.5. Principal Component Analysis—PCA

Multivariate PCA (Figure 5) was employed to determine the nature and strength of correlations among the soil microbiological and chemical parameters studied (B, Ac, Ff, Co, Ce, Ol, pH, TOC, TN, available Mg, P, and K), hillslope location, and genetic level. Two principal components (PC1 and PC2) were extracted from the available data, explaining a total variance of 71.63%. PC1 accounted for the majority of the variance at 55.14%, while PC2 explained 16.49%. The PCA revealed that the first component (PC1) was significantly negatively associated with TOC (−0.823), TN (−0.751), and assimilated forms of K (−0.710) and P (−0.694). The second component, PC2, can be interpreted as describing microbial abundance, as it was significantly positively correlated with Ff, B, Co, Ac, Ol, and Ce.

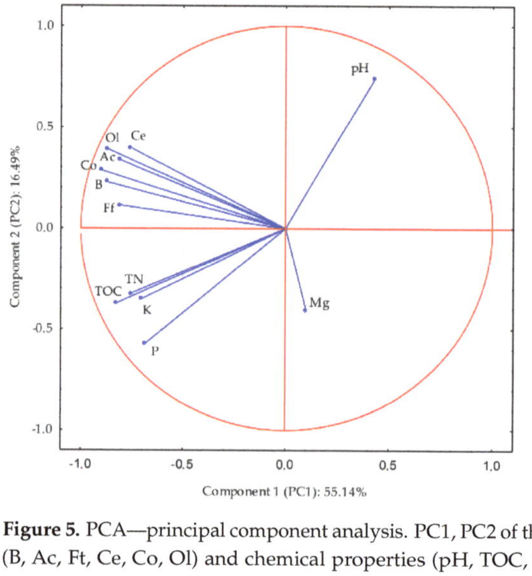

Figure 5. PCA—principal component analysis. PC1, PC2 of the studied soil microbiological properties (B, Ac, Ft, Ce, Co, Ol) and chemical properties (pH, TOC, TN, content of available P, K, and Mg). Abbreviations: see Tables 2 and 4.

The application of Cluster Analysis (CA) to the discussed parameters allowed for the determination of similarities (differences) between the sampling point's location and depth (Figure 6). Based on the dendrogram generated from the obtained data, three clusters were identified. Cluster 1, in the majority of cases (3 out of 5), grouped locations where soil samples were taken from the summit of the hills. These locations (RA1B1, RA1B2, and TA1B1) were characterized by a higher abundance of actinobacteria and the concentration of available phosphorus on the summit, compared to location S. Soil with the highest pH values and a lower overall bacterial count in the subsurface layer at the summit (locations SA1B2 and TA1B2) was grouped in Cluster 2. Cluster 3 distinguished 5 objects (RA2B1, TA2B1, SA2B1, TA2B2, SA1B1) where soil samples from the footslope were characterized by the highest content of TOC, TN, assimilated forms of P and K, and a higher abundance of selected microbial groups (Co, Ac).

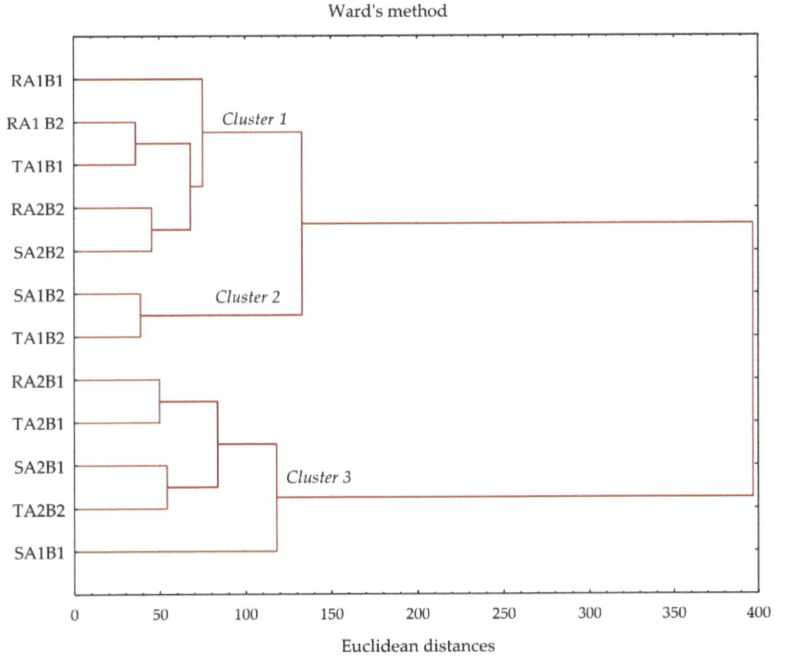

Figure 6. Dendrogram analysis of physicochemical and microbial soil characteristics. Abbreviations: see Table 3.

4. Discussion

Erosion and geomorphic processes are primarily associated with the impact of water, glaciers, and ice sheets on the Earth's surface. Currently, in the Anthropocene, the predominant geomorphic factor is the direct and indirect influence of human activities [39,40]. To date, the majority of studies on soil erosion have been dedicated to assessing changes in physicochemical parameters, with fewer focusing on the analysis of bacterial and fungal communities, as well as individual trophic groups. Addressing this gap in the literature has the potential to contribute to the formulation of a comprehensive approach for the early detection and prevention of anthropogenic erosion. In addition, this paper analyzes Luvisols, which, due to their widespread occurrence, impact food security and the stability of crop production [41,42].

In this study, all examined groups of microorganisms exhibited significant differences in abundance between the cultivation layer and the subsurface layer, regardless of the location on the eroded slope. This aligns with the scientific consensus, indicating that individual horizons should be treated as distinct ecological habitats [43]. The decrease in the abundance of microorganisms along the depth gradient is influenced by several factors. Among the main factors is the increase in bulk density, which reduces the oxygen level, thereby limiting the developmental conditions for aerobic microorganisms [44]. Additionally, it is noteworthy that the arable layer contains more labile organic compounds, thereby promoting the development of heterotrophic microbiota. Another factor directly shaping the levels of microbial abundance is the impact of rainfall on the arable layer. Intense precipitation leads to soil moisture and the release of carbon from aggregates, contributing to increased availability of energy substrates [38,39].

The conducted research revealed variations in the microbial abundance in soil sampled from the summit compared to that at the foot of eroded slopes. While a significant increase in the overall bacterial count was observed in field T (LSD = 10.868), with a rise in count between the foot and summit in the plow layer by 666.77%, correlating with other reports [45–47], a similar phenomenon was not observed for oligotrophs and cellulolytic microorganisms. The higher level of oligotrophic abundance in the plow layer at the summit of the slope can be justified by lower carbon availability, creating an ecological niche for organisms with lower environmental requirements and enhancing the competitive advantage of this trophic group [48]. Although all soil samples were taken from post-wheat harvest sites, and changes in physicochemical parameters favored locations at the slope's base, cellulolytic microorganisms predominated at the summit. As highlighted by Huang et al. [49], sunlight intensity on the slope significantly influences microbial abundance, pointing to a relatively higher count of Gram-negative bacteria and aerobic microorganisms in shaded positions. The results obtained also revealed the prevalence of actinomycetes over the fungal population. This trend aligns with the evolutionary strategy of actinomycetes, as their capacity to metabolize a broad spectrum of hydrolytic enzymes and bioactive compounds enables them to exert fungistatic effects [50]. Furthermore, it has been demonstrated that the abundance of fungi in the soil is dependent on the use of organic fertilizers, which promote the proliferation of mycorrhizal fungi [51]. The soil pH is also of significance, indicating better environmental adaptation of fungi to lower pH levels [52]. In the case of our own research on two cultivation fields, the pH value at the hill's summit was alkaline and decreased to neutral with the depth of the soil samples. This trend may also indicate less favorable conditions for the development of fungi, as the optimal pH for their environment is slightly acidic.

All tested sites exhibited statistically significant differences in pH values, which increased with a decrease in elevation, confirming findings from previous studies [53]. The higher pH values, persisting at the summit on two sites, result from the presence of $CaCO_3$, indicating a shallowing of the soil profile and mixing with calcium carbonate-rich bedrock [54]. Changes in organic carbon content are also a crucial indicator of advancing soil degradation, contributing to a reduction in soil productivity. The decrease in TOC content in the upper part of the hill aligns with the trend associated with the displacement of substrates along the elevation gradient. Moreover, as noted by Jia et al., the slope gradient plays a significant role in influencing changes in soil organic carbon content, particularly concerning water erosion [55]. The phenomenon of TOC transport, as a result of tillage erosion, leads to increased concentration at the deposition site at the foot of the hill, thereby hindering its decomposition and reducing bioavailability [56]. Observations regarding the dynamics of TN content levels on the slope are similar to the changes occurring in TOC. The connection between the redistribution of these two important elements has already been noted in previous studies [57,58]. Following a meta-analysis of 204 studies, Li et al. highlighted that the conversion of severely degraded fields into grassland contributes to enhancing the overall soil TOC and TN content, aligning with the principles of sustainable agricultural management [59].

It is worth noting that this trend also applies to the concentrations of assimilable forms of phosphorus and potassium. The presence of these elements is controlled by the amount of organic matter in the soil, which, due to precipitation and cultivation practices, moves along the slope gradient and accumulates at the foot. Subsequently, as a result of the degradation of organic components by microorganisms, there is a release of elements and their reintegration into the nutrient cycle [53]. In the case of plant-available magnesium, no statistically significant changes were found in the levels of the altitude gradient, but it is noticeable that two fields showed higher levels of this element at the summit of the slope, which was also noted by other authors [60]. There is a possibility that, similar to the higher pH level at the summit of the hill, the element is released by the bedrock due to soil erosion.

Nutrients removed from cultivated fields along with plant yields must be adequately balanced by applied fertilizers. This is particularly important in precision agriculture, which focuses on improving the efficiency of agricultural production. Precision agriculture technologies allow for optimizing fertilization, the use of plant protection products, and the amount of fuel needed for cultivation operations [61,62]. This enables better utilization of agroecosystem capabilities, especially concerning non-renewable soil cover exposed to anthropogenic denudation.

5. Conclusions

In moraine areas subjected to intense agricultural anthropopression, the slope effect plays a significant role in shaping soil microbiota and changes in chemical parameters. The conducted research indicates a clear tendency to accumulate higher levels of organic carbon, total nitrogen, and plant-available forms of P and K at the foot of the slope, likely resulting from the transfer of soil material due to tillage and water erosion. Also, for most of the studied microbial groups, more stimulating conditions for their development were observed at the foot than at the summit, associated with the accumulation of larger quantities and easier access to nutritional substrates in these areas.

Microbial analysis indicates significant differences in microecosystems, depending on soil conditions and location. Variances in microbial composition between surface and depth, as well as among different locations, suggest the presence of specific microenvironments in the study area. These environments are strongly linked to the supply of nutrients and chemical elements, which are directly shaped by erosion processes. Recognizing the role of microorganisms could be a crucial aspect in better understanding, monitoring, and safeguarding biodiversity and the sustainable functioning of ecosystems. Monitoring microbiological and chemical changes in soil quality can be utilized as a prognostic tool to assess further erosive changes. Determining the impact of tillage erosion on the content of organic carbon and nitrogen, as well as shaping pH values and nutrient content in the soil is essential for targeted protection of agrocenosis.

The study found that over-intensification of agriculture negatively affects soil in moraine areas, leading to erosion. Cultivation operations move soil components from the tops of hills to the foothills. The findings underscore the need to adapt agricultural practices to the terrain, especially in hilly areas, and to use sustainable agriculture for long-term soil and environmental protection.

Managing the cultivation system and counteracting the effect of anthropogenic denudation is crucial both from an agricultural and environmental protection standpoint. By better understanding these processes and their consequences, one can contribute to the sustainable use of soil resources and the protection of agrocenosis.

Author Contributions: Conceptualization, P.K. and M.P.; methodology, B.B.-B. and M.K.; formal analysis, P.K. and M.P.; data curation, P.K., M.P., B.B.-B. and M.K.; writing—original draft preparation, P.K. and B.B.-B.; writing—review and editing, B.B.-B. and M.K.; visualization, P.K.; supervision, B.B.-B. and M.K.; funding acquisition, P.K. All authors have read and agreed to the published version of the manuscript.

Funding: The APC was funded by Bydgoszcz University of Science and Technology, as part of the "Działania Naukowe Młodych" program, grant number: DNM/22/2023.

Institutional Review Board Statement: Not applicable.

Data Availability Statement: Data is contained within the article.

Conflicts of Interest: The authors declare no conflict of interest.

References

1. Alewell, C.; Egli, M.; Meusburger, K. An Attempt to Estimate Tolerable Soil Erosion Rates by Matching Soil Formation with Denudation in Alpine Grasslands. *J. Soils Sediments* **2015**, *15*, 1383–1399. [CrossRef]
2. Conforti, M.; Buttafuoco, G. Assessing Space–Time Variations of Denudation Processes and Related Soil Loss from 1955 to 2016 in Southern Italy (Calabria Region). *Environ. Earth Sci.* **2017**, *76*, 457. [CrossRef]
3. Przewoźna, B. Changes of Bulk Density, Air-Water Properties and Morphology of Soils in Basins without Outlets as an Effect of Erosion and Anthropogenic Denudation (a Study from Northwestern Poland). *Soil Sci. Plant Nutr.* **2014**, *60*, 30–37. [CrossRef]
4. Kabała, C.; Musztyfaga, E. Clay-Illuvial Soils in the Polish and International Soil Classifications. *Soil Sci. Annu.* **2015**, *66*, 204–213. [CrossRef]
5. Gray, J.M.; Leys, J.F.; Yang, X.; Zhang, M. Monitoring of Sustainable Land Management Using Remotely Sensed Vegetation Cover and Variable Tolerable Soil Erosion Targets across New South Wales, Australia. *Soil Use Manag.* **2023**, *39*, 849–866. [CrossRef]
6. Świtoniak, M.; Bednarek, R. Denudacja Antropogeniczna. In *Antropogeniczne Przekształcenia Pokrywy Glebowej Brodnickiego Parku Krajobrazowego*; Wyd. Nauk.: Toruń, Poland, 2014; pp. 57–84. ISBN 978-83-231-3280-6.
7. Tomlinson, I. Doubling Food Production to Feed the 9 Billion: A Critical Perspective on a Key Discourse of Food Security in the UK. *J. Rural Stud.* **2013**, *29*, 81–90. [CrossRef]
8. Al-Kaisi, M.M.; Elmore, R.W.; Guzman, J.G.; Hanna, H.M.; Hart, C.E.; Helmers, M.J.; Hodgson, E.W.; Lenssen, A.W.; Mallarino, A.P.; Robertson, A.E.; et al. Drought Impact on Crop Production and the Soil Environment: 2012 Experiences from Iowa. *J. Soil Water Conserv.* **2013**, *68*, 19A–24A. [CrossRef]
9. Jin, F.; Yang, W.; Fu, J.; Li, Z. Effects of Vegetation and Climate on the Changes of Soil Erosion in the Loess Plateau of China. *Sci. Total Environ.* **2021**, *773*, 145514. [CrossRef]
10. Yao, Y.; Liu, Y.; Zhou, S.; Song, J.; Fu, B. Soil Moisture Determines the Recovery Time of Ecosystems from Drought. *Glob. Change Biol.* **2023**, *29*, 3562–3574. [CrossRef]
11. Pandey, S.; Kumar, P.; Zlatic, M.; Nautiyal, R.; Panwar, V.P. Recent Advances in Assessment of Soil Erosion Vulnerability in a Watershed. *Int. Soil Water Conserv. Res.* **2021**, *9*, 305–318. [CrossRef]
12. Boardman, J.; Poesen, J.; Evans, R. Socio-Economic Factors in Soil Erosion and Conservation. *Environ. Sci. Policy* **2003**, *6*, 1–6. [CrossRef]
13. Montanarella, L.; Pennock, D.J.; McKenzie, N.; Badraoui, M.; Chude, V.; Baptista, I.; Mamo, T.; Yemefack, M.; Singh Aulakh, M.; Yagi, K.; et al. World's Soils Are under Threat. *Soil* **2016**, *2*, 79–82. [CrossRef]
14. Panagos, P.; Borrelli, P.; Robinson, D. FAO Calls for Actions to Reduce Global Soil Erosion. *Mitig. Adapt. Strateg. Glob. Change* **2020**, *25*, 789–790. [CrossRef]
15. Qiu, L.; Zhang, Q.; Zhu, H.; Reich, P.B.; Banerjee, S.; van der Heijden, M.G.A.; Sadowsky, M.J.; Ishii, S.; Jia, X.; Shao, M.; et al. Erosion Reduces Soil Microbial Diversity, Network Complexity and Multifunctionality. *ISME J.* **2021**, *15*, 2474–2489. [CrossRef]
16. Abdul Rahman, N.S.N.; Abdul Hamid, N.W.; Nadarajah, K. Effects of Abiotic Stress on Soil Microbiome. *Int. J. Mol. Sci.* **2021**, *22*, 9036. [CrossRef]
17. Lehman, R.M.; Cambardella, C.A.; Stott, D.E.; Acosta-Martinez, V.; Manter, D.K.; Buyer, J.S.; Maul, J.E.; Smith, J.L.; Collins, H.P.; Halvorson, J.J.; et al. Understanding and Enhancing Soil Biological Health: The Solution for Reversing Soil Degradation. *Sustainability* **2015**, *7*, 988–1027. [CrossRef]
18. Khmelevtsova, L.E.; Sazykin, I.S.; Azhogina, T.N.; Sazykina, M.A. Influence of Agricultural Practices on Bacterial Community of Cultivated Soils. *Agriculture* **2022**, *12*, 371. [CrossRef]
19. Piotrowska-Długosz, A.; Kobierski, M.; Długosz, J. Enzymatic Activity and Physicochemical Properties of Soil Profiles of Luvisols. *Materials* **2021**, *14*, 6364. [CrossRef]
20. Loba, A.; Zhang, J.; Tsukamoto, S.; Kasprzak, M.; Beata Kowalska, J.; Frechen, M.; Waroszewski, J. Multiproxy Approach to the Reconstruction of Soil Denudation Events and the Disappearance of Luvisols in the Loess Landscape of South-Western Poland. *Catena* **2023**, *220*, 106724. [CrossRef]
21. Radziuk, H.; Świtoniak, M. The Effect of Erosional Transformation of Soil Cover on the Stability of Soil Aggregates within Young Hummocky Moraine Landscapes in Northern Poland. *Agronomy* **2022**, *12*, 2595. [CrossRef]
22. Kobierski, M. Morphology, Properties and Mineralogical Composition of Eroded Luvisols in Selected Morainic Areas of the Kujavian and Pomeranian Province. *Univ. Technol. Life Sci. Bydg.* **2013**, *Monograph No. 166*, 1–121.
23. Świtoniak, M. Use of Soil Profile Truncation to Estimate Influence of Accelerated Erosion on Soil Cover Transformation in Young Morainic Landscapes, North-Eastern Poland. *Catena* **2014**, *116*, 173–184. [CrossRef]
24. Świtoniak, M.; Mroczek, P.; Bednarek, R. Luvisols or Cambisols? Micromorphological Study of Soil Truncation in Young Morainic Landscapes—Case Study: Brodnica and Chełmno Lake Districts (North Poland). *Catena* **2016**, *137*, 583–595. [CrossRef]
25. Pindral, S.; Świtoniak, M. The Usefulness of Soil-Agricultural Maps to Identify Classes of Soil Truncation. *Soil Sci. Annu.* **2017**, *68*, 2–10. [CrossRef]
26. Solon, J.; Borzyszkowski, J.; Bidłasik, M.; Richling, A.; Badora, K.; Balon, J.; Brzezińska-Wójcik, T.; Chabudziński, Ł.; Dobrowolski, R.; Grzegorczyk, I.; et al. Physico-Geographical Mesoregions of Poland: Verification and Adjustment of Boundaries on the Basis of Contemporary Spatial Data. *Geogr. Pol.* **2018**, *91*, 143–170. [CrossRef]

27. Kostrzewski, A.; Zwoliński, Z.; Andrzejewski, L.; Florek, W.; Mazurek, M.; Niewiarowski, W.; Podgórski, Z.; Rachlewicz, G.; Smolska, E.; Stach, A.; et al. Współczesna ewolucja rzeźby młodoglacjalnej Niżu Polskiego. In *Współczesne Przemiany Rzeźby Polski*; Bogucki Wydawnictwo Naukowe: Kraków, Poland, 2021; pp. 483–571. ISBN 978-83-7986-382-2.
28. IUSS Working Group WRB. *World Reference Base for Soil Resources. International Soil Classification System for Naming Soils and Creating Legends for Soil Maps*, 4th ed.; International Union of Soil Sciences (IUSS): Vienna, Austria, 2022.
29. Crawford, D.L.; Lynch, J.M.; Whipps, J.M.; Ousley, M.A. Isolation and Characterization of Actinomycetes Antagonists of a Fungal Root Pathogen. *Appl. Environ. Microbiol.* **1996**, *59*, 3899–3905. [CrossRef]
30. Atlas, R.M. *Handbook of Microbiological Media*; CRC Press: Boca Raton, FL, USA, 2010; ISBN 978-0-429-13049-6.
31. Gupta, V.; Gulati, P.; Bhagat, N.; Dhar, M.S.; Virdi, J. Detection of Yersinia Enterocolitica in Food: An Overview. European. *J. Clin. Microbiol. Infect. Dis.* **2015**, *34*, 641–650. [CrossRef]
32. Hattori, R.; Hattori, T. Sensitivity to Salts and Organic Coumpounds of Soil Bacteria Isolated on Diluted Media. *J. Gen. Appied Microbiol.* **1980**, *26*, 1–14. [CrossRef]
33. *ISO 10390:2005*; Soil, Treated Biowaste and Sludge—Determination of PH. Polish Committee for Standardizations: Warsaw, Poland, 2013.
34. *PN-R-04020*; Agricultural Analysis. Determination of the Content Available Magnesium. Polish Standards Committee: Warszawa, Poland, 1994.
35. *PN-R-04023*; Agricultural Analysis—Determination of the Content of Available Phosphorus in Mineral Soils. Polish Standards Committee: Warszawa, Poland, 1996.
36. *PN-R-04022*; Agricultural Analysis—Determination of the Content Available Potassium in Mineral Soils. Polish Standards Committee: Warszawa, Poland, 1996.
37. Ward, J.H. Hierarchical Grouping to Optimize an Objective Function. *J. Am. Stat. Assoc.* **1963**, *58*, 236–244. [CrossRef]
38. Statistica. *Data Analysis Software System*, Version 12; TIBCO Software Inc.: Palo Alto, CA, USA, 2019. Available online: https://www.Tibco.Com/Products/Data-Science(accessed on 15 October 2023).
39. Cendrero, A.; Remondo, J.; Beylich, A.A.; Cienciala, P.; Forte, L.M.; Golosov, V.N.; Gusarov, A.V.; Kijowska-Strugała, M.; Laute, K.; Li, D.; et al. Denudation and Geomorphic Change in the Anthropocene; a Global Overview. *Earth Sci. Rev.* **2022**, *233*, 104186. [CrossRef]
40. Zalasiewicz, J.; Waters, C.N.; Williams, M.; Barnosky, A.D.; Cearreta, A.; Crutzen, P.; Ellis, E.; Ellis, M.A.; Fairchild, I.J.; Grinevald, J.; et al. When Did the Anthropocene Begin? A Mid-Twentieth Century Boundary Level Is Stratigraphically Optimal. *Quat. Int.* **2015**, *383*, 196–203. [CrossRef]
41. Mabuhay, J.A.; Nakagoshi, N.; Isagi, Y. Influence of Erosion on Soil Microbial Biomass, Abundance and Community Diversity. *Land Degrad. Dev.* **2004**, *15*, 183–195. [CrossRef]
42. Gapeshin, D.I.; Ivanova, A.E.; Gracheva, T.A.; Pozdnyakov, L.A.; Demidov, V.V. Microbiological Indicators of Erosion Processes in Arable Alluvial Soils. *Mosc. Univ. Soil Sci. Bull.* **2022**, *77*, 374–383. [CrossRef]
43. Michel, H.M.; Williams, M.A. Soil Habitat and Horizon Properties Impact Bacterial Diversity and Composition. *Soil Sci. Soc. Am. J.* **2011**, *75*, 1440–1448. [CrossRef]
44. Alaoui, A.; Diserens, E. Mapping Soil Compaction—A Review. *Curr. Opin. Environ. Sci. Health* **2018**, *5*, 60–66. [CrossRef]
45. Sul, W.J.; Asuming-Brempong, S.; Wang, Q.; Tourlousse, D.M.; Penton, C.R.; Deng, Y.; Rodrigues, J.L.M.; Adiku, S.G.K.; Jones, J.W.; Zhou, J.; et al. Tropical Agricultural Land Management Influences on Soil Microbial Communities through Its Effect on Soil Organic Carbon. *Soil Biol. Biochem.* **2013**, *65*, 33–38. [CrossRef]
46. Denef, K.; Six, J.; Paustian, K.; Merckx, R. Importance of Macroaggregate Dynamics in Controlling Soil Carbon Stabilization: Short-Term Effects of Physical Disturbance Induced by Dry–Wet Cycles. *Soil Biol. Biochem.* **2001**, *33*, 2145–2153. [CrossRef]
47. Jamshidi, A.H.; Sun, L.; Niu, Y.; Liu, X.; Fan, Z.; Yi, W.; Zhang, S. Slope Gradient Altered Microbial Community Composition in the Sloping Cropland in Black Soil. *Catena* **2023**, *232*, 107416. [CrossRef]
48. Naylor, D.; McClure, R.; Jansson, J. Trends in Microbial Community Composition and Function by Soil Depth. *Microorganisms* **2022**, *10*, 540. [CrossRef]
49. Huang, Y.-M.; Liu, D.; An, S.-S. Effects of Slope Aspect on Soil Nitrogen and Microbial Properties in the Chinese Loess Region. *Catena* **2015**, *125*, 135–145. [CrossRef]
50. Selim, M.S.M.; Abdelhamid, S.A.; Mohamed, S.S. Secondary Metabolites and Biodiversity of Actinomycetes. *J. Genet. Eng. Biotechnol.* **2021**, *19*, 72. [CrossRef] [PubMed]
51. Xiang, X.; Liu, J.; Zhang, J.; Li, D.; Xu, C.; Kuzyakov, Y. Divergence in Fungal Abundance and Community Structure between Soils under Long-Term Mineral and Organic Fertilization. *Soil Tillage Res.* **2020**, *196*, 104491. [CrossRef]
52. Rousk, J.; Bååth, E.; Brookes, P.C.; Lauber, C.L.; Lozupone, C.; Caporaso, J.G.; Knight, R.; Fierer, N. Soil Bacterial and Fungal Communities across a PH Gradient in an Arable Soil. *ISME J.* **2010**, *4*, 1340–1351. [CrossRef] [PubMed]
53. Khan, F.; Hayat, Z.; Ramzan, M.; Shah, Z.; Hanif, M. Effect of Slope Position on Physico-Chemical Properties of Eroded Soil. *Soil Environ.* **2013**, *31*, 22–28.
54. Matecka, P.; Świtoniak, M. Delineation, Characteristic and Classification of Soils Containing Carbonates in Plow Horizons within Young Moraine Areas. *Soil Sci. Ann.* **2020**, *71*, 23–36. [CrossRef]
55. Jia, S.; He, X.; Wei, F. Soil Organic Carbon Loss under Different Slope Gradients in Loess Hilly Region. *Wuhan Univ. J. Nat. Sci.* **2007**, *12*, 695–698. [CrossRef]

56. Dungait, J.A.J.; Ghee, C.; Rowan, J.S.; McKenzie, B.M.; Hawes, C.; Dixon, E.R.; Paterson, E.; Hopkins, D.W. Microbial Responses to the Erosional Redistribution of Soil Organic Carbon in Arable Fields. *Soil Biol. Biochem.* **2013**, *60*, 195–201. [CrossRef]
57. Zhang, J.H.; Wang, Y.; Li, F.C.; Zhang, J.H.; Wang, Y.; Li, F.C. Soil Organic Carbon and Nitrogen Losses Due to Soil Erosion and Cropping in a Sloping Terrace Landscape. *Soil Res.* **2015**, *53*, 87–96. [CrossRef]
58. Nie, X.-J.; Zhang, H.-B.; Su, Y.-Y. Soil Carbon and Nitrogen Fraction Dynamics Affected by Tillage Erosion. *Sci. Rep.* **2019**, *9*, 16601. [CrossRef]
59. Li, Y.; Li, Y.; Zhang, Q.; Xu, G.; Liang, G.; Kim, D.-G.; Carmona, C.R.; Yang, M.; Xue, J.; Xiang, Y.; et al. Enhancing Soil Carbon and Nitrogen through Grassland Conversion from Degraded Croplands in China: Assessing Magnitudes and Identifying Key Drivers of Phosphorus Reduction. *Soil Tillage Res.* **2024**, *236*, 105943. [CrossRef]
60. Brubaker, S.C.; Jones, A.J.; Lewis, D.T.; Frank, K. Soil Properties Associated with Landscape Position. *Soil Sci. Soc. Am. J.* **1993**, *57*, 235–239. [CrossRef]
61. Yarashynskaya, A.; Prus, P. Precision Agriculture Implementation Factors and Adoption Potential: The Case Study of Polish Agriculture. *Agronomy* **2022**, *12*, 2226. [CrossRef]
62. Breza-Boruta, B.; Kotwica, K.; Bauza-Kaszewska, J. Effect of Tillage System and Organic Matter Management Interactions on Soil Chemical Properties and Biological Activity in a Spring Wheat Short-Time Cultivation. *Energies* **2021**, *14*, 7451. [CrossRef]

agriculture

Article

The Quantity and Quality of Humic Substances following Different Land Uses in Karst Peak-Cluster Depression in Guangxi, China

Tingting Cui [1], Jianbing Zhang [1,*] and Weiqun Luo [2,3]

1 Guangxi Key Laboratory of Earth Surface Processes and Intelligent Simulation, Nanning Normal University, Nanning 530001, China; cuitt@nnnu.edu.cn
2 Institute of Karst Geology, Chinese Academy of Geological Sciences, Guilin 541004, China; lweiqun@mail.cgs.gov.cn
3 Pingguo National Field Observation and Research Station of Karst Ecosystem in Guangxi, Pingguo 531400, China
* Correspondence: zjb@nnnu.edu.cn

Abstract: There were some ecological and environmental problems in limestone soil in the karst peak-cluster depression in Pingguo City of Guangxi, such as the destruction of soil structure, soil acidification and soil fertility decline, and these problems were closely related to soil organic matter. The soil in this site was classified as Cambisols. Therefore, this project took soil under five major land uses (grassland, afforestation, sugarcane field, corn field and pitaya field) in the karst area of Guangxi as the research object. The contents and molecular structure properties of humic acids, fulvic acids and humins in soils were studied by solid-state ^{13}C nuclear magnetic resonance spectroscopy and elemental composition. From the perspective of the chemical structure of humic substances, the differences in the quantity and structural characteristics of humic acids, fulvic acids and humins in different land uses were revealed. The results showed that the organic carbon content of both afforestation (34.83 g kg^{-1}) and natural restored grassland (31.67 g kg^{-1}) were significantly higher than that of sugarcane field (17.60 g kg^{-1}), corn field (16.35 g kg^{-1}) and pitaya field (14.31 g kg^{-1}) ($p < 0.05$). The contents of three humic fractions in grassland were relatively high, and the contents of three humic fractions in sugarcane field were relatively low ($p < 0.05$). The structural characteristics of humic substances showed that the protein components of the three humic fractions in the afforestation were high, indicating that the humic substances in the afforestation contained more unstable components. The Alkyl C/*O*-alkyl C and Hydrophobic C/hydrophilic C ratios of the three humic fractions of corn field were high, indicating high stability and maturity in humic substances in corn field.

Keywords: karst; different land uses; humic fractions; ^{13}C NMR; elemental compositions

Citation: Cui, T.; Zhang, J.; Luo, W. The Quantity and Quality of Humic Substances following Different Land Uses in Karst Peak-Cluster Depression in Guangxi, China. *Agriculture* **2023**, *13*, 2246. https://doi.org/10.3390/agriculture13122246

Academic Editors: Anetta Siwik-Ziomek and Anna Figas

Received: 7 November 2023
Revised: 29 November 2023
Accepted: 4 December 2023
Published: 6 December 2023

1. Introduction

Soil organic matter (SOM) is essential for agricultural production systems because it regulates the chemical, biological and physical properties of soils and therefore plays an important role in nutrient cycling [1]. Soil organic carbon (SOC) is related to soil function and the sustainability of agroecosystems, and measuring the amount and structural characteristics of SOC is crucial for evaluating soil function and understanding soil carbon sequestration processes [2]. However, most SOC is present in the form of humic substances that are difficult to decompose, namely humic acid (HA), fulvic acid (FA) and humin (Hu), which accounts for about 60–80% of total SOM [3]. And humic substances are still the object of intense discussion and research, even contention about their formation, composition and structure [4,5]. The high molecular weight and complex chemical structure of humic substances indicate that light depletion can produce higher carbon emissions [6].

The 540,000 km^2 karst area in southwest China, including the three provinces of Yunnan, Guizhou and Guangxi, is the core of karst areas in East Asia and one of the three major karst areas in the world [7]. The southwest karst region is also one of the regions of the largest continuously exposed carbonate rocks in China, which is very fragile and sensitive to agricultural activities [8]. Carbonate rocks are highly soluble and cannot produce a large amount of soil, especially on slopes, where the amount of soil is less and the soil layer is shallow, generally less than 30 cm [9]. Therefore, soil resources in karst areas are very valuable. Before the last century, grasslands and forests in many places were degraded into arable land due to the overexploitation of land resources [10]. In most of the world's most fertile soils under long-term cultivation, SOM has been reported to be depleted in large quantities, leading to soil degradation and decreased fertility [11]. Since around 2000, most of the degraded farmland in this region have been restored through natural regeneration and afforestation under the implementation of ecological projects such as returning farmland to forest and closing mountains [12]. How the complexity of humic substances in organic matter in the region changes as vegetation recovers remains not understood. Therefore, it is very important and meaningful to study the influence mechanism of karst vegetation restoration on the decomposition and sequestration of soil humic substances. Studies have shown that land uses play a key role in soil carbon sequestration [13]. Organic carbon accumulation and carbon stock vary with land uses due to changes in the quality (such as species of litter) and quantity (such as the amount of litter and litter C input) of source substances, soil physicochemical properties and biological activity, which in turn change the degree of decomposition of organic matter [14]. Land uses also have a strong influence on the quantity and quality of SOM, for example, the accumulation of soil organic carbon after afforestation in grassland and farmland is a slow and continuous process, and about 30% of the accumulated carbon was stored in an unstable form on the forest surface after afforestation [15]. In addition to the quantity of organic matter input, land uses also affect the quality of organic matter input, which in turn affect the mineralization rate of soil carbon [16]. Therefore, land-use changes have a large impact on global CO_2 emissions by affecting soil carbon storage [17]. Due to the irrational development and utilization of land resources, the change in the molecular structure of soil humic substances leads to an increase in its hydrophilicity, which leads to a decrease in the hydrophobicity capacity of the surface of clay particles of humic substances and a weakening of the structural stability of soil aggregates [18]. But in organic soils, because of mineralization caused by cultivation, the hydrophobicity increases relative to the hydrophilicity, probably due to hydrophilic carbon mineralization and excitation effect are faster than hydrophobic carbon under the action of microorganisms [19]. Certain components of humic substances should be relatively unstable, their quantities and chemistry more sensitive to land uses than the corresponding properties of total SOM. In the southwest China Karst area, high calcium contents in limestone soils may play a key role in SOM protection by forming complexes with HA or surrounding SOM with $CaCO_3$ precipitation [20].

In order to control the rocky desertification of land, close hillsides to facilitate afforestation have been carried out since the 1990s. The phenomenon of reversal of rocky desertification was shown in local areas. Although there are many studies on the effects of different land uses on soil organic carbon, humic substances account for most of the organic matter. Therefore, in this study, soil samples of a 0–10 cm soil layer are collected in five different land uses in the karst peak-cluster depression. Quantitative and qualitative analysis of different humic fractions are carried out. The objectives of this study are: (1) differences in content of humic fractions between different land uses are analyzed; (2) evaluate the molecular composition of humic fractions extracted from subtropical limestone soils under different land uses. Therefore, we hypothesized that vegetation restoration in karst areas leads to improved humic substances within these agro-ecosystems.

2. Materials and Methods

2.1. Site Description and Soil Collection

The study area was located in Pingguo National Field Observation and Research Station of Karst Ecosystem, Guangxi (Figure 1). The landform of this area was a typical karst peak-cluster depression. The climatic conditions were South Asian tropical monsoon climate, rich in heat and abundant rainfall. The average annual temperature was 20.2~22.6 °C, the average annual rainfall was 1322.3 mm and the daily humidity was 72.06%. The rocks were mainly pure limestone and siliceous limestone. The soil in this site was classified as Cambisols (FAO Soil Taxonomy).

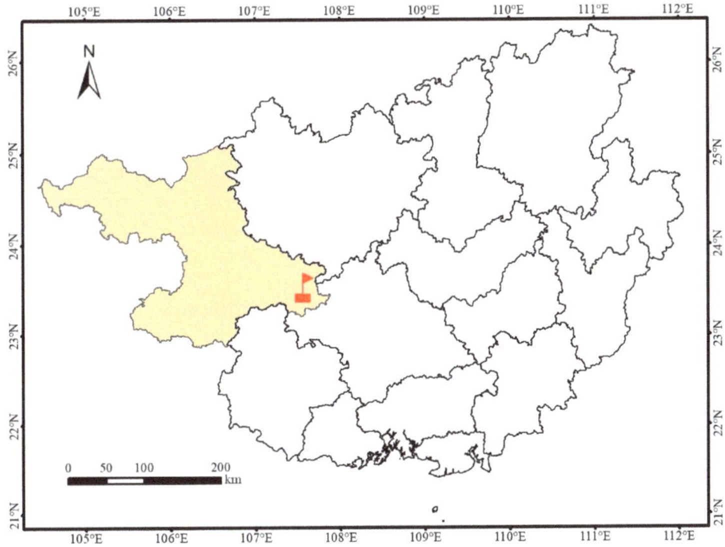

Figure 1. The map of the location of the Pingguo National Field Observation and Research Station of Karst Ecosystem, Guangxi.

Five dominant land uses including grassland (GL), afforestation (AF), sugarcane field (SF), corn field (CF) and pitaya field (PF) were identified and sampled in 2021. Grassland (23°25′57″ N, 107°23′26″ E, altitude of 203 m) soil samples were collected from plots that had been returned to grassland for more than 10 years, and the vegetation of grassland was mainly *Miscanthus, Nephrolepis cordifolia* (L.) *C. Presl, Vitex negundo*, etc. The grass was neither cultivated nor fertilized. Forest soil samples were collected from the plots (23°23′37″ N, 107°23′23″ E, altitude of 279 m) of artificially converted farmland to forest, and the dominant species was *Delavaya toxocarpa* Franch., which was planted in 2006. There was no tillage in afforestation and there was a litter layer on the surface. After the sugarcane seed strips were planted once in the sugarcane plot (23°24′31″ N, 107°24′35″ E, altitude of 136 m), it could be harvested continuously three times and once a year. The sugarcane fields were ploughed once every 2–3 years. The management of sugarcane fields was to spread mineral fertilizer. When the sugarcanes were ripe, the leaves were returned to the field. Corn plots (23°19′55″ N, 107°23′28″ E, altitude of 346 m) were planted with corn for two seasons per year and sown by tilling. The corn field was ploughed twice a year. Corn fields should also be fertilized with mineral fertilizers. Pitaya (*Hylocereus undatus* 'Foo-Lon') was planted in the plot (23°23′11″ N, 107°23′18″ E, altitude of 347 m) for about 10 years. After planting, the land was no longer ploughed and pitaya was harvested four times a year. In addition to the application of mineral fertilizer, the leaf litter was fertilized on the ground around the roots of pitaya.

Soil samples were collected in March 2021. Soil samples were collected randomly from several points at each of the target plot and combined to make a composite sample. There were three replicates per plot. The samples were air dried, ground and passed through a 2 mm sieve for HS extraction and passed through 0.25 mm sieve for determination of the carbon contents.

2.2. *Quantitative Analysis and Extraction of Humic Substances*

Humic substances were analyzed according to the method described by Dou [21]. Briefly, the soil was extracted with a mixed alkali solution and then centrifuged. The supernatant contained humic acid (HA) and fulvic acid (FA), and the insoluble residue was humin (Hu). After acidifying the supernatant to pH 1, the precipitate was HA and the supernatant fraction was FA. The HA was filtered out and dissolved in 0.05 mol L^{-1} NaOH. All fractions were used for concentration analysis. Soil organic carbon (SOC) and the content of humic substances were determined using a $K_2Cr_2O_7$ external heating method [22]. The isolation and purification of FA, HA and Hu were processed using the procedure described by International Humic Substances Society [23]. The three fractions were separated through the classic acid base fractionation. The received HA, FA and Hu residues were required to remove ash. Then three residues were dialyzed to eliminate excess salts. The three fractions were freeze-dried afterwards.

2.3. *Elemental Compositions*

The C, H and N content of all humic fractions were determined by an elemental analyzer (Elementar Vario EL cube, Langenselbold, Germany). The ash content was measured in a muffle furnace by heating the humic substance (HS) at 750 °C for 4 h. The O content was calculated by gravimetric difference and as follows: O% = 100 − C% − H% − N% − ash%. After that C, H, N and O contents were recalculated on an ash-free basis. The H/C, O/C and C/N atomic ratios were determined from elemental analysis [24,25].

2.4. *The Solid State ^{13}C NMR Analysis*

In order to determine the detailed composition of carbon functional groups in the humus component sample, the samples were analyzed using ^{13}C cross polarization/total side-band suppression (CP/TOSS) NMR (Bruker Avance III 400, Billerica, MA, USA) at a ^{13}C frequency of 100.6 MHz. The samples were kept in closed 4 mm NMR MAS rotors. Each spectrum obtained was divided into seven chemical shift regions: 0–45 ppm, alkyl C; 45–60 ppm, OCH_3/NCH; 60–90 ppm, O-alkyl C; 90–110 ppm, Di-O-alkyl C; 110–145 ppm, aromatic C; 145–160 ppm, aromatic C-O; 160–190 ppm, COO/NC=O [26]. The MestReNova software v12.0.1 (Mestrelab Research, A Coruña, Spain) was used to collect and evaluate the spectra. The relative distribution of each signal area was calculated as a percentage of the total spectral range (ppm). Three ratios were used to calculate the structural properties of organic carbon [26,27]. Alkyl C/O-alkyl C (A/O-A): (0–45)/(45–110); aliphatic C/aromatic C: [(0–45) + (45–60) + (60–90) + (90–110)]/[(110–145) + (145–160)]; hydrophobic C/hydrophilic C: [(0–45) + (110–145) + (145–160)]/[(45–60) + (60–90) + (90–110) + (160–190)].

2.5. *Statistical Analysis*

General calculations and data handling were performed with Excel for Mac (Microsoft Corporation, Redmond, WA, USA) and ORIGIN 2021 (OriginLab Corporation, Northampton, MA, USA). All statistical analyses were performed using SPSS 19.0 (IBM Corporation, Armonk, NY, USA). One-way Analysis of Variance (ANOVA) followed by the Least Significant Difference test ($p < 0.05$) was used to assess the significance of differences in soil properties among different land-use types. To further evaluate the chemical structural (relative contribution of different functional groups and indices) differences among the humic fractions and among land use and to identify the key factors of the identified differences, principal component analysis (PCA) was carried out. This analysis was conducted using Canoco for version 5 (Microcomputer Power, Ithaca, NY, USA).

3. Results and Discussion

3.1. Effects of Different Land Uses on Soil Organic Carbon

Data on the contents of soil organic carbon under different land use types are presented in Figure 2. The soil organic carbon contents of forest soil and grassland were 34.83 g kg^{-1} and 31.67 g kg^{-1}, respectively, while the soil organic carbon content of sugarcane, maize and pitaya was 17.60 g kg^{-1}, 16.35 g kg^{-1} and 14.31 g kg^{-1}. It could be seen that the soil organic carbon contents of both AF and GL were significantly higher than that of cultivated agricultural soil (SF, CF and PF). This result indicated that long-term agricultural cultivation resulted in a relatively low level of soil organic carbon content. Forests had greater canopy cover and tree density, so it could help to accumulate more carbon in both aboveground and underground biomass [28]. Higher aboveground and underground biomass could input a large amount of carbon into the soil, thus increasing the content of soil organic carbon [29]. During the natural restoration of grassland, a large root system was formed underground. Grassland underground biomass carbon was the main source of soil organic carbon [30,31]. Therefore, grassland had a greater potential for soil carbon sequestration than farmland [32]. And most of the aboveground litter and underground biomass were removed from farmland, which limited the supply of fresh organic matter and, thus, reduced soil organic carbon content [28]. The soil was disturbed during the planting process [33,34], which also accelerated the decomposition of soil organic carbon. For soil in sugarcane and corn fields, agricultural measures such as tilling could increase the frequency of soil disturbance, accelerate the mineralization and decomposition rate of soil stable organic carbon and lead to the decline of soil organic carbon. Studies have shown that in degraded sub-humid pastures with a low SOM content, it was possible to reduce greenhouse gas emissions by abandoning tillage and ecological restoration [35]. But corn, sugarcane and pitaya were the main sources of income for local farmers, so low or loss of soil organic carbon was inevitable in the region.

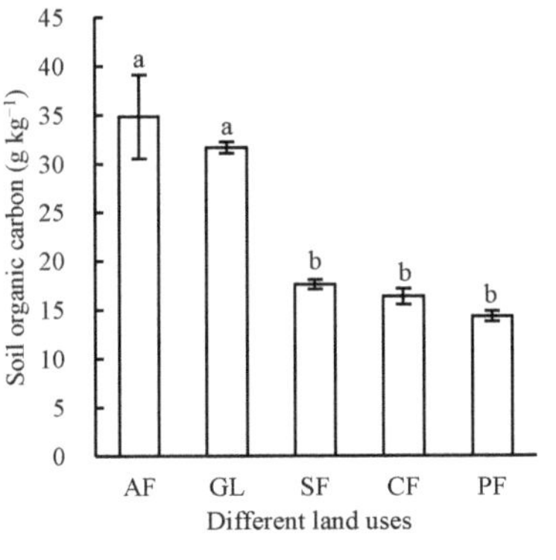

Figure 2. Soil organic carbon content in five different land uses. Error bars represent the standard deviations of the mean (n = 3). Different lowercase letters indicate significant differences among different land uses ($p < 0.05$). AF: afforestation, GL: grassland, SF: sugarcane field, CF: corn field, PF: pitaya field.

3.2. Effects of Different Land Uses on Humic Fractions

Figure 3 shows the content of humic fractions and the proportion of these fractions in soil organic carbon. Among all land uses, the content of Hu (11.43–28.97 g kg^{-1}) was higher than that of HA (1.51–3.49 g kg^{-1}) and FA (1.28–2.40 g kg^{-1}). This was consistent with the results of Guimarães et al. [36]. Hayes and Swift had considered that 30–50% might be a more realistic estimate of the abundance of Hu in SOM [11].

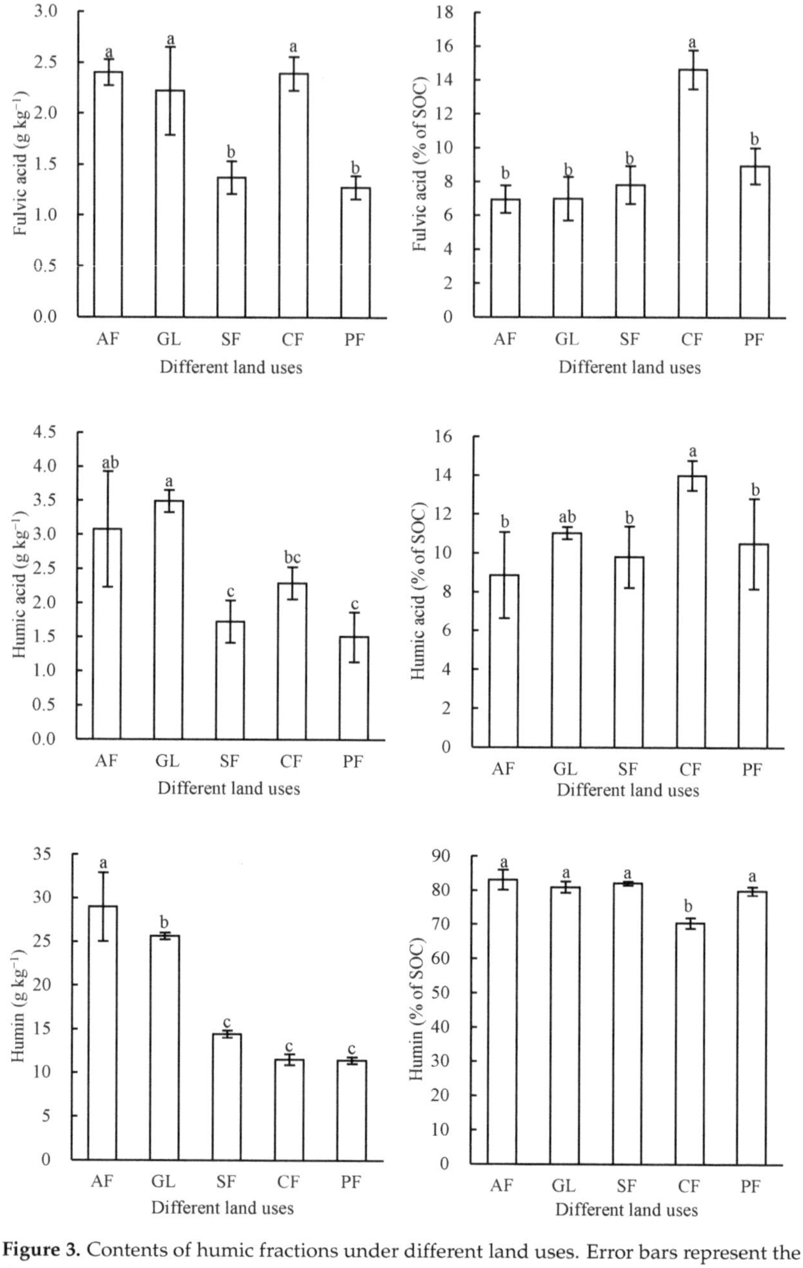

Figure 3. Contents of humic fractions under different land uses. Error bars represent the standard deviations of the mean (n = 3). Different lowercase letters indicate significant differences among different land uses ($p < 0.05$). AF: afforestation, GL: grassland, SF: sugarcane field, CF: corn field, PF: pitaya field.

The FA content in SF and PF was significantly lower than that in AF, GL and CF. It may be because FA in SF and dragon PF is more easily lost. This also indicated that long-term cultivation reduced the FA content in soil [37,38]. The occurrence of more intensive decomposition processes in cultivated soils negatively affected not only the total SOM, but also humic substances [38,39]. Conversely, grassland restoration could promote soil C sequestration by increasing the FA fraction, which was driven by root cellulose [30]. The HA content in SF, CF and PF was significantly lower than in GL. The Hu content in AF and GL was significantly higher than that in SF, CF and PF. These results indicated that different land uses have a certain influence on the amount of humic fraction in the same area. Hu was the most stable and decomposed fractions of organic matter, but it was also affected by tillage. Caravaca et al. had found that the concentration of Hu in cultivated soil was lower than in uncultivated soil [40].

Figure 3 also showed that the percentage of Hu (70.41–83.10%) was higher than FA (6.96–14.64%) and HA (8.86–13.99%). The high contribution of insoluble Hu components to SOC indicated the formation of a broad and strong organic–mineral complex under neutral to alkaline conditions [41]. It can also be seen from Figure 3 that, except for CF, the contribution percentages of FA, HA and Hu to the SOC of the other four land uses were not significantly different. It may be due to the fact that in these five land uses, regular tillage and fertilization practices in corn planting fields affected the contribution of percentage of humus components.

The ratios of HA/FA and HA/(HA + FA) were the most widely used indicators of soil humus quality and the degree of humification, which also reflected the degree of polymerization [42]. As can be seen from Figure 4, the comparison rules of the two ratios were consistent. The two ratios in GL were significantly higher than that in SF, but not significantly difference from other land uses. The results showed that the humic substances extracted from alkali in GL had high degree of polymerization and humification. Some studies have shown that the ratio of HA/FA decreases with grassland restoration [30,42]. However, Loke et al. reported that in one of three semi-arid sites in the grasslands of South Africa, the ratio of HA/FA declined as the primary grassland was converted to farmland and then increased with grassland restoration [38]. Both HA and FA were the fractions of HS in soil, however, HA had a more complex structure and higher degree of polymerization than FA [43,44]. Therefore, the higher the HA/FA ratio, the higher the degree of polymerization and the better the stability of the soil humic substance. Therefore, it could be seen from the figure that the average ratio of HA/FA in GL was higher than in SF, which indicated that the HS formed in GL had stronger aromaticity and condensation, so it had better stability. Except for SF, the ratios of HA/FA were greater than 1, which indicated that HA played a dominant role in the long-term evolution of the other four types of land use [30,45].

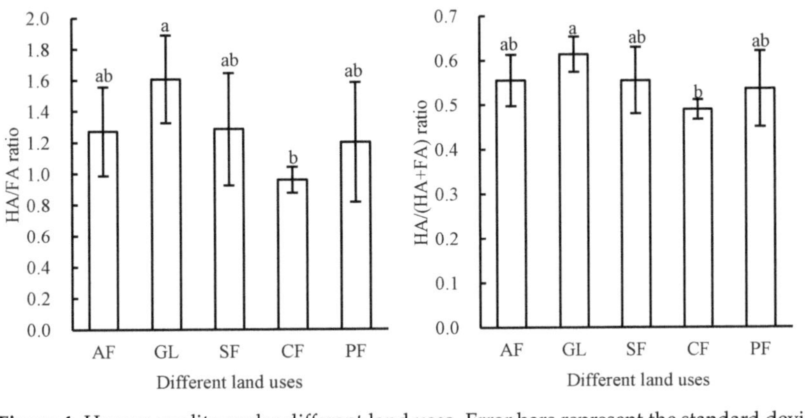

Figure 4. Humus quality under different land uses. Error bars represent the standard deviations of the mean (n = 3). Different lowercase letters indicate significant differences among different land uses ($p < 0.05$). AF: afforestation, GL: grassland, SF: sugarcane field, CF: corn field, PF: pitaya field.

3.3. Effects of Different Land Uses on Elemental Compositions of Humic Fractions

The elemental compositions of all the humic substances isolated from different land uses is shown in Table 1. Different chemical compositions might represent evolutionary state or degrees of humification. In general, the contents of C (35.72–56.24%) and O (37.1–57.8%) were relatively high in all humic substances, while the contents of H (3.32–6.58%) and N (1.32–4.67%) were relatively low. The results showed that the element compositions of FA, HA and Hu in the study area were mainly C and O, which was consistent with previous studies [46]. It had been suggested that the atomic ratio of C/N of SOM reflected the degree of microbial decomposition of plant-derived organic matter [41]. The H/C ratio was an important indicator of the carbon degree enrichment with hydrogen and indicated the type of structure of hydrogen [47]. The O/C ratio represented the degree of oxidation [48]. The C content of FA in SF and CF was higher than that in GL and AF. But SF and CF had a lower SOC content than GL and AF. It might be that fertilization affects the distribution of the C content in FA. Other studies have reported that organic amendments increased the C contents of FA in a Calcic Kastanozem with a low SOC content [49]. The results of this study also showed that the H/C ratio of HA in the five land uses was similar. Previous studies reported that narrower H/C ratios are associated with a greater degree of humification [50]. Therefore, the degrees of humification of HA for the five land uses in this paper were high. And this was consistent with the results of their aromaticity (Table 2). Among them, the H/C ratio of HA in PF was the lowest, and its H content was also the lowest. It indicated that the HA condensation degree of PF was the highest. This might be caused by the deep oxidation of SOM, directly acting on the increase in humification [51]. The C/N ratio of HA in AF was the lowest, indicating that the nitrogen groups in AF were the most, while the C/N ratio of HA in CF was the highest, indicating that HA in CF contained fewer nitrogen groups [52,53].

Table 1. Elemental analyses of humic fractions under different land uses. AF: afforestation, GL: grassland, SF: sugarcane field, CF: corn field, PF: pitaya field, FA: fulvic acid, HA: humic acid, Hu: humin.

	C (%)	H (%)	N (%)	O (%)	C/N	H/C	O/C
GL-Hu	56.24	5.25	4.01	34.5	16.37	1.12	0.46
AF-Hu	55.27	5.49	3.54	35.7	18.24	1.19	0.48
SF-Hu	45.19	6.58	2.04	46.2	25.85	1.75	0.77
CF-Hu	41.51	6.33	2.34	49.8	20.70	1.83	0.90
PF-Hu	36.99	3.32	1.90	57.8	22.74	1.08	1.17
GL-HA	53.73	4.71	4.42	37.1	14.18	1.05	0.52
AF-HA	52.82	4.83	4.67	37.7	13.18	1.10	0.54
SF-HA	51.80	4.62	4.07	39.5	14.86	1.07	0.57
CF-HA	52.56	4.59	3.99	38.9	15.36	1.05	0.55
PF-HA	53.27	4.57	4.03	38.1	15.41	1.03	0.54
GL-FA	37.78	5.51	2.18	54.5	20.26	1.75	1.08
AF-FA	35.72	5.16	1.32	57.8	31.50	1.73	1.21
SF-FA	54.35	6.28	2.21	37.2	28.65	1.39	0.51
CF-FA	47.38	5.32	2.50	44.8	22.11	1.35	0.71
PF-FA	36.57	5.02	1.97	56.4	21.71	1.65	1.16

Higher C/N ratios with HA in SF and Hu in PF indicated that these humic fractions contained fewer nitrogen groups, and the degree of humification was relatively low [54]. Compared with other four land uses, the O/C ratio of HA was the lowest in GL, which might indicate that HA in GL contained more newly formed organic matter [55]. As may be noted from the table, the O/C ratio of Hu in GL was the lowest. The lower oxidation degree corresponds to the lower proportion of O−alkyl C (Table 2), which means that Hu in GL contained few carbohydrates.

Table 2. Chemical composition of organic matter (% organic carbon functional groups) in humic substances extracted from five land uses' soil. AF: afforestation, GL: grassland, SF: sugarcane field, CF: corn field, PF: pitaya field.

	Alkyl C	OCH_3/NCH	O-Alkyl C	Di-O-Alkyl C	Aromatic C	Aromatic C-O	COO/N-C=O	Alkyl C/O-Alkyl C	Aliphatic C/Aromatic C	Hydrophobic C/Hydrophilic C	Aromaticity
	0–45	45–60	60–90	90–110	110–145	145–160	160–190				
						Fulvic acid					
CD	13.57	8.20	30.29	10.10	15.93	3.79	19.25	0.28	3.15	0.49	4.21
LD	12.34	8.27	31.60	8.49	15.40	3.74	20.16	0.26	3.17	0.46	4.12
GZ	14.55	8.12	27.58	9.81	17.60	4.57	17.77	0.32	2.71	0.58	3.85
HL	11.76	7.81	33.46	10.22	14.75	3.66	18.32	0.23	3.43	0.43	4.03
YM	13.29	8.19	26.04	8.59	19.19	5.10	19.60	0.31	2.31	0.60	3.76
						Humic acid					
CD	20.56	6.39	9.78	5.39	33.73	7.78	16.37	0.95	1.01	1.64	4.33
LD	19.83	8.17	11.48	5.91	30.09	7.83	16.52	0.78	1.20	1.37	3.84
GZ	17.48	7.39	11.93	5.55	32.77	8.24	16.64	0.70	1.03	1.41	3.98
HL	18.12	6.34	10.51	5.80	34.96	7.79	16.67	0.80	0.95	1.55	4.49
YM	16.02	6.58	9.27	5.23	39.12	7.93	15.68	0.76	0.79	1.72	4.94
						Humin					
CD	26.25	7.09	11.02	4.99	29.40	7.87	13.65	1.14	1.32	1.73	3.73
LD	25.00	7.73	15.21	7.47	25.52	7.47	11.60	0.82	1.68	1.38	3.41
GZ	31.66	4.70	15.05	6.27	26.96	5.64	9.72	1.22	1.77	1.80	4.78
HL	20.21	4.79	10.21	5.00	39.58	7.71	12.29	1.01	0.85	2.09	5.14
YM	46.45	3.32	18.01	0.47	26.54	1.42	3.79	2.13	2.44	2.91	18.67

3.4. Structural Diversity Analysis of Humic Substances Observed by ^{13}C NMR

The ^{13}C NMR spectrum of all humic substances are shown in Figure 5. The relative distribution of each signal area of soil humic acids, fulvic acids and humins are listed in Table 2. It can be seen that the ^{13}C NMR spectra of all FA were as follows: the contribution proportion of OCH_3/NCH, O-alkyl C, Di-O-alkyl C and COO/N-C=O to FA was higher than that of HA and Hu. The O-alkyl C signal was a methoxy-carbon signal in lignin, an oxygen-containing carbon signal in carbohydrates [56]. The COO/N-C=O group corresponded to the carboxyl group or polypeptide and was characterized by acidity [50]. The carboxyl group was also related to the humification degree of HS [45]. The ^{13}C NMR spectrum of alkyl C and aromatic C for all land uses generally had considerable resonance compared with FA. The ^{13}C NMR is unable to distinguish between polyaromatic carbons and isolated rings due to the three-carbon maximum distance at which nuclear relaxation affects the nucleus in study [57]. There are some differences in the aromatic region, where the two classical well-differentiated regions can be recognized: (110–160 ppm = aromatic/unsaturated (110–145 = unsubstituted, 145–160 = heterosubstituted, vanillyl + syringil lignin units). The predominance of the first type of aromatic carbons is usually attributed to greater maturity, and the second type would indicate a greater amount of lignin, which would indicate less transformation or maturity of the organic matter [27].

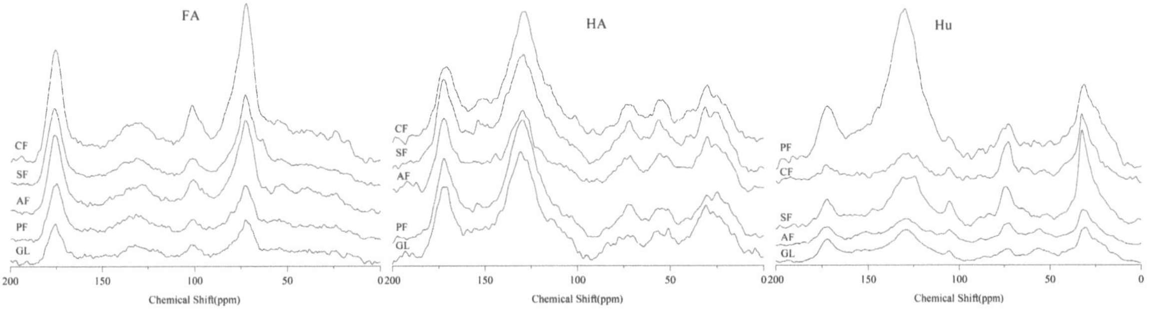

Figure 5. ^{13}C cross polarization/total side-band suppression (CP/TOSS) NMR spectra acquired from humic substances for five land uses. AF: afforestation, GL: grassland, SF: sugarcane field, CF: corn field, PF: pitaya field, FA: fulvic acid, HA: humic acid, Hu: humin.

3.5. The Principal Component Analysis (PCA) for Humic Substances Based on the Organic Carbon Functional Groups as Revealed by ^{13}C NMR

The PCA based on the functional groups of humic fractions extracted two principal components (PC1 and PC2), which described 92.89% of the total variance (Figure 6a). The resulting biplot, consisting of a score plot and a loading plot, is shown in Figure 6a. The PC1 loadings were large and positive for alkyl C and aromatic C groups and large and negative for O-alkyl C, Di-O-alkyl C, OCH_3/NCH and COO/N-C=O groups. The PC2 loading was high and negative for the aromatic C-O group. The biplot also revealed differences in C functional groups of HS between land uses. The PC1 presented a separation between the FA with other humic fractions, and the FA were characterized by large abundant of O-alkyl C and Di-O-alkyl C but low amounts of alkyl C and aromatic C. The HAs formed a cluster in the fourth quadrant of biplot, indicating their enrichment in aromatic C and aromatic C-O. The distribution of Hus indicated that it was rich in alkyl C and aromatic C. As can be seen in Figure 6b–d, sample points of different land use were relatively dispersed in PCA. Therefore, the carbon skeletons of ^{13}C NMR spectra of each humic substance under different land use were similar, but the relative contents of various functional groups were obviously different.

In the biplot of FA (Figure 6b), PF had more O-alkyl C and Di-O-alkyl C on the PC1 axis, but less aromatic C and aromatic C-O. There were more O-alkyl C and Di-O-alkyl C and less aromatic C and aromatic C-O along the PC1 axis under PF (Figure 6b). These results indicated that PF tended to increase the percentages of carbohydrates and decrease those of aromatics. As could also be seen from Table 2, PF had the highest ratio of aliphatic C/aromatic C. The ratio is an important index of humification [58]. As a fruit, pitaya was applied a lot of organic fertilizer during its growth, and withered and pruned leaves in the roots of the pitaya are used as green manure. Thus, carbohydrates, proteins, fatty acids, etc. were incorporated into the soil, resulting in the formation of FAs in pitaya soil with more aliphatic compounds. In the biplot of HA (Figure 6c), CF was significantly different from other land uses along the PC1 axis. It might be that the HA of CF had a high percentage of aromatic carbon and, therefore, had a high degree of humification. Ferrari et al. studied soils under corn cultivation fertilized with different organic and mineral fertilizers and observed that high levels of the typical aromatics and methoxy-functional groups in lignin [59]. These results were also consistent with those of Tadini et al. [52]. In the axis of PC2, there were more alkyl C of HA in GL (Figure 6c), indicating that HA of GL had a prominent aliphatic. These aliphatic compounds might be derived from microbial secretions, since microbial metabolism was one of the possible pathways of formation of humic substance described in the literature [5,60]. In the PCA plot of humin (Figure 6d), there was a significant difference in PC1 between YM treatment and other treatments, characterized by high contents of alkyl C and O-Alkyl C, and low contents of COO/N-C=O and aromatic C-O. Studies showed that alkyl C was the most stable form of SOM [61]. Hu in CF contained a large amount of alkyl C, indicating that Hu in CF was stable [62]. The aromatic C-O content of CF treatment was low, indicating a low lignin content and high maturity. Previous studies have reported a decrease in the aromatization of Hu in agricultural or forest soils after the long-term application of mineral fertilizers [63,64]. The functional group content of Hu in GL was opposite to that in CF (Figure 6d), indicating that the contents of polypeptides and carboxylic acids in Hu with GL were higher. OCH_3/NCH contents were higher than that of other land uses in the ^{13}C NMR spectra (Figure 5) and PCA plots (Figure 6) of three humic fractions in FL. Studies have shown that NCH was usually attributed to protein fragments [65]. Therefore, the abundance of protein substances in three humic fractions of FL were relatively high. The protein substances were mainly derived from microorganisms and plants, and the decomposition of forest litter might increase the accumulation of original and labile SOM in the soil [66]. The ratios of A/O-A and hydrophobic C/hydrophilic C of the three humic fractions in CF were the highest (Table 2). The A/O-A ratio were indicators of the degree of SOM decomposition and the change in organic matter [26]. The high ratio corresponded to the enhanced stability of C against degradation [67]. The Hu aromaticity index in the CF treatment was the highest, which also indicated that the Hu in the CF treatment was of high maturity and high humification decomposition intensity. CF was the most disturbed among five land uses, and the disturbance of soil might lead to the loss of active organic carbon in soil first, so the carbon components that were more difficult to decompose remained.

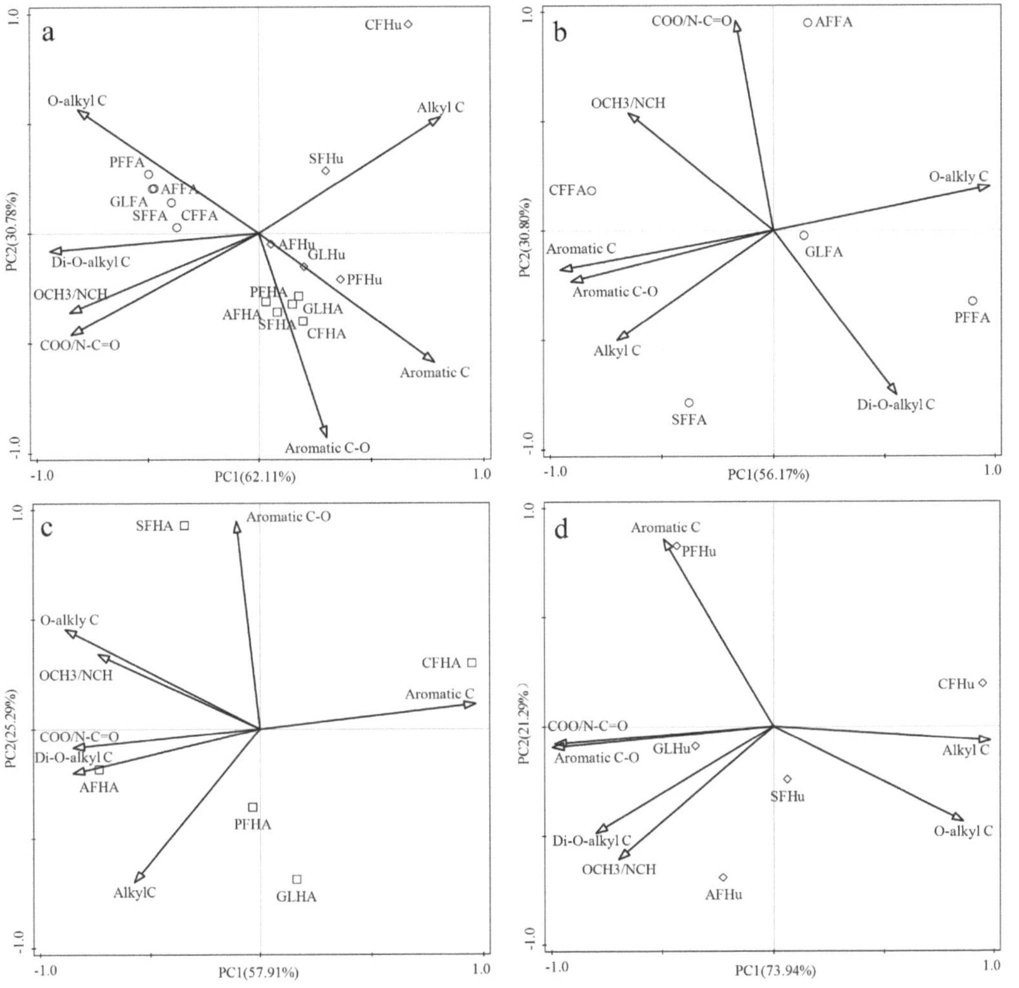

Figure 6. Biplot generated by principal component analysis for (**a**) functional groups of fulvic acids, humic acids and humins, (**b**) functional groups of FA from all soils, (**c**) functional groups of HA from all soils and (**d**) functional groups of Hu from all soils. AF: afforestation, GL: grassland, SF: sugarcane field, CF: corn field, PF: pitaya field, FA: fulvic acid, HA: humic acid, Hu: humin.

4. Conclusions

Our results show that soil organic carbon content in both afforestation and naturally restored grassland are significantly higher than in the sugarcane field, corn field and pitaya field. Different land uses in the same area also have a certain impact on the amount of soil humic fractions. The structural characteristics of humic fractions of five different land uses show that different land uses directly affect the structure of humic fractions. Three humic fractions in afforestation contain more unstable components. HA and Hu in grassland contain more newly formed substances and carboxylic acid compounds. The HA and

Hu of corn field show high maturity. Therefore, factors such as crops grown in different land use types, field management practices, use of organic fertilizers and surface litter are incorporated into the structure of formed of humic substances in the soil. Moreover, the results of this study have important guiding significance for improving the storage and stability of soil organic carbon in karst areas.

Author Contributions: Conceptualization, investigation, methodology, validation, writing—original draft preparation, funding acquisition, T.C.; supervision, project administration, writing—review and editing, J.Z.; resources, W.L. All authors have read and agreed to the published version of the manuscript.

Funding: This research was funded by the Guangxi Natural Science Foundation (No.: 2020GXNSFBA297134), the Start−up Scientific Foundation of Nanning Normal University (No.: 602021239169) and the Opening Foundation of Key Laboratory of Earth Surface Processes and Intelligent Simulation of Nanning Normal University (No.: GTEU-KLOP-X1807).

Institutional Review Board Statement: Not applicable.

Informed Consent Statement: Not applicable.

Data Availability Statement: The data presented in this study are available in article.

Conflicts of Interest: The authors declare no conflict of interest.

References

1. Chowdhury, S.; Bolan, N.; Farrell, M.; Sarkar, B.; Sarker, J.R.; Kirkham, M.B.; Hossain, M.Z.; Kim, G.-H. Chapter Two—Role of cultural and nutrient management practices in carbon sequestration in agricultural soil. In *Advances in Agronomy*; Sparks, D.L., Ed.; Academic Press: New York, NY, USA, 2021; Volume 166, pp. 131–196.
2. Lal, R. Soil carbon sequestration to mitigate climate change. *Geoderma* **2004**, *123*, 1–22. [CrossRef]
3. Tinoco, P.; Almendros, G.; González-Vila, F.J.; Sanz, J.; González-Pérez, J.A. Revisiting molecular characteristics responsive for the aromaticity of soil humic acids. *J. Soils Sediments* **2015**, *15*, 781–791. [CrossRef]
4. De Nobili, M.; Bravo, C.; Chen, Y. The spontaneous secondary synthesis of soil organic matter components: A critical examination of the soil continuum model theory. *Appl. Soil. Ecol.* **2020**, *154*, 103655. [CrossRef]
5. Dou, S.; Shan, J.; Song, X.; Cao, R.; Wu, M.; Li, C.; Guan, S. Are humic substances soil microbial residues or unique synthesized compounds? A perspective on their distinctiveness. *Pedosphere* **2020**, *30*, 159–167. [CrossRef]
6. Berg, B.; Mcclaugherty, C. *Plant Litter, Decomposition, Humus Formation, Carbon Sequestration*, 4th ed.; Springer Nature Switzerland AG: Cham, Switzerland, 2020.
7. Ding, Y.; Nie, Y.; Chen, H.; Wang, K.; Querejeta, J.I. Water uptake depth is coordinated with leaf water potential, water-use efficiency and drought vulnerability in karst vegetation. *New Phytol.* **2021**, *229*, 1339–1353. [CrossRef] [PubMed]
8. Wang, K.; Zhang, C.; Chen, H.; Yue, Y.; Zhang, W.; Zhang, M.; Qi, X.; Fu, Z. Karst landscapes of China: Patterns, ecosystem processes and services. *Landsc. Ecol.* **2019**, *34*, 2743–2763. [CrossRef]
9. Jiang, Z.; Lian, Y.; Qin, X. Rocky desertification in Southwest China: Impacts, causes, and restoration. *Earth Sci. Rev.* **2014**, *132*, 1–12. [CrossRef]
10. Liu, J.; Zhang, Z.; Xu, X.; Kuang, W.; Zhou, W.; Zhang, S.; Li, R.; Yan, C.; Yu, D.; Wu, S.; et al. Spatial patterns and driving forces of land use change in China during the early 21st century. *J. Geogr. Sci.* **2010**, *20*, 483–494. [CrossRef]
11. Hayes, M.H.B.; Swift, R.S. Vindication of humic substances as a key component of organic matter in soil and water. In *Advances in Agronomy*; Academic Press: New York, NY, USA, 2020; Volume 163, pp. 1–37.
12. Wang, Z.; Yue, F.; Xue, L.; Wang, Y.; Qin, C.; Zeng, J.; Ding, H.; Fu, Y.; Li, S. Soil nitrogen transformation in different land use and implications for karst soil nitrogen loss controlling. *Catena* **2023**, *225*, 107026. [CrossRef]
13. Gupta, M.K.; Sharma, S.D.; Kumar, M. Sequestered organic carbon stock in the soils under different land uses in western region of Haryana. *Indian. For.* **2015**, *141*, 718–725.
14. Martí-Roura, M.; Hagedorn, F.; Rovira, P.; Romanyà, J. Effect of land use and carbonates on organic matter stabilization and microbial communities in Mediterranean soils. *Geoderma* **2019**, *351*, 103–115. [CrossRef]
15. Poeplau, C.; Don, A.; Vesterdal, L.; Leifeld, J.; Van Wesemael, B.; Schumacher, J.; Gensior, A. Temporal dynamics of soil organic carbon after land-use change in the temperate zone–carbon response functions as a model approach. *Glob. Chang. Biol.* **2011**, *17*, 2415–2427. [CrossRef]
16. Han, X.; Zhao, F.; Tong, X.; Deng, J.; Yang, G.; Chen, L.; Kang, D. Understanding soil carbon sequestration following the afforestation of former arable land by physical fractionation. *Catena* **2017**, *150*, 317–327. [CrossRef]
17. Zhong, Z.; Han, X.; Xu, Y.; Zhang, W.; Fu, S.; Liu, W.; Ren, C.; Yang, G.; Ren, G. Effects of land use change on organic carbon dynamics associated with soil aggregate fractions on the Loess Plateau, China. *Land. Degrad. Dev.* **2019**, *30*, 1070–1082. [CrossRef]

18. Volikov, A.B.; Kholodov, V.A.; Kulikova, N.A.; Philippova, O.I.; Ponomarenko, S.A.; Lasareva, E.V.; Parfyonova, A.M.; Hatfield, K.; Perminova, I.V. Silanized humic substances act as hydrophobic modifiers of soil separates inducing formation of water-stable aggregates in soils. *Catena* **2016**, *137*, 229–236. [CrossRef]
19. Deng, S.; Zheng, X.; Chen, X.; Zheng, S.; He, X.; Ge, T.; Kuzyakov, Y.; Wu, J.; Su, Y.; Hu, Y. Divergent mineralization of hydrophilic and hydrophobic organic substrates and their priming effect in soils depending on their preferential utilization by bacteria and fungi. *Biol. Fert. Soils* **2021**, *57*, 65–76. [CrossRef]
20. Ma, L.; Xiao, B.; Di, X.; Huang, W.; Wang, S. Characteristics and distributions of humic acids in two soil profiles of the southwest China Karst area. *Acta Geochim.* **2016**, *35*, 85–94. [CrossRef]
21. Dou, S. *Soil Organic Matter*; Science Press: Beijing, China, 2010.
22. Cui, T.; Li, Z.; Wang, S. Effects of in-situ straw decomposition on composition of humus and structure of humic acid at different soil depths. *J. Soils Sediments* **2017**, *17*, 2391–2399. [CrossRef]
23. Kuwatsuka, S.; Watanabe, A.; Itoh, K.; Arai, S. Comparison of two methods of preparation of humic and fulvic acids, IHSS method and NAGOYA method. *Soil. Sci. Plant Nutr.* **1992**, *38*, 23–30. [CrossRef]
24. Zhang, Z.; Liu, S.; Wang, X.; Huang, S.; Sun, K.; Xia, X. Differences in structure and composition of soil humic substances and their binding for polycyclic aromatic hydrocarbons in different climatic zones. *Environ. Pollut.* **2023**, *322*, 121121. [CrossRef]
25. dos Santos, J.V.; Fregolente, L.G.; Mounier, S.; Hajjoul, H.; Ferreira, O.P.; Moreira, A.B.; Bisinoti, M.C. Fulvic acids from Amazonian anthropogenic soils: Insight into the molecular composition and copper binding properties using fluorescence techniques. *Ecotoxicol. Environ. Safe* **2020**, *205*, 111173. [CrossRef] [PubMed]
26. Ji, H.; Han, J.; Xue, J.; Hatten, J.A.; Wang, M.; Guo, Y.; Li, P. Soil organic carbon pool and chemical composition under different types of land use in wetland: Implication for carbon sequestration in wetlands. *Sci. Total Environ.* **2020**, *716*, 136996. [CrossRef] [PubMed]
27. de Aguiar, T.C.; de Oliveira Torchia, D.F.; van Tol de Castro, T.A.; Tavares, O.C.H.; de Abreu Lopes, S.; de Souza da Silva, L.; Castro, R.N.; Berbara, R.L.L.; Pereira, M.G.; García, A.C. Spectroscopic–chemometric modeling of 80 humic acids confirms the structural pattern identity of humified organic matter despite different formation environments. *Sci. Total Environ.* **2022**, *833*, 155133. [CrossRef]
28. Gogoi, A.; Sahoo, U.K.; Saikia, H. Vegetation and ecosystem carbon recovery following shifting cultivation in Mizoram-Manipur-Kachin rainforest eco-region, Southern Asia. *Ecol. Process* **2020**, *9*, 21. [CrossRef]
29. Ahirwal, J.; Gogoi, A.; Sahoo, U.K. Stability of soil organic carbon pools affected by land use and land cover changes in forests of eastern Himalayan region, India. *Catena* **2022**, *215*, 106308. [CrossRef]
30. Bai, X.; Guo, Z.; Huang, Y.; An, S. Root cellulose drives soil fulvic acid carbon sequestration in the grassland restoration process. *Catena* **2020**, *191*, 104575. [CrossRef]
31. Baer, S.G.; Kitchen, D.J.; Blair, J.M.; Rice, C.W. Changes in Ecosystem Structure and Function along a Chronosequence of Restored Grasslands. *Ecol. Appl.* **2002**, *12*, 1688–1701. [CrossRef]
32. Lal, R.; Negassa, W.; Lorenz, K. Carbon sequestration in soil. *Curr. Opin. Environ. Sustain.* **2015**, *15*, 79–86. [CrossRef]
33. Mishra, G.; Sarkar, A. Studying the relationship between total organic carbon and soil carbon pools under different land management systems of Garo hills, Meghalaya. *J. Environ. Manag.* **2020**, *257*, 110002. [CrossRef]
34. Lungmuana; Singh, S.B.; Vanthawmliana; Saha, S.; Dutta, S.K.; Rambuatsaiha; Singh, A.R.; Boopathi, T. Impact of secondary forest fallow period on soil microbial biomass carbon and enzyme activity dynamics under shifting cultivation in North Eastern Hill region, India. *Catena* **2017**, *156*, 10–17. [CrossRef]
35. Chu, L.; Grafton, R.Q.; Nguyen, H. A global analysis of the break-even prices to reduce atmospheric carbon dioxide via forest plantation and avoided deforestation. *Policy Econ.* **2022**, *135*, 102666. [CrossRef]
36. Guimarães, D.V.; Gonzaga, M.I.S.; da Silva, T.O.; da Silva, T.L.; da Silva Dias, N.; Matias, M.I.S. Soil organic matter pools and carbon fractions in soil under different land uses. *Soil. Tillage Res.* **2013**, *126*, 177–182. [CrossRef]
37. Sun, C.; Liu, J.; Wang, Y.; Zheng, N.; Wu, X.; Liu, Q. Effect of long-term cultivation on soil organic carbon fractions and metal distribution in humic and fulvic acid in black soil, Northeast China. *Soil. Res.* **2012**, *50*, 562–569. [CrossRef]
38. Loke, P.F.; Kotzé, E.; Preez, C.C.; Twigge, L. Dynamics of Soil Carbon Concentrations and Quality Induced by Agricultural Land Use in Central South Africa. *Soil. Sci. Soc. Am. J.* **2019**, *83*, 366–379. [CrossRef]
39. Kotzé, E.; Loke, P.F.; Akhosi-Setaka, M.C.; Du Preez, C.C. Land use change affecting soil humic substances in three semi-arid agro-ecosystems in South Africa. *Agric. Ecosyst. Environ.* **2016**, *216*, 194–202. [CrossRef]
40. Caravaca, F.; Lax, A.; Albaladejo, J. Aggregate stability and carbon characteristics of particle-size fractions in cultivated and forested soils of semiarid Spain. *Soil. Tillage Res.* **2004**, *78*, 83–90. [CrossRef]
41. Stevenson, F.J. *Humus Chemistry: Genesis, Composition, Reactions*, 2nd ed.; John Wiley & Sons: New York, NY, USA, 1994.
42. Raiesi, F. The quantity and quality of soil organic matter and humic substances following dry-farming and subsequent restoration in an upland pasture. *Catena* **2021**, *202*, 105249. [CrossRef]
43. Godlewska, P.; Schmidt, H.P.; Ok, Y.S.; Oleszczuk, P. Biochar for composting improvement and contaminants reduction. A review. *Bioresour. Technol.* **2017**, *246*, 193–202. [CrossRef]
44. Yu, Z.; Liu, X.; Chen, C.; Liao, H.; Chen, Z.; Zhou, S. Molecular insights into the transformation of dissolved organic matter during hyperthermophilic composting using ESI FT-ICR MS. *Bioresour. Technol.* **2019**, *292*, 22007. [CrossRef]

45. Abakumov, E.V.; Cajthaml, T.; Brus, J.; Frouz, J. Humus accumulation, humification, and humic acid composition in soils of two post-mining chronosequences after coal mining. *J. Soils Sediments* **2013**, *13*, 491–500. [CrossRef]
46. He, S.; Zheng, Z.; Zhu, R. Long-term tea plantation effects on composition and stabilization of soil organic matter in Southwest China. *Catena* **2021**, *199*, 105132. [CrossRef]
47. Polyakov, V.; Loiko, S.; Istigechev, G.; Lapidus, A.; Abakumov, E. Elemental and molecular composition of humic acids isolated from soils of tallgrass temperate rainforests (Chernevaya taiga) by ^{1}H-^{13}C HECTCOR NMR spectroscopy. *Agronomy* **2021**, *11*, 1998. [CrossRef]
48. Mielnik, L.; Hewelke, E.; Weber, J.; Oktaba, L.; Jonczak, J.; Podlasiński, M. Changes in the soil hydrophobicity and structure of humic substances in sandy soil taken out of cultivation. *Agric. Ecosyst. Environ.* **2021**, *319*, 107554. [CrossRef]
49. Jindo, K.; Hernández, T.; García, C.; Sánchez-Monedero, M.A. Influence of Stability and Origin of Organic Amendments on Humification in Semiarid Soils. *Soil. Sci. Soc. Am. J.* **2011**, *75*, 2178–2187. [CrossRef]
50. Tan, K.H. *Humic Matter in Soil and the Environment: Principles and Controversies, 2nd ed*; CRC Press: Boca Raton, FL, USA, 2014.
51. Tadini, A.M.; Bernardi, A.C.C.; Milori, D.M.B.P.; Oliveira, P.P.A.; Pezzopane, J.R.M.; Martin-Neto, L. Spectroscopic characteristics of humic acids extracted from soils under different integrated agricultural production systems in tropical regions. *Geoderma Reg.* **2022**, *28*, e00476. [CrossRef]
52. Tadini, A.M.; Xavier, A.A.P.; Milori, D.M.B.P.; Oliveira, P.P.A.; Pezzopane, J.R.; Bernardi, A.C.C.; Martin-Neto, L. Evaluation of soil organic matter from integrated production systems using laser-induced fluorescence spectroscopy. *Soil. Till Res.* **2021**, *211*, 105001. [CrossRef]
53. Tadini, A.M.; Goranov, A.I.; Martin-Neto, L.; Bernardi, A.C.C.; Oliveira, P.P.A.; Pezzopane, J.R.M.; Hatcher, P.G. Structural characterization using 2D NMR spectroscopy and TMAH-GC × GC-MS: Application to humic acids from soils of an integrated agricultural system and an Atlantic native forest. *Sci. Total Environ.* **2022**, *815*, 152605. [CrossRef] [PubMed]
54. Dou, S.; Chen, E.; Xu, X.; Tan, S.; Zhang, J. Effect of improving soil fertility by organic material application (ISFOMA) on structural characteristics of humic acids in soil. *Acta Pedol. Sin.* **1992**, *29*, 199–207.
55. Aranda, V.; Macci, C.; Peruzzi, E.; Masciandaro, G. Biochemical activity and chemical-structural properties of soil organic matter after 17 years of amendments with olive-mill pomace co-compost. *J. Environ. Manag.* **2015**, *147*, 278–285. [CrossRef]
56. Spaccini, R.; Piccolo, A. Molecular characteristics of humic acids extracted from compost at increasing maturity stages. *Soil. Biol. Biochem.* **2009**, *41*, 1164–1172. [CrossRef]
57. Vinci, G.; Cangemi, S.; Bridoux, M.; Spaccini, R.; Piccolo, A. Molecular properties of the Humeome of two calcareous grassland soils as revealed by GC/qTOF-MS and NMR spectroscopy. *Chemosphere* **2021**, *279*, 130518. [CrossRef]
58. Xu, J.; Zhao, B.; Li, Z.; Chu, W.; Mao, J.; Olk, D.C.; Zhang, J.; Xin, X.; Wei, W. Demonstration of chemical distinction among soil humic fractions using quantitative solid-state ^{13}C NMR. *J. Agric. Food Chem.* **2019**, *67*, 8107–8118. [CrossRef]
59. Ferrari, E.; Francioso, O.; Nardi, S.; Saladini, M.; Ferro, N.D.; Morari, F. DRIFT and HR MAS NMR characterization of humic substances from a soil treated with different organic and mineral fertilizers. *J. Mol. Struct.* **2011**, *998*, 216–224. [CrossRef]
60. Guo, X.; Liu, H.; Wu, S. Humic substances developed during organic waste composting: Formation mechanisms, structural properties, and agronomic functions. *Sci. Total Environ.* **2019**, *662*, 501–510. [CrossRef] [PubMed]
61. Gao, L.; Wang, B.; Li, S.; Han, Y.; Zhang, X.; Gong, D.; Ma, M.; Liang, G.; Wu, H.; Wu, X.; et al. Effects of different long-term tillage systems on the composition of organic matter by ^{13}C CP/TOSS NMR in physical fractions in the Loess Plateau of China. *Soil. Tillage Res.* **2019**, *194*, 104321. [CrossRef]
62. Mylotte, R.; Verheyen, V.; Reynolds, A.; Dalton, C.; Patti, A.F.; Chang, R.R.; Burdon, J.; Hayes, M.H.B. Isolation and characterisation of recalcitrant organic components from an estuarine sediment core. *J. Soils Sediments* **2015**, *15*, 211–224. [CrossRef]
63. Zhang, J.; Dou, S.; Song, X. Effect of long-term combined nitrogen and phosphorus fertilizer application on ^{13}C CPMAS NMR spectra of humin in a typic hapludoll of northeast China. *Eur. J. Soil. Sci.* **2009**, *60*, 966–973. [CrossRef]
64. Preston, C.M.; Newman, R.H. A long-term effect of N fertilization on the 13C CPMAS NMR of de-ashed soil humin in a second-growth Douglas-fir stand of coastal British Columbia. *Geoderma* **1995**, *68*, 229–241. [CrossRef]
65. Zhou, Z.; Chen, N.; Cao, X.; Chua, T.; Mao, J.; Mandel, R.D.; Bettis, E.A.; Thompson, M.L. Composition of clay-fraction organic matter in Holocene paleosols revealed by advanced solid-state NMR spectroscopy. *Geoderma* **2014**, *223–225*, 54–61. [CrossRef]
66. Hayes, M.H.B.; Mylotte, R.; Swift, R.S. Humin: Its Composition and Importance in Soil Organic Matter. In *Advances in Agronomy*; Academic Press: New York, NY, USA, 2017; Volume 143, pp. 47–138.
67. Sun, H.; Jiang, J.; Cui, L.; Feng, W.; Wang, Y.; Zhang, J. Soil organic carbon stabilization mechanisms in a subtropical mangrove and salt marsh ecosystems. *Sci. Total Environ.* **2019**, *673*, 502–510. [CrossRef]

agriculture

MDPI

Article

Soil Attributes and Their Interrelationships with Resistance to Root Penetration and Water Infiltration in Areas with Different Land Uses in the Apodi Plateau, Semiarid Region of Brazil

Pollyana Mona Soares Dias [1], Jeane Cruz Portela [1,*], Joaquim Emanuel Fernandes Gondim [1], Rafael Oliveira Batista [1], Leticia Sequinatto Rossi [2], Jonatan Levi Ferreira Medeiros [1], Phâmella Kalliny Pereira Farias [1], Paulo Jardel Mota [1], Diego José da Costa Bandeira [1], Luís Cesar de Aquino Lemos Filho [1], Geisiane Xavier de Matos [1], Francisco de Oliveira Mesquita [3], Felipe Ozias de Oliveira [1], Antonio Genilson Rodrigues Araújo [1] and Claudeone Manoel do Nascimento [1]

[1] Department of Agronomic and Forest Science, Federal Rural University of the Semi-Arid-UFERSA, Mossoró 59625-900, RN, Brazil; rafaelbatista@ufersa.edu.br (R.O.B.); phamellakalliny@hotmail.com (P.K.P.F.); diego.bandeira@alunos.ufersa.edu.br (D.J.d.C.B.)
[2] Department of Soils and Nature Resources, State University of Santa Catarina-UDESC, Lages 88520-000, SC, Brazil
[3] PPGEC Graduate Program in Ecology and Conservation, State University of Paraíba, Campina Grande 58429-500, PB, Brazil
* Correspondence: jeaneportela@ufersa.edu.br

Abstract: Studies on soils and their interrelationships with land use in the context of the semi-arid region of Brazil are still scarce, even though they have the potential to assist in understanding the use and management of soil and agricultural crops. From this perspective, this study investigated four land uses in different locations of the Apodi Plateau, an elevated area in semi-arid region of northeastern Brazil. The different soils were analyzed for their resistance to root penetration, water infiltration, inorganic fractions, soil density, total porosity, potential of hydrogen, electrical conductivity, total organic carbon, potential acidity, and sum of bases. The soil resistance to root penetration and water infiltration were determined in the field. The results obtained were interpreted using multivariate and geostatistical analysis. The resistance data were subjected to the Shapiro–Wilk test at 5% of probability and expressed in maps, whereas infiltration data curves were constructed to estimate the amount of infiltrated water at the different time intervals. The textural classification was an important factor for the analysis of soil resistance to root penetration (Q) and the infiltration rate, being evidenced in the cluster analysis and allowing the formation of two groups, one for the surface layers of the areas and another for the subsurface layers, with the inorganic sand and clay fractions standing out with the greatest dissimilarity. The establishment of conservation practices for soil management is suggested to correct the pore space problems and the degradation of agroecosystems in areas with soils whose conditions are similar to the ones of this study.

Keywords: conservation agriculture; multiple soil classes; tillage practices; geostatistics; kriging; dry forest

Citation: Dias, P.M.S.; Portela, J.C.; Gondim, J.E.F.; Batista, R.O.; Rossi, L.S.; Medeiros, J.L.F.; Farias, P.K.P.; Mota, P.J.; Bandeira, D.J.d.C.; Filho, L.C.d.A.L.; et al. Soil Attributes and Their Interrelationships with Resistance to Root Penetration and Water Infiltration in Areas with Different Land Uses in the Apodi Plateau, Semiarid Region of Brazil. *Agriculture* **2023**, *13*, 1921. https://doi.org/10.3390/agriculture13101921

Academic Editors: Vito Armando Laudicina and Bettina Eichler-Löbermann

Received: 6 July 2023
Revised: 21 September 2023
Accepted: 27 September 2023
Published: 30 September 2023

1. Introduction

Soil health is a parameter that cannot be measured directly, requiring information on structural attributes such as water infiltration and soil resistance to root penetration, which can be used to interpret the effects of soil and water degradation processes that ultimately compromise biodiversity [1–5], attributes that are significantly impacted by land use. Therefore, evaluating the influence of the inter-relation of different factors on physical, chemical, and structural characteristics can assist in identifying the physical forces that govern soil structure in the field [6–8].

Several studies have aimed to assess soil resistance to penetration [9–14] and infiltration [15–18] given the impact of these variables on plant growth, crop performance, and the sustainable development of agroecosystems. These attributes are mutually associated and can serve as parameters related to water–structural functions of the soil. For example, soils with less resistance to root penetration are associated with higher water infiltration and structural and environmental functionalities [19,20]. The relationship of these attributes with soil production capacity is even more important in family farming areas, which are more dependent on natural resources [21]. However, in arid and semi-arid lands, water shortage restricts agricultural development [18]. Therefore, it is essential to compare soil attributes with different land uses, as these attributes can be changed due to environmental and anthropic actions [22], which justifies the need to understand how different soil management practices can modify the soil attributes in order to decrease degradation and ensure sustainable land use [23].

In this scenario, kriging is an advanced geostatistical procedure used by several researchers [24–26] that considers spatial dependence, data treatment, and inferential procedures. Furthermore, although it is common to use geostatistics and multivariate analysis separately, they can clarify the dynamics of water in soils and be decisive for the proper planning of agricultural practices when used in association.

In the semi-arid region of northeastern Brazil, the Apodi Plateau is an outstanding region in the context of agricultural production, with expressive irrigated, rainfed, and livestock areas and the predominance of ultisols, cambisols, and oxisols [27]. However, inadequate human action has reduced the production capacity of the region's soils.

From this perspective, the innovative character of this research refers to the use of geostatistics associated with a multivariate tool to discriminate environments and soil classes in agroecosystems in an interrelated manner, which can be used on a global scale. The novelty of this study consists in exploring data on the soil's resistance to root penetration and water infiltration, parameters used to recognize physical and water restrictions. As a result, this information can contribute to other research aimed at the conservation of environmental and ecosystem services, as well as to new actions on the subject.

The importance of these physical properties for the growth and development of agricultural crops that consequently influence soil quality and, when necessary, reorient and replace inadequate techniques of soil and crop management is highlighted in this study. In addition, clarifying these issues could benefit regional agriculture by providing useful information for more adequate soil and crop management, not only under semi-arid conditions, but also in a global context.

Our main hypotheses are (i) restrictions with regard to resistance to root penetration and water infiltration in the soil are clear in soils with clayey textures, with a history of intensive soil preparation and consolidation of the soil surface, and in lower locations of the landscape. (ii) Kriging is a regression method used in geostatistics to approximate or interpolate data. It is believed that kriging can improve the performance of the quantitative estimates of soil attributes, especially with regard to root penetration, in association with multivariate statistics.

We also evaluate the importance of geostatistical analysis, through kriging, to complement the multivariate statistical analysis to provide accuracy in our findings, especially concerning root penetration.

From this perspective, this study aimed to evaluate the interrelationships between the physical attributes of the soil related to water infiltration and resistance to root penetration in soils of the Moacir Lucena Settlement Project in the Apodi Plateau, a semi-arid region of Brazil, using multivariate statistics and geostatistical analyses through the kriging method.

2. Materials and Methods

2.1. Study Area

This study was conducted in the Moacir Lucena Settlement Project (Figure 1), Apodi (Brazil). Apodi is in the semi-arid region of the State of Rio Grande do Norte (RN), in the micro-region of the Apodi Plateau, in the Oeste Potiguar mesoregion of RN. The climate of the region is classified as *BSh* (hot semi-arid) according to Köppen [28], with a mean annual rainfall between 500 and 600 mm. The natural vegetation belongs to the Caatinga Phytogeographic Domain.

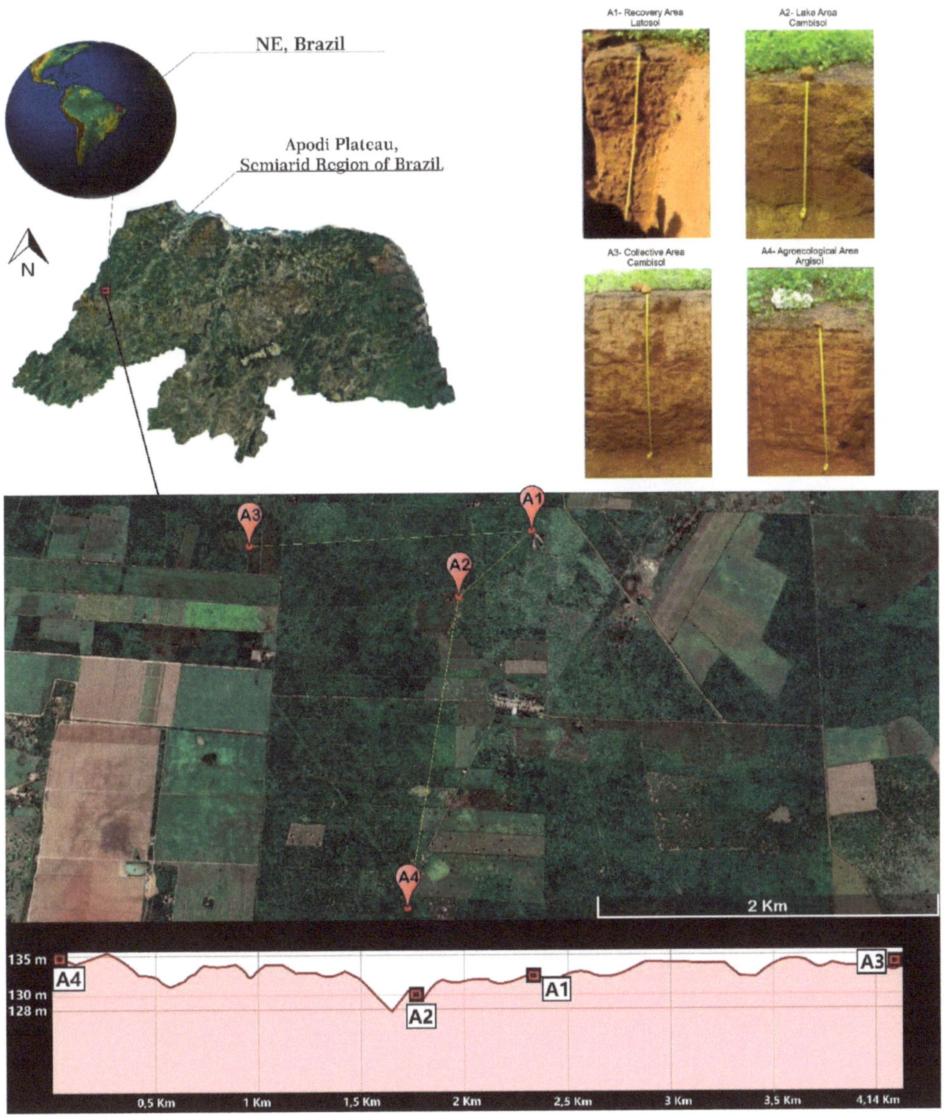

Figure 1. Location map of the Moacir Lucena Settlement Project in the Apodi Plateau, a semi-arid region of Brazil.

The soils of the agricultural areas studied were classified according to World Reference Base-WRB [29] published by the Food and Agriculture Organization of the United Nations—FAO and the Brazilian Soil Classification System [30]. The research was developed in four agroecosystems (Table 1) used as study sites: Recovery Area (A1); Lake Area (A2); Collective Area (A3); and Agroecological Area (A4).

Table 1. Characteristics of the areas studied in the Apodi Plateau, a semi-arid region of Brazil.

Land Use	Soil Classification		Environmental History	
	SBCS [3]	WRB/FAO	Characteristics	Area (ha)
Recovery Area (LATOSOL) [1]	LATOSOL	FERRALSOL	The area has rested for 16 years to recover its native forest and soil, where cotton was previously cultivated.	2.5
Lake Area (CAMBISOL) [1]	CAMBISOL	CAMBISOL	The area is flooded in the rainy season (temporary lake) due to its position in the landscape (moderate depression). Presence of sediment deposition.	4–5
Collective Area (CAMBISOL)	CAMBISOL	CAMBISOL	Area used for crop sowing (dry season) and sorghum cultivation (rainy season), with the presence of grazing animals for 2 months/year. Intensive soil preparation.	35 [2]
Agroecological Area (ARGISOL) [1]	ARGISOL	ACRISOL	Area cultivated with short-cycle crops.	3

Note: [1] Land use: Inserted within the limits of the Permanent Preservation Area (APP); Within the area, 8 hectares (ha) of native forest are maintained, separating the area used to grow short-cycle crops and the area with cashew trees; [2] Environmental History; and [3] Brazilian Society of Soil Science.

The chosen areas lack information about their limitations and the potential necessary for the sustainable and efficient management of natural resources. As a result, extensive lowland areas, with the potential for grazing, for example, tend to be underutilized, which affects the quality of life of farmers.

The study areas also show significant space–time variations in soil attributes, in addition to having a distinctive history of uses and management. The places selected for the study include areas of higher agricultural aptitude, intended for the cultivation of short-cycle crops and fruit trees (Agroecological and Community Areas), as well as areas that, due to their intensive use and location in the landscape, have been transformed into permanent preservation areas (Recovery and Lake Areas).

Leaving soil fallow is a strategy commonly used in the Brazilian semi-arid region. However, few studies that reflect soil response to fallow periods in the long term have been developed, especially in arid or semi-arid environments.

2.2. Sample Collection

Disturbed and undisturbed samples were collected from the soil layers for physical and chemical analysis. The disturbed samples were collected from the 0.00–0.10 and 0.10–0.20 m layers to evaluate soil resistance to root penetration and soil moisture. Water infiltration was measured in the 0.00–0.10 m layer.

Disturbed and undisturbed samples were collected from the following soil horizons: Area: 1—Recovery Area (LATOSOL): A: (0.00–0.04 m); AB: (0.04–0.17 m);

Area 2—Lake Area (CAMBISOL): A: (0.00–0.03 m); BA: (0.03–0.15 m);

Area 3—Collective Area (CAMBISOL): Ap: (0.00–0.06 m) BA: (0.06–0.18 m);

Area 4—Agroecological Area (ARGISOL): A: (0.00–0.03 m) BA: (0.03–0.16 m).

The disturbed samples were collected using a tray and shovel, after which they were identified and packed in plastic bags. Subsequently, the samples were air-dried, ground, and passed through 2 mm sieves to obtain air-dried fine earth. Ten unformed samples were collected per layer in each class (80 samples in the four environments) using volumetric rings 0.05 high and 0.05 m wide and an Uhland-type apparatus. After sampling, the rings were coated with aluminum foil to maintain the structure and moisture of the original soil and taken to the laboratory.

2.3. *Soil Analyses*

2.3.1. Soil Resistance to Root Penetration

The evaluation of soil resistance to root penetration (Q) was performed using PenetroLOG equipment (Falker—USA) with a support capacity of 90 kgf (198 lb). The data were read at every centimeter by an automatic measurement system, according to ASABE S.313.3 [31], until the 40 cm layer, with a reference area of 0.5 hectares in each environment (25 readings per area). At the time of the test, deformed samples were collected to evaluate the gravimetric moisture content in the 0.00–0.10 and 0.10–0.20 m layers, with ten different sampling points per layer and area, totaling 80 points, thus obtaining the mean water content values in the soil.

2.3.2. Water Infiltration into the Soil

The infiltration rate was determined by the concentric ring method developed by Bernardo and collaborators [32], using two metallic cylinders coupled to the inner cylinder, with diameters of 30 cm in the inner ring and 50 cm in the outer ring (Figure 2a). The cylinders, measuring 40 cm in height, were positioned at a depth of 10 cm into the soil. The water height was measured inside the inner cylinder at 0, 1, 2, 3, 4, 5, 10, 15, 20, 30, 45, 60, 90, and 120 min (Figure 2b). Infiltration was considered constant when the reading value was repeated at least three times, according to the recommendations of the authors mentioned before.

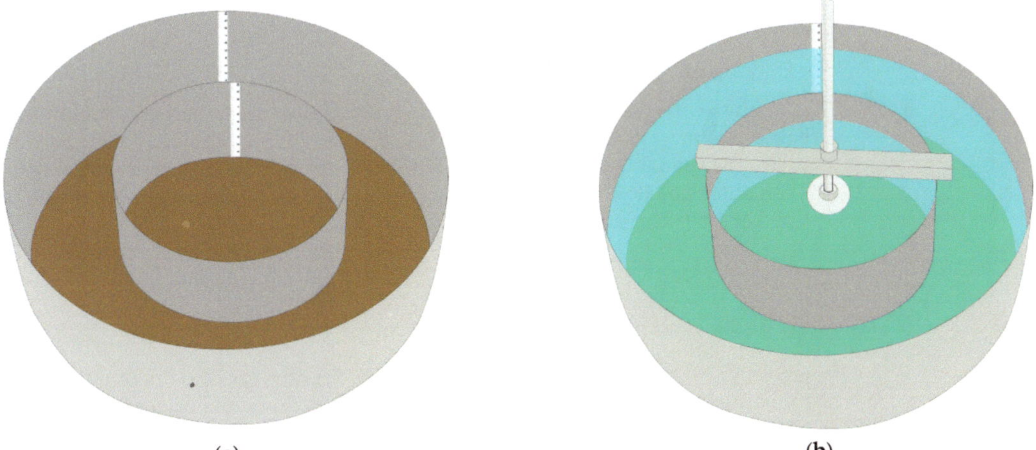

(a) (b)

Figure 2. Water infiltration test: (**a**) by using the concentric ring method; (**b**) by performing readings of the infiltrated water height.

In order to determine the water infiltration rate into the soil, three replications were performed in each soil class by collecting disturbed soil samples to quantify the gravimetric moisture [33]. The cumulative infiltration curves were obtained by Equation (1), and the Basic Infiltration Speed (BIS) was calculated by Equation (2), which allowed the categorization of soil classes according to Bernardo and collaborators [32].

$$I = a * T^{n} \quad (1)$$

$$BIS = 60 * a * n * T^{n-1} \quad (2)$$

where:

I—cumulative infiltration (cm); BIS—Basic Infiltration Speed (cm/h);
a—constant;

T—infiltration time (min);
n—constant that ranges from 0 to 1.

2.3.3. Supplementary Physicochemical Analyses

The physical and chemical attributes were determined to complement the soil analysis. This was performed to quantify the properties of soil resistance to root penetration and water infiltration. The physical analysis consisted of determining textural parameters, soil bulk density, total porosity, and the gravimetric moisture content. The granulometry was obtained by the pipette method, using the chemical dispersant sodium hexametaphosphate. The sand fraction (2 to 0.05 mm) was obtained by sieving; the clay fraction (<0.002 mm) by sedimentation; and the silt fraction (0.05 to 0.002 mm) was obtained by the difference between the two previous fractions [34]. Soil density was obtained by calculating the ratio of dry soil mass to the total volume of the ring [35]. Total porosity (TP) was obtained by saturating undisturbed samples for 48 h [36]. The gravimetric moisture content was obtained by the difference between the mass of air-dried samples and the mass of the samples after 3 days at 105 °C on the oven.

The chemical analyses were carried out according to Teixeira et al. [33] and consisted of determining the following attributes: potential of hydrogen (pH); electrical conductivity (EC) in water; total organic carbon (TOC) by organic matter digestion [37]; calcium (Ca^{2+}) and exchangeable magnesium (Mg^{2+}), determined with a potassium chloride extractor; sodium (Na^+) and potassium (K^+), determined with the Mehlich-1 extractor; and potential acidity ($H + Al^{3+}$), determined with calcium acetate (after which the sum of bases—SB was calculated).

2.4. Statistical Analysis—Geostatistical Analysis

The geostatistical analysis was performed to subsidize the semivariogram modeling based on the measurement data. The ordinary kriging interpolation method was used to verify the spatial dependence of the studied variables using the software Vesper 1.6 [38]. Thematic maps were generated with the interpolated values using the software Quantum GIS 2.18 [39]. The estimate was made using Equation (3) [40]:

$$\gamma^*(h) = \frac{1}{2.(h)} * \sum_{\substack{Z.(S_i) \\ J=1}}^{N_i.(h)} [(S_i) - h]^2 \tag{3}$$

where γ (h) are the semivariances, and (h) is the number of pairs of points z (S_i) and z (S_i + h) separated by a distance h, informing how different the values become as a function of h.

Using the semivariogram generated by geostatistics as a reference, maps were made for the variation of the maximum depth reached by the equipment in centimeters (cm) and the mechanical resistance to soil penetration in kPa. This variation was demonstrated by means of a color gradient that varied from red to green. For layer variation, red was used for the smallest values and green for the largest values. For the variation in Q, the opposite occurs. The following semivariogram models were tested: spherical (Equations (2) and (3)), exponential (Equation (4)), and Gaussian (Equation (5)). The adjustment of the models was made using the Root Mean Square Error (RMSE) and the Akaike Information Criterion (AIC) [41].

$$\gamma^*(h) = C_0 + C_1 * \left[1.5 * \left(\frac{h}{a}\right) - 0.5 * \left(\frac{h}{a}\right)^3\right],\ 0 < h < a \tag{4}$$

$$\gamma^*(h) = C_0 + C_1,\ h \geq a \tag{5}$$

$$\gamma^*(h) = C_0 + C_1 * \left[1 - e^{\left(\frac{-3.h}{a}\right)}\right],\ 0 < h < d \tag{6}$$

$$\gamma^*(h) = C_0 + C_1 * \left[1 - e^{\left(\frac{-3.h^2}{a^2}\right)}\right], 0 < h < d \quad (7)$$

where γ^* (h)—semivariances; d—the maximum distance at which the semivariogram is defined; h—distance; a—soil-dependent constant.

The scale proposed by Ribeiro [42] (Table 2) was used to interpret data related to resistance to root penetration. The degree of spatial dependence (DSD) of the semivariograms was obtained by Equation (6) and evaluated according to intervals proposed by Cambardella [43]: DSD < 25%—strong spatial dependence; 25% < DSD < 75%—moderate spatial dependence; and DSD > 75%—weak spatial dependence.

$$DSD = \left(\frac{C_0}{C_0 + C_1}\right) * 100 \quad (8)$$

where GD—degree of dependence; C_0—nugget effect; C_1—structural variance.

Table 2. Evaluation criteria for the maps of resistance to root penetration in the Moacir Lucena Settlement areas, Apodi Plateau, a semi-arid region according to Ribeiro (2010).

Soil Root Penetration Resistance (kPa)	Compaction Level	Impediment Level to Root Growth
0–2000	Low	No impediment
2000–4000	Moderate	Slight impediment
4000–6000	High	Reduced development
6000–8000	Very high	Minimum development

2.5. Multivariate Analysis

The results for the resistance to root penetration and the physical and chemical attributes (Section 2.3.3) were interpreted by multivariate statistics (principal component analysis and factor analysis) using the software Statistica 7.0 [44]. The correlation matrix was used to standardize the data, considering correlations equal to or higher than 0.70 [45], and verify the similarities and distinctions between the studied areas depending on the potential or restrictions of the environments.

3. Results

Soil Physical and Chemical Attributes

There was variation in the textural classification in the different soil samples (surface and subsurface) (Table 3). This variation ranged from light sandy-clay (A1, A2, and A3) to sandy loam (A4) on the surface, and from sandy loam to loamy in the subsurface, with emphasis on the clayey texture of A2 (Lake Area—CAMBISOL) and A4 (Agroecological Area—ARGISOL). The mean bulk density ranged from 1.32 to 1.71 g.cm^{-3} in the studied soil classes and was lower in the cambisol of the Lake Area.

Table 3. Physical analysis of the soil classes of the Moacir Lucena Settlement Project, Apodi Plateau, a semi-arid region of Brazil.

Layer (cm)	Particle Size Distribution			Bulk Density	Total Porosity	Gravimetric Soil Moisture	Textural Classification
	Sand	Silt	Clay				
		g.kg^{-1}		g.cm^{-3}		%	
			Area 1—Recovery Area (LATOSOL)				
A (0–4)	660	87	253	1.60	39.81	2.4	light sandy-clay
AB (4–17)	492	92	416	1.61	36.01	5.7	sandy loam

Table 3. *Cont.*

Layer (cm)	Particle Size Distribution			Bulk Density	Total Porosity	Gravimetric Soil Moisture	Textural Classification
	Sand	Silt	Clay				
		g.kg^{-1}		g.cm^{-3}		%	
				Area 2—Lake Area (CAMBISOL)			
A (0–3)	653	96	251	1.71	31.81	3.7	light sandy-clay
BA (3–15)	415	135	450	1.32	47.69	3.0	loamy/clay
				Area 3—Collective Area (CAMBISOL)			
Ap (0–6)	720	69	211	1.55	41.11	0.9	light sandy-clay
BA (6–18)	525	47	428	1.52	38.83	3.7	sandy loam
				Area 4—Agroecological Area (ARGISOL)			
A (0–3)	660	200	140	1.43	43.79	3.2	sandy
BA (3–16)	435	145	420	1.49	42.85	3.2	loamy/clay

Water infiltration changed according to the texture, with the A3 area showing the highest infiltration rate on the surface, as well as a predominant sand fraction. The areas showed constant infiltration after 60 min of evaluation (Figure 3a), except for A3, which showed greater oscillation in the infiltration rate (Figure 3b). High water infiltration values are expected in plateau soils due to their location at higher elevations and the flat relief. However, the same did not occur in the A4 area (ARGISOL), which showed a strong physical impediment and a gradual reduction in the infiltration rate due to clay accumulation in depth.

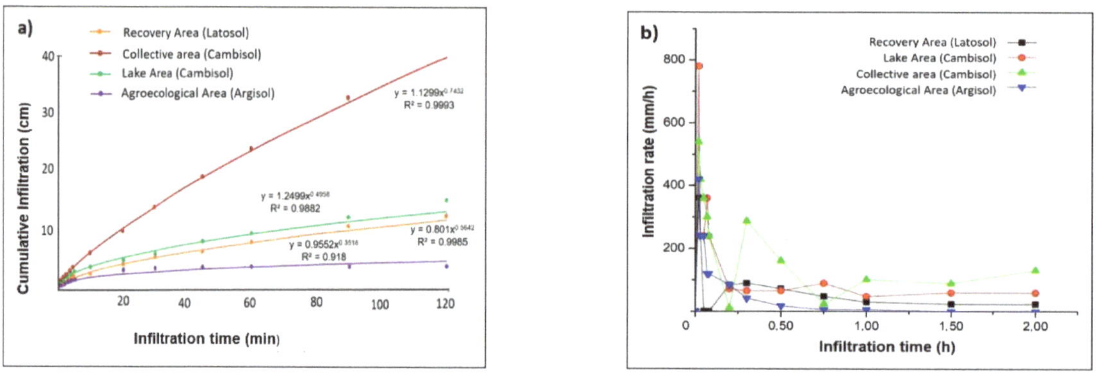

Figure 3. Infiltration curve of soil samples illustrating: (**a**) infiltration; (**b**) the infiltration rate.

Areas 4 (Agroecological Area—ARGISOL) and 2 (Lake Area—CAMBISOL) showed the highest Q (Table 4) on the surface, whereas Areas 1 (Recovery Area—LATOSOL) and 3 (Collective Area—CAMBISOL) had the lowest Q in the subsurface. The greater the accumulation of the clay fraction, the more pronounced the gravimetric moisture and Q. Thus, clay accumulation in all soil classes in the subsurface caused an increase in Q in-depth, with texture being a decisive factor in the Q of the studied soils.

Table 4. Descriptive statistics parameters for maximum depth and soil resistance to root penetration in the soils of the Moacir Lucena Settlement Project, in Apodi Plateau, a semi-arid region of Brazil.

Attribute	Mean	Median	Maximum	Minimum	SD	CV%	Classification	Ck	Ca	W
A1(Layer)	5.00	5.00	7.00	3.00	1.33	26.57	high	−0.87	−0.86	NS
A1 (Q)	3923.54	3918.92	4298.20	3621.13	191.54	4.88	low	0.00	0.22	NS
A2 (Layer)	11.00	7.00	30.00	4.00	8.57	**77.92**	**very high**	0.34	1.28	*
A2 (Q)	4147.59	3994.31	5051.92	3658.44	420.15	10.13	medium	−0.31	0.79	NS
A3 (Layer)	8.29	8.00	13.00	4.00	2.41	29.10	high	0.22	0.15	NS
A3 (Q)	3405.20	3409.70	4240.50	2400.64	543.44	15.96	medium	−0.83	−0.35	NS
A4 (Layer)	12.07	7.00	34.00	5.00	9.34	**77.39**	**very high**	2.02	1.72	*
A4 (Q)	4016.37	4041.33	4832.83	3382.00	486.01	12.10	medium	−0.74	0.53	NS

Notes: Layer—Maximum depth; Q—Soil resistance to root penetration; SD—Standard Deviation; CV—Coefficient of Variation; Ck—Kurtosis Coefficient; Ca—Asymmetry Coefficient; W—Shapiro–Wilk Test; *—Non-normal distribution by the Shapiro–Wilk test (p-value < 0.05); NS—Normal distribution by Shapiro–Wilk test (p-value > 0.05). There was intensive conventional soil preparation over time.

The soils showed pH values close to neutrality, with aluminum levels below the criteria required by the Brazilian Soil Classification System to identify them as aluminic or alithic. Potential acidity levels ranged from 1.02 to 2.66 $cmol_c.kg^{-1}$. These values can be justified by the absence of aluminum (Table 5).

Table 5. Chemical attributes of the soils of the Moacir Lucena Settlement, Apodi Plateau, a semi-arid region of Brazil.

Layer (cm)	pH in Water	EC ($dS.m^{-1}$)	TOC ($g.kg^{-1}$)	(H + Al)	SB
				$cmol_c.kg^{-1}$	
Area 1—Recovery Area (LATOSOL)					
A (0–4)	7.24	0.37	4.80	1.96	8.61
AB (4–17)	6.68	0.17	3.73	2.66	6.54
Area 2—Lake Area (CAMBISOL)					
A (0–3)	7.64	0.98	6.23	1.26	10.90
BA (3–15)	6.75	0.27	3.20	2.17	8.55
Area 3—Collective Area (CAMBISOL)					
Ap (0–6)	7.10	0.41	4.83	1.32	6.33
BA (6–18)	6.89	0.20	3.80	1.02	4.52
Area 4—Agroecological Area (ARGISOL)					
A (0–3)	7.02	0.75	6.73	2.49	11.49
BA (3–16)	6.93	0.57	3.80	2.18	9.80

Note: pH—Potential of hydrogen; EC—Electrical Conductivity; TOC—Total Organic Carbon; (H + Al)—Potential acidity; SB—Sum of Bases.

In all areas, the TOC values were lower than 1%, with the most expressive ones corresponding to the surface soil layers (where there is a greater quantity of organic matter), especially in Areas 2 and 4. The values of calcium and magnesium were the most representative of the sum of bases, especially in the less weathered soils (cambisol and argisol) (Table 5). A2 showed the highest electrical conductivity value (0.98 $dS.m^{-1}$). However, the values were generally low.

The variability in soil attributes can be classified according to the coefficient of variation (CV) [46], a statistical measure that relatively quantifies how far the values are moving away from the mean; thus, higher values indicate distancing from the mean [47]. In this study, the CV (Table 4) ranged from high (A1 and A3) to very high (A2 and A4) for the layers, and from low (A1) to medium (A2–A4) for Q. The Shapiro–Wilk test at 5% of probability showed that most parameters have a normal distribution, except for the depths of A2 and

A3, which showed a greater distance from zero for the coefficients of kurtosis (Ck) and asymmetry (Ca).

The range (A) (Table 6) is the main spatial correlation parameter provided by geostatistics; thus, from this distance, the variable starts to show random spatial variability [48]. The range of variable Q was 29.30 (A1) to 80.01 (A2), whereas the range of the layer was 10.54 (A1) to 37.72 (A3). Furthermore, the degree of spatial dispersion of the semivariogram demonstrates a predominance of strong spatial dependence between the parameters evaluated (DSD < 25%), except for the Q of A1, which showed a weak spatial dependence (DSD > 75%).

Table 6. Models and estimated parameters of the semivariogram.

Attribute	N. Lags	Tolerance (%)	Co	Co + C1	A	Model	Q	AIC	DSD	Classification
A1(Layer)	20	30	0.279	1.743	10.54	Exponential	0.59	47.78	16.007	FDE
A1 (Q)	15	20	0	38,247	29.30	Spherical	13,973	355.8	0	FDE
A2 (Layer)	10	20	47.1	75.53	30.34	Spherical	16.07	84.56	62.359	MSD
A2 (Q)	15	25	67,827	282,918	34.47	Gaussian	70,618	329.6	23.974	SSD
A3 (Layer)	20	30	0.660	7.828	37.72	Exponential	2.838	83.79	8.431	SSD
A3 (Q)	20	30	63,568	345,595	80.01	Spherical	68,558	406.7	18.394	SSD
A4 (Layer)	20	30	0	81.68	31.5	Spherical	28.73	96.19	0	SSD
A4 (Q)	20	35	119,141	290,666	27.30	Exponential	69,557	329.2	40.989	MSD

Notes: N—number of lags; C0—nugget effect; Co + C1—sill; A—range (m); Q—soil resistance to root penetration; AIC—Akaike Information Criterion; DSD—degree of dependence; SSD—Strong Spatial Dependence; MSD—Moderate Spatial Dependence.

We can correlate the resistance values with the adopted soil management, soil class, and the maximum depth reached. In the Collective Area (CAMBISOL), in which there was an increase in the clay fraction, the depth ranged from 5 to 11 cm (Figure 4). In the Lake Area, in which there was sediment accumulation, the variation was from 8 to 15 cm. In the Agroecological Area (ARGISOL), there was a variation from 5 to 27 cm since this soil is deeper and more weathered. Finally, the Recovery Area (LATOSOL) showed little variation in depth.

In general, the resistance to root penetration showed high values in the subsurface for all studied soils, with uniform variations between the layers of all classes, corresponding to approximately 100 kPa for A1 and 200–300 kPa for the other classes (Figure 5).

A high negative correlation was found for the clay and Q variables with the sand inorganic fraction, as well as a high positive correlation between sand and TOC. The silt fraction was positively correlated with SB. The clay fraction showed an inversely proportional correlation with the sand fraction. The pH, TOC, and EC showed a negative correlation only with potential acidity, which had a positive correlation with Q. The SB was positive for all variables. BD and TP showed an inverse correlation. Furthermore, there was a negative correlation between TP and BD, Q and pH, and Q and TOC, and a positive correlation of EC with two other variables, pH and TOC (Table 7).

Table 7. Correlation matrix of physical soil attributes in the Moacir Lucena Settlement Project, Apodi Plateau, a semi-arid region of Brazil.

	Sand	Silt	Clay	BD	TP	U	Q	pH	TOC	SB	EC	(H + Al)
Sand	1.00											
Silt	−0.14	1.00										

Table 7. *Cont.*

	Sand	Silt	Clay	BD	TP	U	Q	pH	TOC	SB	EC	(H + Al)
Clay	−0.92	−0.27	1.00									
BD	0.49	−0.53	−0.26	1.00								
TP	−0.35	0.51	0.13	−0.95	1.00							
U	−0.51	0.06	0.47	0.19	−0.40	1.00						
Q	−0.83	−0.35	0.95	−0.19	0.07	0.55	1.00					
pH	0.68	−0.12	−0.61	0.65	−0.56	−0.33	−0.72	1.00				
TOC	0.78	0.37	−0.90	0.35	−0.32	−0.18	−0.90	0.71	1.00			
SB	0.15	0.79	−0.47	−0.04	0.07	−0.03	−0.63	0.47	0.64	1.00		
EC	0.43	0.22	−0.51	0.46	−0.50	0.04	−0.65	0.84	0.76	0.70	1.00	
(H + Al)	−0.36	0.69	0.07	−0.36	0.37	0.43	0.15	−0.52	−0.09	0.38	−0.22	1.00

Note: BD—Bulk density; TP—Total Porosity; U—Gravimetric moisture, Q—Soil resistance to root penetration; pH—Potential of hydrogen; TOC—Total Organic Carbon; SB—Sum of Bases, EC—Electric Conductivity; (H + Al)—Potential acidity.

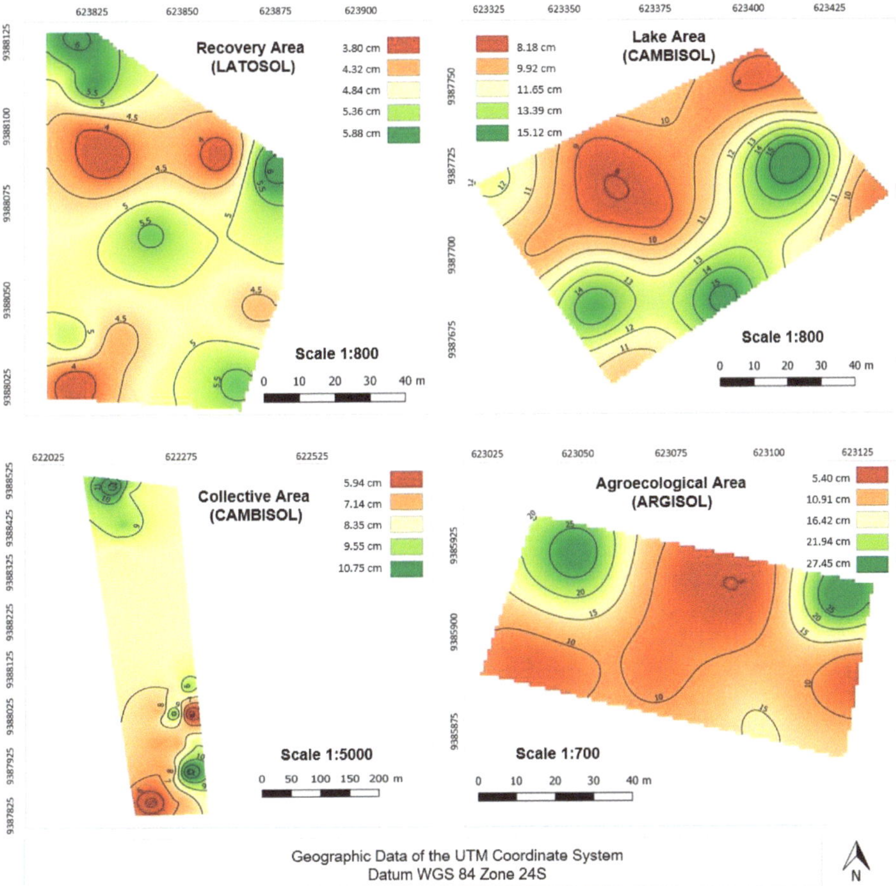

Figure 4. Maps of layers of the Moacir Lucena Settlement Project, Apodi Plateau, a semi-arid region of Brazil.

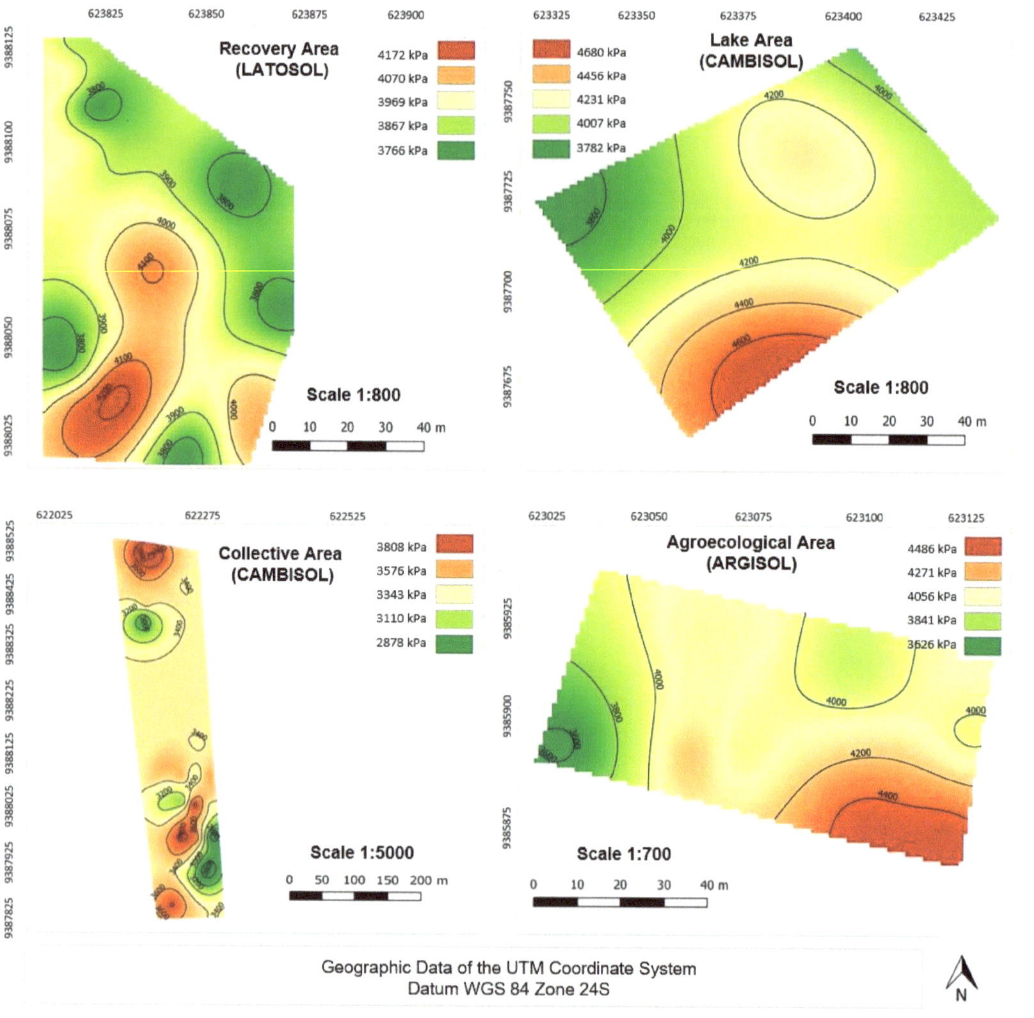

Figure 5. Maps of mechanical resistance to root penetration in the soils of the Moacir Lucena Settlement Project, Apodi Plateau, a semi-arid region of Brazil.

In the Cluster Analysis, two groups were formed at 20% dissimilarity (Figure 6). The first was represented by the surface horizons and showed dissimilarity with the clay, silt, and sand inorganic fractions and with variables TP, U, pH, TOC, SB, EC, and (H + Al^{3+}). The second group was defined according to the subsurface horizon of all areas, with dissimilarity for variable Q.

Factors 1–3 explained 90.12% of data variation (Table 8). Factor 1 made it possible to estimate the variables of sand, clay, Q, pH, and TOC. Factor 2 highlighted silt, SB, and potential acidity (H + Al). Factor 3 only highlighted the TP. The cumulative variance obtained for factors 1 and 2 was 73.78%, showing great representativeness for the studied environments.

Table 8. Factor loads corresponding to the 12 physical attributes of the soils analyzed and their respective eigenvalues, total variances observed, and cumulative variances.

Attributes	Factor 1	Factor 2	Factor 3
Sand	0.88	−0.26	0.18
Silt	0.20	0.91	−0.33
Clay	−0.93	−0.12	−0.04
BD	0.22	−0.30	0.87
TP	−0.09	0.24	−0.96
U	−0.66	0.41	0.56
Q	−0.98	−0.18	−0.02
pH	0.76	−0.10	0.54
TOC	0.85	0.31	0.31
SB	0.50	0.80	0.11
EC	0.60	0.33	0.61
(H + Al)	−0.31	0.78	−0.21
Eigenvalues (%)	5.70	3.15	1.96
Total Variance (%)	47.54	26.25	16.34
Cumulative Variance (%)	47.54	73.78	90.12

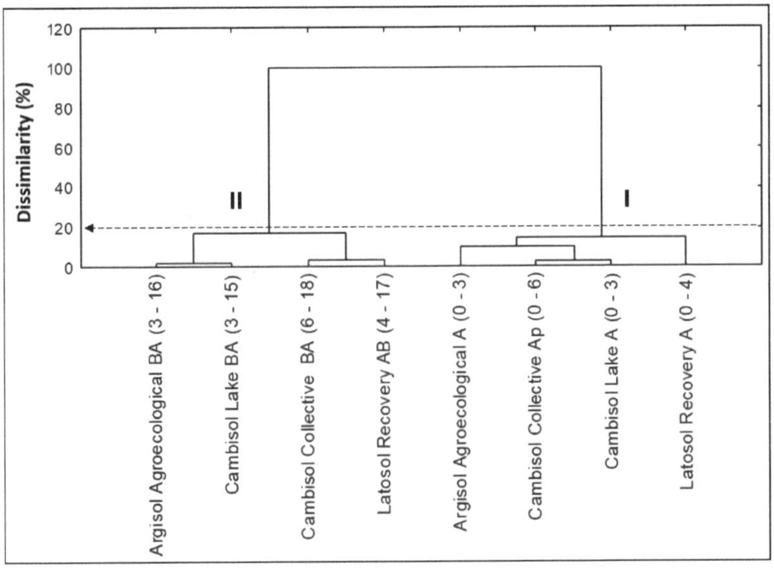

Figure 6. Vertical dendrogram of the distance matrix by the single bond grouping method.

The correlation circles (Figure 7a,c) and clouds of variables (Figure 7b,d) highlighted the influence of physical attributes to differentiate the studied environments. The inorganic fractions were not clustered close to each other. Thus, we infer that the areas show variability in texture, reflecting the predominance of the variables discriminated for environments and portraying the existing interrelationships for each local particularity. The discriminating variables were TP, BD, SB, pH, Q, and silt for the cambisol; SB, TOC, silt, and EC for the argisol; and Q, U, sand, and clay for the latosol.

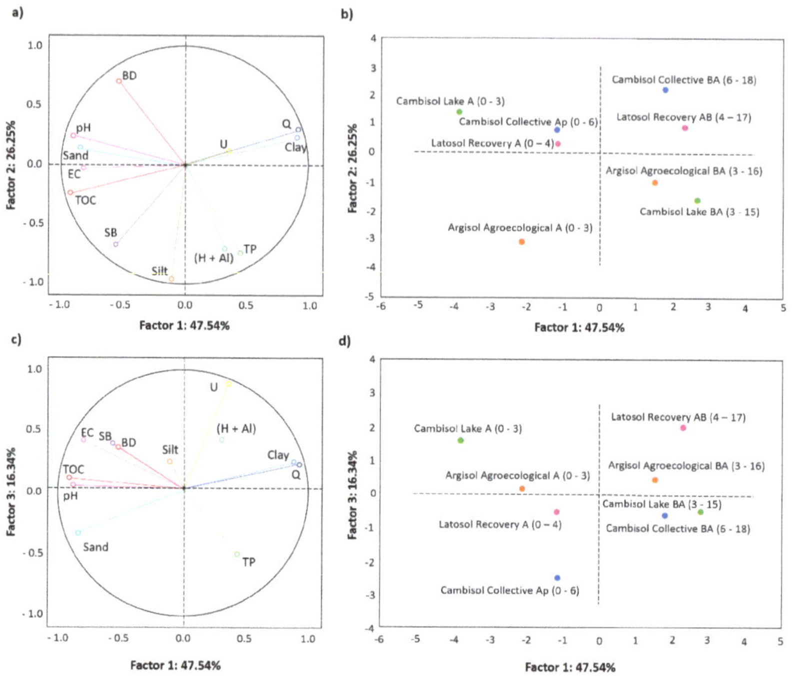

Figure 7. Cloud distribution of: (**a**) variables in the correlation circle (factors 1–2); (**b**) points representing the relationships between factors 1–2; (**c**) variables in the correlation circle (factors 1–3); (**d**) points representing the relationships between factors 1–3 of the studied environments. Note: BD—Bulk Density; TP—Total Porosity; U—Gravimetric moisture, Q—Soil resistance to root penetration; pH—Potential of hydrogen; TOC—Total Organic Carbon; SB—Sum of Bases, EC—Electric Conductivity; (H + Al)—Potential acidity.

4. Discussion

Soil Physical and Chemical Attributes

Smaller BD values were observed in the subsurface of A2 (Lake Area) and on the surface of A4 (Agroecological Area). In A2, the greater BD on the surface can be explained by the accumulation of sediments coming from higher elevations (slope), as found in the study by Quan et al. [49]. In A4, the BD reduction on the surface can be justified by the supply of organic matter [50,51].

The surface structure of the soil was compromised by the BD increase and TP reduction, constituting an impediment to its recovery and the restructuring of the porous system in all layers [52]. The highest bulk density (Table 3) found on the surface of A2 was due to clays with high colloidal activity, a characteristic of cambisols, with cohesion forces contributing to the consolidation of the surface. Fertile soils with clays of high colloidal activity (2:1 clay) compromise the gas exchange parameters, thus influencing the soil's structural parameters.

Cambisols are considered young (little weathered) and show expressive silt fraction values, contributing to their compactness [53]. Freddi et al. [54], in their study with a latosol (oxisol), found that soil density had a positive influence on the soil water content due to reduced macroporosity and the redistribution of pore sizes, corroborating the results achieved for the same soil class in this study (Figure 3a). The hydraulic functionality differs for soils with different textures, with microporosity playing an important role in this parameter due to the presence of materials with fine textures, e.g., clay and SOM, showing higher retention capacity when the soil is moistened [55,56].

With regard to the local climate, there are two well-defined periods in the region: dry and rainy [57]. In the investigation of Q, the water content was at low levels (U < 10% in all soil layers) due to the study being carried out in the dry period. Thus, the soil water content is reduced by the climatic characteristics of the semi-arid region, which has high temperatures and evaporation rates, and by the inherent characteristics of the soils. This directly influences the gas exchange dynamics and favors the cohesion forces provided by the clay fraction (Table 3), reducing the infiltration rate (Figure 3b), and increasing the Q, according to the study by Souza et al. [58]. Otherwise, the increase in the soil water content reduces cohesion, compactness, and the soil shear resistance [59].

The greater infiltration shown by Area 3 occurred due to the predominance of the sand fraction and greater macroporosity resulting from intensive soil preparation, as verified in the study carried out by Hlaváčiková et al. [60]. The similarity in the proportion of inorganic fractions in the studied layers for areas A1 (Recovery Area) and A2 (Lake Area) explains the proximity of the water infiltration curves (Figure 2a). In turn, the gradual reduction in the infiltration speed seen in Area 4 (Agroecological Area) occurred due to the densification promoted by the clayey soil texture.

The maximum speed observed at the beginning of the test in A2 (Figure 3a) stood out from the other areas due to sediment deposition (Table 1), which facilitated water entry into the soil. Despite this, A3 (Collective Area) showed the highest infiltration rate. The irregularity in the infiltration speed of this area can be attributed to the compactness at the surface due to the transit of animals. In the subsurface, this is explained by the soil management history and the characteristics of the area, due to the conventional intensive preparation and planting of annual crops (Table 1). Practices such as the use of cover plants and the maintenance of plant residues in orchards avoid compaction and favor water infiltration, consequently controlling erosion processes [61,62].

The pH ranged from 6.68 to 7.64 between soil layers, ranging from slightly acidic to neutral (Table 5). These values tending to alkalinity found in the cambisol soil class were due to the limestone material present in the Apodi Plateau region, which is rich in bases such as calcium and magnesium, justifying the observed high values of sum of bases [63]. The pH values in argisol and latosol soils were due to the characteristic climatic pattern of the region (low rainfall), which reduces chemical weathering (Section 2.1).

EC showed values below 4 $dS.m^{-1}$ in all areas, implying low salt concentrations in the soil solution, with low potential risks posed by salinity. According to Richards [64], soils are considered saline only when the electrical conductivity (EC) of the saturation extract is higher than or equal to 4 $dS.m^{-1}$ and when the percentage of exchangeable sodium is lower than 15%. A similar study was conducted by Sparks [65] and Zaman et al. [66], also using the pH (lower than 8.5) and sodium adsorption rate (lower than 13), finding that the studied soil posed no restrictions with regard to salinization and sodification. The most expressive value was observed on the surface layer (0–3 cm) of the A2 area (Lake Area) due to area's position in the landscape (sediment deposition).

The Caatinga Domain, which is representative of the Brazilian semi-arid region, shows low carbon accumulation due to edaphoclimatic conditions, resulting in a reduction in the input of senescent plant material on the soil surface and intense radiation, which favors rapid mineralization as a result of microbial respiration [67,68]. Oliveira et al. [69] observed that the interaction between the semi-arid climate of northeastern Brazil, extensive pasture, and poorly conducted occupation rates have caused soil degradation, reducing the soil contents of nitrogen and carbon. As a result, conservation practices associated with polycultures are important for adding residues to the soil surface and improving structural and chemical attributes in agroecosystems [70], as verified in A4. This was also observed in A2 due to the soil water content, which remains saturated temporarily in the rainy season, favoring the maintenance of TOC.

Ferrari et al. [71] analyzed the spatial variability of soil resistance to penetration in different layers and observed spatial dependence in the first 20 cm of the soil class, with high variability in the reach for different depths, corroborating our study. The authors claim that this was mainly due to variation in TOC. However, variables with strong spatial dependence can also be influenced by intrinsic soil attributes, e.g., texture, in addition to being altered by different soil uses and management and agricultural crops, which contribute to a weak spatial dependence [43]. The degree of spatial dependence was considered moderate or strong for the layers in the three soil classes studied (oxisol, argisol, and cambisol), as observed by Cortez et al. [72] in a latosol (oxisol), by Souza et al. [58] in an argisol, and by Campos et al. [73] in a cambisol.

Alonso et al. [74] highlighted the importance of micrometric and decimetric sampling, which are representative of the structural functionality of the soil, related to the resistance to root penetration. Arshad et al. [75] stressed that the definition of an adequate planning for land use and the adoption of appropriate practices regarding local particularities require the understanding of spatial variability, which is potentiated by the landscape that influences the water dynamics. The higher the length and degree of the slope, the more susceptible the environment to soil and water loss, compromising the production capacity of the soil [76].

Soil attributes with high variability are less accurate and more difficult to manage in specific locations [77]. However, in open systems, it is common and acceptable to find these values, as in the studies carried out by Sağlam and Dengiz and Souza et al. [25,78]. The range parameter showed greater spatial variability in the Collective Area in relation to the Q attribute, corroborating Aquino et al. [79]. The range values obtained for Q in this study are greater than those observed by Campos et al. [73] in a Haplic cambisol in the State of Amazonas. In another study, Lima et al. [80] added that the study of the spatial variability of soil attributes, especially the resistance to root penetration, is important as it directly influences the root development of agricultural crops.

The variograms with medium to strong GD generate maps with a more accurate dependency structure than those with a weak GD [78]. This allows us to infer that the maps prepared show the local reality of the areas. The spatial variability maps of soil resistance to root penetration show that all environments had higher surface Q values (Figures 4 and 5), as observed by Schjønning et al. [10]. In the subsurface, although the Q values were lower than on the surface, these were still high and considered restrictive [12,14,81]. These results are mainly due to the fact that the analysis was carried out in a dry period, in which soils with higher clay content had cohesive particles and provided higher Q values [82]. BD showed slight differences in the layers, except for the sediment deposition area (A2). The larger BD on the A2 surface contributed to the increase in Q, as observed in the study by Wang et al. [13] and Xing et al. [1], in which the Q values ranged from 0.08 to 1.57 MPa when the density increased from 1.01 to 1.43 $Mg.m^{-3}$. The subsurface Q values can be explained by the increase in the inorganic particles of silt and clay, which reduced the macropores.

The Q values were high in all layers and land uses, being above the limits established (2 MPa) by Guimarães et al. [83]. The degree of impediment varied between soil layers 'impaired for root penetration' and layers with reduced crop development (Tables 2 and 6, Figures 4 and 5), whereas the level of compaction ranged from moderate to high. Thus, the studied soils are restrictive to root growth [13], requiring adequate management for the development of root systems.

The Lake Area (A2) was the agroecosystem with the highest surface Q value, followed by the Agroecological (A4), Recovery (A1), and Collective (A3) areas. Souza et al. [58] also reported an increase in surface Q values in the dry period in cambisols in the semi-arid region of Brazil. In their study, the authors stressed that two conditions limit the growth of the root system: low soil water contents and rapid Q increases. These prevent roots from exploring deep layers. Thus, the cambisol areas in the study only have the potential for short-cycle crops, as seen in the study of Mota et al. [84].

Therefore, the spatial variability of Q occurred as a function of the textural and structural variation of the soil and the management adopted in the areas. Some agroecosystems (A1 and A3) had a history of intensive machinery use in the past (Table 1), and despite the care taken with the conservation of the areas, they have not yet had time to recover their structural condition based on the high values observed for Q. Mohieddinne et al. [85] highlighted the average duration of recovery for clayey soils (54 years), acidic sandy soils—

Podzol (70 years), and neutral sandy soils (20 years), whereas Schäffer et al. [86] evaluated a time period of almost four decades for the recovery of silty soils.

From this perspective, the authors mentioned before show that soil recovery is also associated with biological activity in the soil (presence of organisms such as earthworms) and crops with an aggressive pivoting root system to disrupt dense layers [87,88]. Due to changes in soil attributes arising from inadequate management, which compromises the soil's production capacity [89], the conservation of agricultural lands is the main solution to guarantee ecological stability [90]. Socio-ecological principles should guide the planning of integrated approaches between the agricultural suitability of lands and appropriate and sustainable supportive conservation practices in order to enhance land potential and mitigate climate change and biodiversity loss [91,92].

The negative correlation between clay and Q with the sand fraction (Table 7) is justified by the distinct nature of these fractions. The sand fraction has a higher proportion of macropores and a smaller proportion of micropores compared to clay, whereas the clay fraction has electrical charges that provide physicochemical phenomena such as flocculation and particle aggregation [93]. This distinction influences the porous arrangement of the soil and Q [94], and the negative correlation of Q with clay allows us to infer that soils with higher clay contents are more sensitive to compaction [95].

The negative correlation of clay with TOC (Table 7) indicates that the maintenance of organic matter in sandy soils is important because these soils have less natural fertility in relation to clayey ones. Thus, the land cover improves the physical and structural attributes, especially in semi-arid soils where organic matter is more easily decomposed due to weather patterns [17,96,97].

The negative correlation of TOC with Q was due to the fact that soils with higher TOC contents are less dense and structured (Table 7), facilitating the development and penetration of the root system into the soil. In this study, this effect was observed on the surface layer of the Agroecological Area, mainly due to the maintenance of the soil cover and less disturbance in relation to other agroecosystems. Marinho et al. [96] added that soil matter is essential in the maintenance and preservation of agroecosystems. Carus et al. [97] and Kosmallaa et al. [98] pointed out that a higher density of vegetation cover mitigates the shear resistance of the soil and controls active agents in erosion processes.

Gabriel et al. [99] stated that TOC can be maintained in the soil through the use of cover provided by the polyculture practice, which helps maintain the soil water content and contributes to reducing the Q. Koudahe et al. and Mondal et al. [62,100] stressed that conservation practices such as the use of cover crops and lighter agricultural machinery tend to improve the physical condition of the soil, reducing compactness and improving root growth, thus corroborating the data obtained in A4. Furthermore, studies that investigated the benefits caused by biomass incorporation into the soil have disclosed positive results between agricultural practices and water retention and carbon sequestration by the soil, in addition to promoting improvements in the chemical, physical, and structural attributes [101,102].

In the cluster analysis, Group I was formed by physical and structural variables that expose surface phenomena, e.g., the accentuated presence of TOC, especially in A2 (CAMBISOL) and A4 (ARGISOL), which provide improvements in other physical attributes, e.g., TP and the maintenance of U (Figure 6). In the case of chemical attributes, the pH, SB, and EC variables are specifically associated with the cambisol class (representative of the Apodi Plateau region). These soils are derived from the limestone rock of the Jandaíra formation. They are rich in exchangeable bases, e.g., calcium and magnesium, which raise the pH and EC of the soil, making it alkaline [63,103]. Furthermore, the climate pattern of the semi-arid region contributes to the permanence of bases in the system [104].

Group II discriminates the Q variable, linking it with the subsurface of agroecosystems. Q showed values above the established limits (2 MPa or 2000 kPa) in all soil layers. Benevenute et al. [9] and Lima et al. [105] reported that the pressure from the passage of machinery associated with agricultural implements used for sowing, cultivation, and harvesting results in increased resistance to root penetration and soil degradation on the surface and subsurface layers. This action causes a rearrangement and then the packing of clay particles, which raises the Q.

Vaz and collaborators obtained a positive correlation of clay on Q increase in their study with Brazilian latosols [106]. The authors reported that clay values above 35% raised the Q parameter. On the other hand, Sobucki et al. [107] reported the interrelationships of soil attributes that interfere with critical Q values, e.g., clay, soil density, mineralogy, and total organic carbon. This is attributed to the arrangement of clay particles in the subsurface compared to the arrangement of other inorganic soil fractions (silt and sand). Thus, the damage caused to the functionality of the porous network by inadequate management alters the physical attributes of the soil [94], compromising its drainage [60,108,109].

The argisol had TOC, SB, silt, and EC as discriminant variables (Figure 7). The high levels of TOC on the surface came from the addition of plant residues. This occurred even under a semi-arid climate, which has little primary biomass, and weather conditions accelerate the process of plant decomposition. Similar results were achieved by Sousa et al. [110], Singh et al. [111], and Sulieman et al. [112]. The weather pattern was also responsible for the EC values on the surface, but with no restrictions on salinity and sodium concentrations, according to Santos [30]. The silt fraction showed higher surface values and an intermediate degree of pedogenetic development. The study carried out by Rêgo et al. [113] corroborates the results pointed out in this study.

The main limitations refer to the resistance to root penetration into the soil, a parameter used to estimate the mechanical impediment that the soil provides to roots and is a physical attribute highly related to plant growth (compaction), negatively interfering with root growth and consequently affecting the natural development of plants.

The scientific merit of this study allows for an integrative and multidisciplinary understanding of the factors involved and their interrelationships with field and laboratory information, being perfectly reproducible in strategic areas on both regional and global scales. Furthermore, this study encourages new conservation practices and actions that complement other areas of soil science.

This study also encourages the establishment of a soil science database containing information regarding physical, structural, and chemical properties, as well as integrating geostatistical tools in the strategic areas of food production and the conservation of natural resources. The main practical implications of this study refer to decision-making regarding the best manner of using and managing natural resources in different environments while assessing their potentials and limitations.

5. Conclusions

The study evaluated the interrelationships between the water–structural and chemical attributes of soils and the properties of soil resistance to root penetration and water infiltration in areas with four land uses: Recovery (LATOSOL); Lake (CAMBISOL); Collective use (CAMBISOL); and Agroecological use (ARGISOL) in the semi-arid region of northeastern Brazil. Our results suggest that geostatistics through kriging complemented the multivariate analysis, reinforcing the accuracy of our findings, especially with regard to soil resistance to root penetration.

There was variability in our findings, with restrictions regarding resistance to root penetration and water infiltration into the soil, mainly in the Lake Area (CAMBISOL) at a lower elevation, with the analysis successfully discriminating the clay texture of the soil, the density, and the total porosity, factors associated with the water deficit in the region and contributing to restrictions regarding water–structural attributes.

The Agroecological Area (ARGISOL) was the only land use in which soil resistance to root penetration was not a discriminating variable through the multivariate analysis due to the high value of total organic carbon resulting from the conservation practices carried out in the area. However, the geostatistical analysis using kriging identified values of resistance to root penetration above the standard limit for crop development (2 MPa) in all land uses studied, demonstrating the importance of the complementary tool and corroborating one of the hypotheses of the study.

In the Collective Area (CAMBISOL) and Recovery Area (LATOSOL), the history of land uses with intensive soil preparation, involving compaction caused by animals and agricultural machinery traffic, contributed to water–structural restrictions (resistance to root penetration and water infiltration into the soil), compromising the porous arrangement.

In general, the results indicate variations in soil resistance as a function of physicochemical attributes (texture, total organic carbon, and sum of bases), with higher critical values of Q in the Lake Area (4.7 MPa) compared to the Collective Area (3.8 MPa) on the surface. Thus, a soil management plan with mitigating actions is suggested to conserve and recover the areas.

Among the existing conservation practices, minimum preparation, maintenance of vegetation cover, and the maintenance of biological diversity are alternatives that can be highlighted for the studied agroecosystems. These measures are necessary in order not to compromise the porous arrangement of the soil and allow the adequate growth and development of the plant root system.

Author Contributions: P.M.S.D.: conceptualization, data curation, formal analysis, investigation, methodology, visualization, validation, project administration, writing—original draft. J.C.P.: conceptualization, investigation, data curation, methodology, visualization, validation, project administration, resources, supervision, writing—original draft. J.E.F.G.: data curation, formal analysis, investigation, software, writing—original draft, visualization, validation, writing—review and editing. R.O.B.: conceptualization, investigation, data curation, methodology, visualization, validation, project administration, resources, supervision, writing—original draft. L.S.R.: methodology, visualization, validation, resources, supervision, writing—original draft. J.L.F.M.: data curation, methodology, visualization, validation, writing—review and editing. P.K.P.F.: data curation, methodology, visualization, validation, resources, writing—review and editing. P.J.M.: data curation, formal analysis, investigation, software, visualization, validation, writing—original draft, writing—review and editing. D.J.d.C.B.: data curation, formal analysis, investigation, writing—review and editing. L.C.d.A.L.F.: data curation, methodology, visualization, validation, writing—review and editing. G.X.d.M.: visualization, validation, data curation, formal analysis. F.d.O.M.: methodology, visualization, validation, writing—review and editing. F.O.d.O.: methodology, visualization, validation, resources, writing—review and editing. A.G.R.A.: data curation, formal analysis, investigation. C.M.d.N.: visualization, validation, data curation. All authors have read and agreed to the published version of the manuscript.

Funding: This research was financed by the CAPES institutional support program—coordination for the improvement of higher education personnel.

Institutional Review Board Statement: Not applicable.

Data Availability Statement: The datasets generated during and/or analyzed during the current study are available from the corresponding author on reasonable request.

Acknowledgments: We thank Pro Rectory Research and Graduate Studies (PROPPG), the Graduate Program in Soil and Water Management (PPGMSA) of the Federal Rural University of the Semi-Arid (UFERSA) and the 'Coordenação de Aperfeiçoamento de Pessoal de Nível Superior' (CAPES), for the financial support in order to carry out the research.

Conflicts of Interest: The authors declare no conflict of interest.

References

1. Xing, S.; Zhang, G.; Zhu, P.; Wang, L.; Wang, Z.; Wang, C. Variation in shear strength of soil-root system under five typical land use types on the Loess Plateau of China. *CATENA* **2023**, *222*, 106883. [CrossRef]
2. Gürsoy, S.; Türk, Z. Effects of land rolling on soil properties and plant growth in chickpea production. *Soil Tillage Res.* **2019**, *195*, 104425. [CrossRef]
3. Chen, S.; Zhang, G.; Zhu, P.; Wang, C.; Wan, Y. Impact of land use type on soil erodibility in a small watershed of rolling hill northeast China. *Soil Tillage Res.* **2023**, *227*, 105597. [CrossRef]
4. Cui, Z.; Huang, Z.; Luo, J.; Qiu, K.; Vicente, M.L.; Wu, G.L. Litter cover breaks soil water repellency of biocrusts, enhancing initial soil water infiltration and content in a semi-arid sandy land. *Agric. Water Manag.* **2021**, *255*, 107009. [CrossRef]

5. Liu, Z.; Ma, D.; Hu, W.; Li, X. Land use dependent variation of soil water infiltration characteristics and their scale-specific controls. *Soil Tillage Res.* **2018**, *178*, 139–149. [CrossRef]
6. Basset, C.; Najm, M.A.; Ghezzehei, T.; Hao, X.; Daccache, A. How does soil structure affect water infiltration? A meta-data systematic review. *Soil Tillage Res.* **2023**, *226*, 105577. [CrossRef]
7. Zhang, Y.; Hartermink, A.E.; Huang, J.; Minasny, B. Digital Soil Morphometrics. In *Encyclopedia of Soils in the Environment*; Academic Press: Cambridge, MA, USA, 2023.
8. Dexter, A.R. Advances in Characterization of Soil Structure. *Soil Tillage Res.* **1988**, *11*, 199–238. [CrossRef]
9. Benevenute, P.A.N.; de Morais, E.G.; Souza, A.A.; Vasques, I.C.F.; Cardoso, D.P.; Sales, F.R.; Severiano, E.C.; Homem, B.G.C.; Casagrande, D.R.; Silva, B.M. Penetration Resistance: An Effective Indicator for Monitoring Soil Compaction in Pastures. *Ecol. Indic.* **2020**, *117*, 106647. [CrossRef]
10. Schjønning, P.; Lamandé, M.; Munkholm, L.J.; Lyngvig, H.S.; Nielsen, J.A. Soil Precompression Stress, Penetration Resistance and Crop Yields in Relation to Differently-Trafficked, Temperate-Region Sandy Loam Soils. *Soil Tillage Res.* **2016**, *163*, 298–308. [CrossRef]
11. da Silva, W.M.; Bianchini, A.; da Cunha, C.A. Modeling and Correction of Soil Penetration Resistance for Variations in Soil Moisture and Soil Bulk Density. *Eng. Agrícola* **2016**, *36*, 449–459. [CrossRef]
12. Vaz, C.M.P.; Manieri, J.M.; de Maria, I.C.; Van Genuchten, M. Scaling the Dependency of Soil Penetration Resistance on Water Content and Bulk Density of Different Soils. *Soil Sci. Soc. Am. J.* **2013**, *77*, 1488–1495. [CrossRef]
13. Wang, H.; Wang, L.; Huang, X.; Gao, W.; Ren, T. An empirical model for estimating soil penetrometer resistance from relative bulk density, matric potential, and depth. *Soil Tillage Res.* **2021**, *208*, 104904. [CrossRef]
14. Catania, P.; Badalucco, L.; Laudicina, V.A.; Vallone, M. Effects of tilling methods on soil penetration resistance, organic carbon and water stable aggregates in a vineyard of semiarid Mediterranean environment. *Environ. Earth Sci.* **2018**, *77*. [CrossRef]
15. Al-Qinna, M.; Abu-Awwad, A. Infiltration Rate Measurements in Arid Soils 706 with surface crust. *Irrig. Sci.* **1998**, *18*, 83–89. [CrossRef]
16. Cui, Z.; Wu, G.L.; Huang, Z.; Liu, Y. Fine roots determine soil infiltration potential than soil water content in semi-arid grassland soils. *J. Hydrol.* **2019**, *578*, 124023. [CrossRef]
17. Li, J.; Wang, W.; Guo, M.; Kang, H.; Wang, Z.; Huang, J.; Sun, B.; Wang, K.; Zhang, G.; Bai, Y. Effects of soil texture and gravel content on the infiltration and soil loss of spoil heaps under simulated rainfall. *J. Soils Sediments* **2020**, *20*, 3896–3908. [CrossRef]
18. Liu, Y.; Guo, L.; Huang, Z.; López-Vicente, M.; Wu, G.L. Root morphological characteristics and soil water infiltration capacity in semi-arid artificial grassland soils. *Agric. Water Manag.* **2020**, *235*, 106153. [CrossRef]
19. Oda, M.; Rasyid, B.; Omae, H. Root Mass May Affect Soil Water Infiltration More Strongly than the Incorporated Residue. *F1000Res* **2020**, *7*, 1523. [CrossRef]
20. Stone, L.R.; Schlegel, A.J. Tillage and Crop Rotation Phase Effects on Soil Physical Properties in the West-Central Great Plains. *Agron. J.* **2010**, *102*, 483–491. [CrossRef]
21. Mendes, K.R.; Portela, J.C.; Gondim, J.E.F.; Ribeiro, M.A.; Medeiros, J.F.; Queiroz, G.C.M. Physical, chemical and structural attributes of soil in agroecosystems in the Brazilian Semi-arid region. *Rev. Ciência Agronômica* **2022**, *53*, 1–10. [CrossRef]
22. Gan, F.; Shi, H.; Gou, J.; Zhang, L.; Liu, C. Effects of bedrock strata dip on soil infiltration capacity under different land use types in a karst trough valley of Southwest China. *CATENA* **2023**, *230*, 107253. [CrossRef]
23. Nuralykyzy, B.; Nurzhan, A.; Li, N.; Huang, Q.; Zhu, Z.; An, S. Influence of land use types on soil carbon fractions in the Qaidam Basin of the Qinghai-Tibet Plateau. *CATENA* **2023**, *231*, 107273. [CrossRef]
24. Júnior, V.V.; Carvalho, M.P.; Dafonte, J.; Freddi, O.S.; Vidal Vázquez, E.; Ingaramo, O.E. Spatial Variability of Soil Water Content and Mechanical Resistance of Brazilian Ferralsol. *Soil Tillage Res.* **2006**, *85*, 166–177. [CrossRef]
25. Sağlam, M.; Dengiz, O. Spatial variability of soil penetration resistance in na alluvial delta plain under different land uses in middle Black Sea Region of Turkey. *Arch. Agron. Soil Sci.* **2017**, *63*, 60–73. [CrossRef]
26. Souza, M.D.C.; de Oliveira, F.P.; Silva, J.R.; Martins, A.F.; da Silva, P.L.F. Spatial variability of resistance to penetration in soil under sugarcane crops with different harvest Methods. *Rev. Caatinga* **2020**, *33*, 480–489. [CrossRef]
27. Mota, J.C.A.; Assis, J.; Amaro, F.J.; Libardi, P.L. Some physical and water properties of three soils in Chapada do Apodi, RN, cultivated with melon. *Rev. Bras. Ciência Solo* **2008**, *32*, 49–58. [CrossRef]
28. Dubreuil, V.; Fante, K.P.; Planchon, O.; Neto, J.L.S. Os Tipos de Climas Anuais No Brasil: Uma Aplicação Da Classificação de Köppen de 1961 a 2015. *Rev. Fr.-Bras. Geogr.* **2018**, *37*. [CrossRef]
29. IUSS Working Group. *World Reference Base for Soil Resources 2014 International Soil Classification System for Naming Soils and Creating Legends for Soil Maps*; FAO—Food and Agriculture Organization of the United Nations: Rome, Italy, 2014; p. 192.
30. Dos Santos, H.G.; Jacomine, P.K.T.; Dos Anjos, L.H.C.; Oliveira, V.A.; Lumbreras, J.F.; Coelho, M.R.; Almeida, J.A.; Filho, J.C.A.; Oliveira, J.B.; Cunha, T.J.F. *Sistema Brasileiro de Classificação de Solos (SiBCS)*, 5th ed.; EMBRAPA: Brasília, Brazil, 2018; p. 356.
31. *ASAE Standard: ASAE EP542*; AMERICAN SOCIETY OF AGRICULTURAL ENGINEERS. Procedures for using and reporting data obtained with the soil cone penetrometer. St Joseph. ASAE: Washington, DC, USA; p. 9646.
32. Bernardo, S.; Soares., A.A.; Mantovani, E.C. *Manual de Irrigação*, 8th ed.; Editora UFV: Viçosa, Brazil, 2008; p. 622.
33. Teixeira, P.C.; Donagemma, G.K.; Fontana, A.; Teixeira, W.G. *Manual de Métodos de Análise de Solo*, 3rd ed.; Embrapa Solos: Brasília, Brazil, 2017.

34. Gee, G.W.; Or, D. Particle Size Analysis. In *Methods of Soil Analysis: Part 4 Physical Methods*, 1st ed.; Dane, J.H., Topp, G., Eds.; Soil Science Society of America, Inc.: Madison, WI, USA, 2002; pp. 255–293. [CrossRef]
35. Blake, G.R.; Hartge, K.H. Bulk Density. In *Methods of Soil Analysis: Part 1 Physical and Mineralogical Methods*, 2nd ed.; Klute, A., Ed.; American Society of Agronomy, Inc.: Madison, WI, USA; Soil Science Society of America, Inc.: Madison, WI, USA, 1986; pp. 363–375. [CrossRef]
36. Kiehl, J. *Manual de Edafologia-Relações Solo-Planta*; Agronômica Ceres: São Paulo, Brazil, 1979; p. 272.
37. Yeomans, J.C.; Bremner, J.M. A rapid and precise method for routine determination of organic carbon in soil. *Commun. Soil Sci. Plant Anal.* **2008**, *19*, 1467–1476. [CrossRef]
38. Minasny, B.; Mcbratney, A.B.; Whelan, B.M. *Vesper Version 1.62*; Centre for Precision Agriculture: Sydney, Australia, 2005.
39. Sherman, L.W.; Strang, H.; Barnes, G.; Woods, D.J.; Bennett, S.; Inkpen, N.; Newbury-Birch, D.; Rossner, M.; Angel, C.; Mearns, M.; et al. Twelve experiments in restorative justice: The Jerry Lee program of randomized trials of restorative justice conferences. *J. Exp. Criminol.* **2015**, *11*, 501–540. [CrossRef]
40. Couto, E.G.; Klamt, E.; Stein, A. Estimativa do teor de argila e do potássio trocável em solos esparsamente amostrados no sul do estado de Mato Grosso. *Rev. Bras. Ciência Solo* **2000**, *24*, 129–140. [CrossRef]
41. Akaike, H. Information Theory and an Extension of the Maximum Likelihood Principle. In *Selected Papers of Hirotugu Akaike*, 1st ed.; Parzen, E., Tanabe, K., Kitagawa, G., Eds.; Springer—International Publisher: New York, NY, USA, 1997; pp. 199–213.
42. Ribeiro, C.A. Variabilidade Espacial da Resistência Mecânica do solo à Penetração em áreas Mecanizadas de cana-de-açúcar (Saccharum Officinarum). xiv, 69 f. Master's Thesis, São Paulo State University, Faculty of agricultural and veterinary sciences, Jaboticabal, Brazil, 2010.
43. Cambardella, C.A.; Moorman, T.B.; Novak, J.M.; Parkin, T.B.; Karlen, D.L.; Turco, R.F.; Konopka, A.E. Field-Scale Variability of Soil Properties in Central Iowa Soils. *Soil. Sci. Soc. Am. J.* **1994**, *58*, 1501–1511. [CrossRef]
44. Statsoft, I.N.C. STATISTICA (sistema de software de análise de dados). *Versão* **2004**, *7*, 1984–2004.
45. Hair, J.F.; Hult, G.T.M.; Ringle, C.M.; Sarstedt, M.A. *Primer on Partial Least Squares Structural Equation Modeling (PLS-SEM)*, 2nd ed.; SAGE Publications, Inc.: Thousand Oaks, CA, USA, 2017; p. 328.
46. Dalchiavon, F.C.; de Carvalho, M.P.E.; Andreotti, M.; Montanari, R. Variabilidade espacial de atributos da fertilidade de um Latossolo Vermelho Distroférrico sob Sistema Plantio Direto. *Rev. Ciência Agronômica* **2012**, *43*, 453–461. [CrossRef]
47. Oliveira Junior, J.C.; de Souza, L.C.P.; de Melo, V.F.; Rocha, H.O. Variabilidade espacial de atributos mineralógicos de solos da formação Guabirotuba, 907 Curitiba (PR). *Rev. Bras. Ciência Solo* **2011**, *35*, 1481–1490. [CrossRef]
48. Lemos Filho, L.C.A.; de Oliveira, E.L.; de Faria, M.A.; de Andrade, L.A.B. Spatial variability of soil density and organic matter content in area cultivated with sugar cane (Saccharum officinarum L.) Variação espacial da densidade do solo e matéria orgânica em área cultivada com cana-de-açúcar (*Saccharum officinarum* L.). *Rev. Cienc. Agrônomica* **2008**, *39*, 193–202.
49. Quan, X.; He, J.; Cai, Q.; Sun, L.; Li, X.; Wang, S. Soil erosion and deposition characteristics of slope surfaces for two loess soils using indoor simulated rainfall experiment. *Soil Tillage Res.* **2020**, *204*, 104714. [CrossRef]
50. Kebebew, S.; Bedadi, B.; Erkossa, T.; Yimer, F.; Wogi, L. Effect of Different Land-Use Types on Soil Properties in Cheha District, South-Central Ethiopia. *Sustainability* **2022**, *14*, 1323. [CrossRef]
51. Topa, D.; Cara, I.G.; Jităreanu, G. Long term impact of different tillage systems on carbon pools and stocks, soil bulk density, aggregation and nutrients: A field meta-analysis. *Catena* **2021**, *199*, 105102. [CrossRef]
52. Fernández, R.; Frasier, I.; Quiroga, A.; Noellemeye, E. Pore morphology reveals interaction of biological and physical processes for structure formation in soils of 782 the semi-arid Argentinean Pampa. *Soil Tillage Res.* **2019**, *191*, 256–265. [CrossRef]
53. Da Costa, J.D.; Portela, J.C.; Farias, P.K.P.; Sobrinho, F.E.; Souza, C.M.M.; de Lopes, T.C.S.; Silva, F.W.A. Characterization and Classification of Soils of the Terra da Esperança Settlement Project in Chapada do Apodi, Brazil. *J. Agric. Sci.* **2019**, *11*, 235. [CrossRef]
54. da Freddi, O.S.; Centurion, J.F.; Duarte, A.P.; Leonel, C.L. Compactação do solo e produção de cultivares de milho em latossolo vermelho: I—Características de planta, solo e índice S. *Rev. Bras. Ciência Solo* **2009**, *33*, 793–803. [CrossRef]
55. Oro, L.A.; Colazo, J.C.; Avecilla, F.; Buschiazzo, D.E.; Asensio, C. Relative soil water content as a factor for wind erodibility in soils with different texture and aggregation. *Aeolian Res.* **2019**, *37*, 25–31. [CrossRef]
56. Yang, C.; Wu, J.; Li, P.; Wang, Y.; Yang, N. Evaluation of soil-water characteristic curves for different textural soils using fractal analysis. *Water* **2023**, *15*, 772. [CrossRef]
57. da Corado Neto, F.C.; de Sampaio, F.M.T.; da Veloso, M.E.C.; Matias, S.S.R.; Andrade, F.R.; Lobato, M.G.R. Variabilidade Espacial da Resistência à Penetração em Neossolo Litólico Degradado. *Rev. Bras. Ciência Solo* **2015**, *39*, 1353–1361. [CrossRef]
58. Souza, R.; Hartzell, S.; Freire Ferraz, A.P.; de Almeida, A.Q.; de Sousa Lima, J.R.; Dantas Antonino, A.C.; de Souza, E.S. Dynamics of soil penetration resistance in water-controlled environments. *Soil Tillage Res.* **2021**, *205*, 104768. [CrossRef]
59. Sion, B.D.; Shoop, S.A.; McDonald, E.V. Evaluation of in-situ relationships between variable soil moisture and soil strength using a plot-scale experimental design. *J. Terramech.* **2022**, *103*, 33–51. [CrossRef]
60. Hlaváčiková, H.; Holko, L.; Danko, M.; Novák, V. Estimation of macropore flow characteristics in stony soils of a small mountain catchment. *J. Hydrol.* **2019**, *574*, 1176–1187. [CrossRef]
61. Ramos, M.F.; da Silva Almeida, W.R.; do Amaral, R.D.L.; Suzuki, L.E.A.S. Degree of compactness and soil quality of peach orchards with different production ages. *Soil Tillage Res.* **2022**, *219*, 105324. [CrossRef]

62. Koudahe, K.; Allen, S.C.; Djaman, K. Critical review of the impact of cover crops on soil properties. *Int. Soil Water Conserv. Res.* **2022**, *10*, 343–354. [CrossRef]
63. Paiva, C.A.; de Medeiros, J.F.; de Almeida, J.G.L.; Aroucha, E.M.M. Desempenho do melão Gália cultivado sob diferentes relações catiônicas do solo. *Semin. Ciências Agrárias* **2021**, *42*, 3089–3106. [CrossRef]
64. Allison, L.E.; Bernstein, L.; Bower, C.A.; Brown, J.W.; Fireman, M.; Hatcher, J.T.; Hayward, H.E.; Pearson, G.A.; Reeve, R.C.; Richards, L.A.; et al. *Diagnosis and Improvement of Saline and Alkaline Soils*; US Department of Agriculture: Washington, DC, USA, 1954; pp. 1–147.
65. Sparks, D.L. *Environmental Soil Chemistry*, 2nd ed.; Academic Press: San Diego, CA, USA, 2003; p. 352.
66. Zaman, M.; Shahid, S.A.; Heng, L. *Guideline for Salinity Assessment, Mitigation and Adaptation Using Nuclear and Related Techniques*, 1st ed.; Springer International Publishing: Cham, Switzerland, 2018; p. 183. [CrossRef]
67. Zhao, C.; Miao, Y.; Yu, C.; Zhu, L.; Wang, F.; Jiang, L.; Hui, D.; Wan, S. Soil microbial community composition and respiration along an experimental precipitation gradient in a semi-arid steppe. *Sci. Rep.* **2016**, *6*, 24317. [CrossRef] [PubMed]
68. Guerra, M.D.F.; de Souza, J.J.L.L.; Schaefer, C.E.G.R.; de Souza, M.J.N. Remnant wetlands under palm swamps in the Araripe Plateau, Brazilian semi-arid. *Catena* **2023**, *226*, 107074. [CrossRef]
69. de Souza Oliveira Filho, J.; da Silva Barrozo, M.V.; de Holanda Bastos, F.; Pereira, M.G. Effects of climate and land-use on physicochemical properties of gneiss-derived soils in tropical drylands: Evidence from northeastern Brazil. *Geoderma Reg.* **2023**, *33*, e00639. [CrossRef]
70. Zhang, Z.; Qiang, H.; McHugh, A.D.; He, J.; Li, H.; Wang, Q.; Lu, Z. Effect of conservation farming practices on soil organic matter and stratification in a mono- cropping system of Northern China. *Soil Tillage Res.* **2016**, *156*, 173–181. [CrossRef]
71. Ferrari, J.M.S.; Gabriel, C.P.C.; Silva, T.B.G.; Mota, F.D.; Gabriel Filho, L.R.A.; Tanaka, E.M. Análise da variabilidade espacial da resistência à penetração do solo em diferentes profundidades/analysis of the spatial variability of resistance to soil penetration at diferente depths. *Rev. Bras. Eng. Biossistemas* **2018**, *12*, 164. [CrossRef]
72. Cortez, J.W.; Cavassini, V.H.; de Motomiya, A.V.A.; Orlando, R.C.; Valente, I.Q.M. Spatialization of soil resistance to penetration for localized management by precision agriculture Tools. *Eng. Agrícola* **2018**, *38*, 690–696. [CrossRef]
73. Campos, M.C.C.; de Oliveira, I.A.; dos Santos, L.A.C.; Aquino, R.E.; Soares, M.R.D. Variabilidade espacial da resistência do solo à penetração e umidade em áreas 738 cultivadas com mandioca na região de Humaitá, AM. *Rev. Agro@Mbiente Line* **2012**, *6*, 9. [CrossRef]
74. Alonso, A.; Froidevaux, M.; Javaux, M.; Laloy, E.; Mattern, S.; Roisin, C.; Vanclooster, M.; Bielders, C. A hybrid method for characterizing tillage-induced soil physical quality at the profile scale with fine spatial details. *Soi Tillage Res.* **2022**, *216*, 00124. [CrossRef]
75. Arshad, M.; Ranamukhaarachchi, S.L.; Ahmad, S.; Nawaz, R.; Qayyum, M.M.N.; Razaq, A.; Faiz, F. Variability and correlation of selected soil attributes and maize yield influenced by tillage systems in mountainous agroecosystem. *J. Soil Water Conserv.* **2022**, *77*, 466–481. [CrossRef]
76. Bettoni, M.; Maerker, M.; Bosino, A.; Conedera, M.; Simoncelli, L.; Vogel, S. Land use effects on surface runoff and soil erosion in a southern Alpine valley. *Geoderma* **2023**, *435*, 116505. [CrossRef]
77. Schmidt, J.P.; Taylor, R.K.; Milliken, G.A. Evaluating the Potential for Site-Specific Phosphorus Applications Without High-Density Soil Sampling. *Soil Sci. Soc. Am. J.* **2002**, *66*, 276–283. [CrossRef]
78. de Souza, Z.M.; de Souza, G.S.; Marques Júnior, J.; Pereira, G.T. Número de amostras na análise geoestatística e na krigagem de mapas de atributos do solo. *Ciência Rural* **2014**, *44*, 261–268. [CrossRef]
79. Aquino, R.E.; Campos, M.C.C.; Oliveira, I.A.; Siqueira, D.S.; Soares, M.D.R.; Freitas, L. Técnicas geoestatísticas na avaliação de atributos químicos em Cambissolo com agrofloresta e cana-de-açúcar em Humaitá, Amazonas. *Brazilian J. Agric. Sci.* **2015**, *10*, 544–552. [CrossRef]
80. de Lima, C.L.R.; da Silva, A.P.; Imhoff, S.; Leão, T.P. Estimativa da capacidade de suporte de carga do solo a partir da avaliação da resistência à penetração. *Rev. Bras. Ciência Solo* **2006**, *30*, 217–223. [CrossRef]
81. Gao, W.; Watts, C.W.; Ren, T.; Whalley, W.R. The effects of compaction and soil drying on penetrometer resistance. *Soil Tillage Res.* **2012**, *125*, 14–22. [CrossRef]
82. Tang, C.S.; Gong, X.P.; Shen, Z.; Cheng, Q.; Inyang, H.; Lv, C.; Shi, B. Soil micro-penetration resistance as an index of its infiltration processes during rainfall. *J. Rock Mech. Geotech. Eng.* **2022**, *14*, 1580–1587. [CrossRef]
83. Guimarães, R.M.L.; Ball, B.C.; Tormena, C.A.; Giarola, N.F.B.; da Silva, Á.P. Relating visual evaluation of soil structure to other physical properties in soils of contrasting texture and management. *Soil Tillage Res.* **2013**, *127*, 92–99. [CrossRef]
84. Mota, J.C.A.; Assis, J.R.N.; Amaro F., J.; Libardi, P.L. Algumas propriedades físicas e hídricas de três solos na Chapada do Apodi, RN, cultivados com melão. *Rev. Bras. Ciência Solo* **2008**, *32*, 49–58. [CrossRef]
85. Mohieddinne, H.; Brasseur, B.; Spicher, F.; Gallet-Moron, E.; Buridant, J.; Kobaissi, A.; Horen, H. Physical recovery of forest soil after compaction by heavy machines, revealed by penetration resistance over multiple decades. *For. Ecol. Manag.* **2019**, *449*, 117472. [CrossRef]
86. Schäffer, J. Recovery of soil structure and fine root distribution in compacted forest soils. *Soil Syst.* **2022**, *6*, 49. [CrossRef]
87. Ambus, J.V.; Reichert, J.M.; Gubiani, P.I.; de Faccio Carvalho, P.C. Changes incomposition and functional soil properties in long-term no-till integrated crop-livestock system. *Geoderma* **2018**, *330*, 232–243. [CrossRef]

88. Beylich, A.; Oberholzer, H.R.; Schrader, S.; Höper, H.; Wilke, B.M. Evaluation of soil compaction effects on soil biota and soil biological processes in soils. *Soil Tillage Res.* **2010**, *109*, 133–143. [CrossRef]
89. Chellappa, J.; Sagar, K.L.; Sekaran, U.; Kumar, S.; Sharma, P. Soil organic carbon, aggregate stability and biochemical activity under tilled and no-tilled agroecosystems. *J. Agric. Res.* **2021**, *4*, 100139. [CrossRef]
90. Jiang, L.; Wang, Z.; Zuo, Q.; Du, H. Simulating the impact of land use change on ecosystem services in agricultural production areas with multiple scenarios considering ecosystem service richness. *J. Clean. Prod.* **2023**, *397*, 136485. [CrossRef]
91. Anusha, B.N.; Babu, K.R.; Kumar, B.P.; Sree, P.P.; Veeraswamy, G.; Swarnapriya, C.; Rajasekhar, M. Integrated studies for land suitability analysis towards sustainable agricultural development in semi-arid regions of AP, India. *Geosyst. Geoenviron.* **2023**, *2*, 100131. [CrossRef]
92. Roy, P.S.; Ramachandran, R.M.; Paul, O.; Thakur, P.K.; Ravan, S.; Behera, M.D.; Sarangi, C.; Kanawade, V.P. Anthropogenic land use and land cover changes—A review on its environmental consequences and climate change. *J. Indian Soc. Remote Sens.* **2022**, *50*, 1615–1640. [CrossRef]
93. dos Santos, P.G.; de Almeida, J.A.; Sequinatto, L. Mineralogy of the Clay Fraction and Chemical Properties of Soils Developed from Sedimentary Lithologies of Pirambóia, Sanga the Cabral and Guará Geological Forma tions in Southern Brazil. *Rev. Bras. Ciência Solo* **2017**, *41*, 1–19. [CrossRef]
94. Reichert, J.M.; da Rosa, V.T.; Vogelmann, E.S.; da Rosa, D.P.; Horn, R.; Reinert, D.J.; Sattler, A.; Denardin, J.E. Conceptual framework for capacity and intensity physical soil properties affected by short and long-term (14 years) continuous no-tillage and controlled traffic. *Soil Tillage Res.* **2016**, *158*, 123–136. [CrossRef]
95. Mileusnić, Z.I.; Saljnikov, E.; Radojević, R.L.; Petrović, D.V. Soil compaction due to agricultural machinery impact. *J. Terramech.* **2022**, *100*, 51–60. [CrossRef]
96. Marinho, A.C.S.; Portela, J.C.; Silva, E.; Silva Dias, N.; Sousa, F.; Silva, A.; Silva, J. Organic matter and physicochemical attributes of a cambisol under different agricultural uses in a semi-arid region of Brazil. *Aust. J. Crop. Sci.* **2016**, *10*, 32–41.
97. Carus, J.; Paul, M.; Schröder, B. Vegetation as self-adaptive coastal protection: Reduction of current velocity and morphologic plasticity of a brackish marsh pioneer. *Ecol. Evol.* **2016**, *6*, 1579–1589. [CrossRef] [PubMed]
98. Kosmalla, V.; Keimer, K.; Schürenkamp, D.; Lojek, O.; Goseberg, N. Erosion resistance of vegetation-covered soils: Impact of different grazing conditions in salt marshes and analysis of soil-vegetation interactions by the novel DiCoastar method. *Ecol. Eng.* **2022**, *181*, 106657. [CrossRef]
99. Gabriel, J.L.; García-González, I.; Quemada, M.; Martin-Lammerding, D.; Alonso-Ayuso, M.; Hontoria, C. Cover crops reduce soil resistance to penetration by preserving soil surface water content. *Geoderma* **2021**, *386*, 114911. [CrossRef]
100. Mondal, S.; Chakraborty, D.; Das, T.K.; Shrivastava, M.; Mishra, A.K.; Bandyopadhyay, K.K.; Aggarwal, P.; Chaudhari, S.K. Conservation agriculture had a strong impact on the sub-surface soil strength and root growth in wheat after a 7-year transition period. *Soil Tillage Res.* **2019**, *195*, 104385. [CrossRef]
101. Vendig, I.; Guzman, A.; De La Cerda, G.; Esquivel, K.; Mayer, A.C.; Ponisio, L.; Bowles, T.M. Quantifying direct yield benefits of soil carbon increases from cover cropping. *Nat. Sustain.* **2023**, 1–10. [CrossRef]
102. Qian, R.; Guo, R.; Naseer, M.A.; Zhang, P.; Chen, X.; Ren, X. Long-term straw incorporation regulates greenhouse gas emissions from biodegradable film farmland, improves ecosystem carbon budget and sustainable maize productivity. *Field Crops Res.* **2023**, *295*, 108890. [CrossRef]
103. Lopes, T.C.d.S.; Portela, J.C.; Batista, R.O.; Bandeira, D.J.d.C.; Leite, I.d.O.; Ramalho, L.B.; Gondim, J.E.F.; Costa, J.D.d.; Gurgel, M.T.; Souza, C.M.M.; et al. Clay Fraction Mineralogy and Structural Soil Attributes of Two Soil Classes under the Semi-Arid Climate of Brazil. *Land* **2022**, *11*, 2192. [CrossRef]
104. Matos, L.V.; Donato, S.L.R.; Kondo, M.K.; Lani, J.L.; Aspiazú, I. Soil attributes and the quality and yield of 'Gigante' cactus pear in agroecosystems of the semi-arid region of Bahia. *J. Arid Environ.* **2021**, *185*, 104325. [CrossRef]
105. Lima, C.C.; De Maria, I.C.; Silva, G.J.W.; Figueiredo, G.C.; Dechen, S.C.F.; Bolonhezi, D. ROOT parameters of sugarcane and soil compaction indicators under deep strip tillage and conventional tillage. *Sci. Rep.* **2022**, *12*, 18537. [CrossRef]
106. Vaz, C.M.P.; Manieri, J.M.; de Maria, I.C.; Tuller, M. Modeling and correction of soil penetration resistance for varying soil water content. *Geoderma* **2011**, *166*, 92–101. [CrossRef]
107. Sobucki, V.; Holthusen, D.; Batistão, A.C.; Mota, J.C.A.; Reichert, J.M. Potential of rheometry in detecting cohesive soils in Brazil as an additional tool to morphological field description and tensile resistance quantification. *Geoderma Reg.* **2022**, *30*, e00553. [CrossRef]
108. Cortez, J.W.; Armando, E.J.; de Motomiya, A.V.A.; Rodrigues, S.A.; Gonçalves, E.A. Soil physical Attributes before and after management under sowing forward speed. *Eng. Agrícola* **2019**, *39*, 607–615. [CrossRef]
109. Hirmas, D.R.; Giménez, D.; Nemes, A.; Kerry, R.; Brunsell, N.A.; Wilson, C.J. Climate-induced changes in continental-scale soil macroporosity may intensify water cycle. *Nature* **2018**, *561*, 100–103. [CrossRef] [PubMed]
110. Sousa, J.E.S.; dos Santos, J.C.B.; Corrêa, M.M.; do Nascimento, A.F.; Schulze, S.M.B.B.; Ferreira, T.O.; de Araújo Filho, J.C.; de Souza Júnior, V.S. Mineralogy and genesis of Planosols under a semi-arid climate, Borborema Plateau, NE Brazil. *CATENA* **2020**, *184*, 104260. [CrossRef]
111. Singh, J.; Ale, S.; DeLaune, P.B.; Barnes, E.M. Simulated effects of cover crops with no-tillage on soil and crop productivity in rainfed semi-arid cotton production systems. *Soil Tillage Res.* **2023**, *230*, 105709. [CrossRef]

112. Sulieman, M.M.; Sallam, A.; Brevik, E.C.; Al-farraj, A.S. Early indicators of pedogenesis at Harrat Khaybar volcano, Saudi Arabia. *Geoderma* **2021**, *383*, 114743. [CrossRef]
113. da Rêgo, L.G.S.; Silva, J.J.A.; Souza, C.M.M.; Portela, J.C.; Moura, I.N.B.M.; Silva, A.C.R.; Miranda, N.O. Pedogenesis in the Barreiras Formation Under Climates of Rio Grande do Norte, Brazil. *J. Agric. Sci.* **2019**, *11*, 19. [CrossRef]

agriculture

Article

Effects of Deep Vertical Rotary Tillage Management Methods on Soil Quality in Saline Cotton Fields in Southern Xinjiang

Zhijie Li [1,2], Hongguang Liu [1,2,*], Haichang Yang [3] and Tangang Wang [4,*]

1 College of Water Conservancy & Architectural Engineering, Shihezi University, Shihezi 832000, China; lzhijie_2023163@163.com
2 Xinjiang Production & Construction Group Key Laboratory of Modern Water-Saving Irrigation, Shihezi 832000, China
3 College of Agronomy, Shihezi University, Shihezi 832000, China; yhc2012@126.com
4 Institute of Agricultural Science of the Third Division of Xinjiang Production and Construction Corps, Tumushuk 843900, China
* Correspondence: liuhongguang-521@163.com (H.L.); ssnkswtg@163.com (T.W.)

Abstract: A long-term high-saline soil environment will limit the improvement of soil quality and cotton yield. Modified tillage management measures can improve soil quality, and the establishment of a soil quality evaluation system will facilitate evaluation of the soil quality and land production potential in southern Xinjiang. The objective of this study was to determine the effects of different tillage management methods on soil quality in saline cotton fields in southern Xinjiang. A three-year experiment was conducted in Tumushuke, Xinjiang, with different deep vertical rotary tillage depths (DTM20, 20 cm; DTM40, 40 cm; DTM60, 60 cm) and conventional tillage (CTM, 20 cm). The soil quality index (SQI) under different tillage management methods was established by using the full dataset (TDS) with a scoring function for eight indicators, including physicochemical properties of the soil from 0 to 60 cm, to evaluate its impact on the soil quality of the saline farmland in southern Xinjiang. The results of the study showed that deep vertical rotary tillage management can effectively optimize soil structure; reduce soil bulk density (BD), soil solution conductivity (EC), and pH; and promote the accumulation of soil organic carbon (SOC) and total nitrogen (TN) in the soil. However, the average diameter of soil water-stable aggregates (MWD) in a 0–60 cm layer becomes smaller with an increasing depth of tillage. This does not reduce crop yields but does promote soil saline leaching. In addition, the significant linear relationship ($p < 0.001$) between seed cotton yield and soil quality indicated that improving soil quality was favorable for crop yield. The principal component analysis revealed BD, MWD, pH, and EC as limiting sensitive indicators for seed cotton yield, while SOC and TN were positive sensitive indicators. The soil quality index (SQI) values of DT40 and DTM60 were significantly higher than that of CTM by 11.02% and 15.27%, respectively. Overall, the results show that DTM60 is the most suitable tillage strategy to improve soil quality and seed cotton yield in this area, and this approach will provide a reliable theoretical basis for the improvement of saline farmland.

Keywords: soil quality; saline farmland; deep vertical rotary tillage; soil tillage management methods; seed cotton yield

Citation: Li, Z.; Liu, H.; Yang, H.; Wang, T. Effects of Deep Vertical Rotary Tillage Management Methods on Soil Quality in Saline Cotton Fields in Southern Xinjiang. *Agriculture* **2023**, *13*, 1864. https://doi.org/10.3390/agriculture13101864

Academic Editors: Anetta Siwik-Ziomek and Anna Figas

Received: 23 August 2023
Revised: 20 September 2023
Accepted: 22 September 2023
Published: 23 September 2023

1. Introduction

Xinjiang in northwestern China is the largest cotton-producing region in the country, with an annual output that can reach 1.8 million t, accounting for 93.6% of China's total production [1]. Cotton production areas located in the southern part of Xinjiang have long been constrained by drought and water scarcity with soil salinization [2]. At the same time, under the long-term cotton cropping and shallow tillage management system based on five-share plowing, soil salinity has been accumulating in the root zone, limiting the improvement of cotton yield [3]. Increasing soil salinity problems and global climate change can lead to degradation of cropland, and rational tillage practices are an effective way to improve soil quality and crop yield potential [4].

Poor soil structure and low soil fertility caused by soil salinization, and high salinity in the soil profile are key factors limiting cotton production in southern Xinjiang [5]. Saline soils can be effectively improved through modified soil tillage management methods to increase soil fertility and improve soil quality [6]. Traditionally, once or twice a year, plowing (spring or autumn plowing) is used in the Xinjiang region to form a loose soil structure, and spring and winter irrigation is used to wash soil saline ions [7]. However, long-term shallow tillage increases soil compaction and shallow tillage layer, which severely limits soil water vapor transmission [8]. Deep tillage is a typical conservation tillage practice aimed at conservation that can achieve soil loosening without disturbing the soil layers at all depths [9]. The improved soil structure and loosened subsoil allow a crop root system to obtain more water and nutrient resources for higher overall crop yields [10–12]. In order to explore the improved soil and tillage management in southern Xinjiang, this work evaluated the use of deep vertical rotary tillage technology. This method can realize different depths of deep tillage by controlling the lifting and lowering of the auger head and rapid rotary grinding and crushing of the soil, which suspends into a ridge and does not disturb the upper soil layer [13]. This tillage method has been widely promoted in China, and it can better harmonize soil water, air, and heat to build a healthy soil tillage layer. Given the water-scarce climate and soil salinization at the southern border, the type of tillage management is closely related to the agro-environmental conditions [14]. The goal of this work was to evaluate the ability of this tillage method to improve soil quality and sustainable agricultural development.

The soil quality index (SQI) is an assessment tool that includes physical, chemical, and biological indicators, and can be used to estimate changes in soil conditions in arable soils over time due to land-use practices and soil management [15]. For effective evaluation, soil quality indicators should be physical, chemical, and biological properties that are sensitive to soil management methods; combining these soil properties into a single indicator can make the assessment more meaningful and practical. But soil quality sensitivity indicators vary considerably from region to region [16]. Thus, for meaningful evaluation of tillage management in southern Xinjiang, we needed to first determine the sensitive indicators affecting soil quality in this region.

Saline soil in Xinjiang is an important reserve arable land resource in China, but agricultural production in this region is limited by water scarcity and soil salinization [17]. Thus, there is significant interest in developing new strategies for improved utilization of saline farmland resources for more sustainable agricultural development. In this study, three different types of deep vertical rotary tillage and one type of conventional tillage were compared for tillage of salinized farmland in southern Xinjiang from 2020 to 2022. By measuring its soil physicochemical properties and yield, analyzing the effects of different tillage management methods on it, and obtaining the soil quality index (SQI) and sensitivity indices under different tillage management methods, the overall functional capacity of the soil in the region was finally determined. The effects of the different tillage management methods on soil physicochemical properties and yield were determined by obtaining the soil quality index (SQI) values to assess the overall functional capacity of the soil in the region. The results of this work should provide a theoretical basis to improve salinized farmland in the southern Xinjiang region.

2. Materials and Methods

Experiments were conducted in a heavily salinized agricultural field in Tumushuke, Xinjiang, China (79°2′5″ E, 40°0′10″ N; altitude 1098 m). The study area has a temperate extremely arid desert climate, with an average annual precipitation of 38.3 mm, an annual evapotranspiration of 1643 to 2202 mm, and an average annual temperature of 11.6 °C. The total precipitation during the experimental period of 2020–2022 was 264 mm, and the average air temperature was 21.61 °C. The total precipitation during the test period was

264 mm, and the average air temperature was 21.61 °C (June 2020–September 2022). The topsoil of the region has a clayey texture and is heavily salinized, with a salt composition dominated by chlorides [18], but there is plenty of sunshine, and cotton is the main cash crop grown in the area. The groundwater table is 7.2–8 m in the irrigated season and 8–10 m in the non-irrigated season. The soil basic information was measured at 0–60 cm in the test area, and the results are shown in Table 1.

Table 1. Soil texture of the soil in the study area.

Soil Depth cm	Soil Fraction/% Sand Particles 0.05–2/mm	Silt Particles 0.05–0.002/mm	Clay Particles <0.002/mm	Soil Texture [1]
0–20	20.80	70.77	8.43	Silty loam
20–40	14.56	76.38	9.06	Silty loam
40–60	9.97	80.49	9.54	Silty loam

[1] Soil particles were graded according to the USDA Soil Taxonomy system of Soil Classification Standards [19].

2.1. Experimental Site and Experimental Design

Twelve experimental plots were established, each with an area of 14 × 15 m^2. Four treatments were tested, and each treatment was replicated three times, with a 10 m isolation zone between neighboring plots to ensure that the treatments were not affected by nearby plots. The treatments were categorized into different deep vertical rotary tillage depths (DTM20, 20 cm; DTM40, 40 cm; DTM60, 60 cm) and conventional tillage (CTM, 20 cm). The deep vertical rotary tillage treatments were started in April 2020 using a loosening machine (Aksu Wufeng Agricultural Machinery Co., Ltd., Aksu, China, Model 1FSGL-230) to achieve different tillage depths, and in 2021 and 2022, a conventional five-share plow was used (Anhui Huaifeng Modern Agricultural Equipment Co., Ltd., Hefei, China, Model 1L-530J). Conventional tillage treatment (CTM) using a five-share plow for tillage in 2020–2022 (Table 2) and spring irrigation (2800 m^3·ha^{-1}) was performed after completion of tillage. The test variety of "Xinluzhong No. 56" cotton was sown in the spring after irrigation, and the straw was crushed and returned to the field after harvesting in September, with a deep turning of 30 cm. Drip irrigation under the membrane was adopted, with the drip irrigation tapes arranged in three tubes and six rows (Figure 1). The drip irrigation tapes had a diameter of 16 mm, a wall thickness of 0.2 mm, a drip head spacing of 30 cm, and a flow rate of 3.2 L·h^{-1}. Irrigation water was mixed with water from wells and canals, with a ratio of irrigation of 1:1. The irrigation and fertilization system was consistent for each year during the experimental period, and the cotton was irrigated 10 times during the season, with an interval of 7–10 d between each irrigation treatment. The total irrigation quota was 47,400–48,000 m^3·ha^{-1}. Water-soluble fertilizers were applied with each irrigation treatment, including 500 kg·ha^{-1} of urea (N mass fraction $\geq$ 46%), 40 kg·ha^{-1} of potassium xanthate (mass fraction of xanthate $\geq$ 50%), 290 kg·ha^{-1} of high-nitrogen and high-phosphorus water-soluble fertilizers ($N + P_2O_5$ mass fraction $\geq$ 74%), and 210 kg·ha^{-1} of high-nitrogen and high-potassium water-soluble fertilizers ($N + K_2O$ mass fraction $\geq$ 70%). Other field management measures were performed according to local implementation.

Table 2. Soil tillage management methods for each treatment in the experimental area.

Soil Tillage Management Methods	Tillage Machinery and Depth 2020	2021–2022
CTM	Five-share plow machinery, 20 cm	Five-share plow machinery, 20 cm
DTM20	Deep vertical rotary tillage, 20 cm	Five-share plow machinery, 20 cm
DTM40	Deep vertical rotary tillage, 40 cm	Five-share plow machinery, 20 cm
DTM60	Deep vertical rotary tillage, 60 cm	Five-share plow machinery, 20 cm

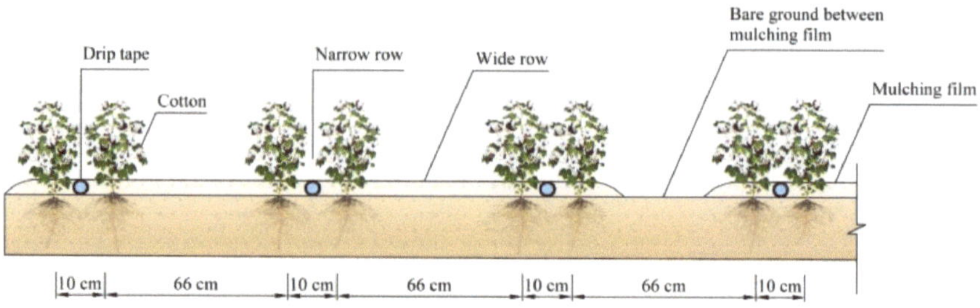

Figure 1. Schematic diagram of cotton planting.

2.2. Collection and Analysis of Soil Samples

On the third day after each irrigation, samples were randomly taken from the wide rows of each treatment to a depth of 60 cm using the "five-point method", and the average value during the reproductive period was taken as the average value of the treatment at a depth of 0–60 cm. A portion of the collected samples was used for pH measurement with a pH meter (Shanghai Yidian Scientific Instrument Co., Ltd., Shanghai, China, Model PHS-3C) and soil EC measurement by a conductivity meter (Shanghai Yidian Scientific Instrument Co., Ltd., Shanghai, China, Model DDS-307). The rest of the soil was used for the determination of soil nutrients, in which soil organic carbon was determined by the potassium dichromate-sulfuric acid external heating method (SOC), total nitrogen was determined by Kjeldahl nitrogen fixation (TN), total potassium was determined by H_2SO_4-H_2O_2 boiling and the flame photometric method (TK), and total phosphorus was determined by H_2SO_4-H_2O_2 boiling and the molybdenum blue method with ascorbic acid (TP) [20].

Cotton harvesting was performed in September each year, and the soil samples were collected and air-dried using the "five-point method." For each soil sample, 100 g of soil were weighed after thorough mixing, and the soil water stability aggregates were determined using the wet sieve method [21]. The average weight diameter (MWD) of the soil water stability aggregates from 0 to 60 cm was calculated by the following formula:

$$\mathrm{MWD} = \frac{\sum_{i=1}^{n} d_i w_i}{\sum_{i=1}^{n} w_i} \tag{1}$$

where w_i is the proportion of agglomerate mass in each particle size range, %; d_i is the average diameter of agglomerates in any level range, mm.

Soil samples were collected using the cutting-ring method between the wide rows of each treatment to a depth of 60 cm during the cotton harvest in September of each year, and each treatment was replicated three times. The soil bulk density (BD) was calculated using the following formula, which was calculated to obtain the mean value of 0–60 cm:

$$d_v = \frac{M \cdot 100}{V(100 + W)} \tag{2}$$

where, dv is the soil bulk density, $g \cdot cm^{-3}$; M is the weight of wet soil, g; V is the volume of the ring knife, cm^3; and W is the soil moisture content in the ring knife, %.

Seed cotton yield was obtained by hand-picking, drying, and weighing 100 randomly selected bolls during the cotton harvest in September each year. Individual boll weight was calculated by dividing the total weight of 100 bolls by the total number of bolls, and the final average was taken and substituted into the following equation as seed cotton yield for each treatment. Each treatment was repeated three times.

$$Y = y \times a \times m \tag{3}$$

where, Y is the seed cotton yield, kg·ha^{-1}; y is the boll weight, g; a is the number of bolls, per plant; m is the planting density, plant·ha^{-1}.

2.3. Soil Quality Assessment

Physical and chemical indicators commonly used in SQI evaluation were initially selected based on previous studies. The selected indicators were used to establish a total dataset (TDS) and included soil physical property indicators (BD, MWD) [22] that can reflect the effects of different tillage management practices on soil structure and soil particle distribution, and soil chemical property indicators (EC, pH, SOC, TN, TP, TK) [22] that can reflect the effects of different tillage management practices on soil ecology and fertility [23]. The soil indicators in TDS were transformed into normalized values between 0.1 and 1.0 using three types of scoring functions. The positive S-type (SSF_1) function was applied to positive slopes, the inverse S-type (SSF_2) function was applied to negative slopes, and the parabolic (SSF_3) function was applied to positive slopes that change to negative slopes at a certain threshold [15]. The standard scoring functions used for normalization of soil quality indicators and their thresholds are shown in Table 3. The score curve equation was used to calculate the soil indicator scores, and the SSF formula shown in Table 3 is as follows [24].

$$\mathrm{TypeS(SFF_1)}:\ f(x)=\begin{cases}1.0 & (x\geq b)\\ \frac{x-a}{b-a} & (a<x<b)\\ 0.1 & (x\leq a)\end{cases} \tag{4}$$

$$\mathrm{TypeS(SFF_2)}:\ f(x)=\begin{cases}0.1 & (x\geq b)\\ \frac{x-a}{b-a} & (a<x<b)\\ 1.0 & (x\leq a)\end{cases} \tag{5}$$

$$\mathrm{TypeS(SFF_3)}:\ f(x)=\begin{cases}0.1 & (x\leq a,\ x\geq b)\\ \frac{x-b}{b_2-b} & (b_2<x<b)\\ \frac{x-a}{b_1-a} & (a<x<b_1)\\ 1.0 & (b_1<x<b_2)\end{cases} \tag{6}$$

where f(x) is the soil quality indicator score; x is the measured value of the soil indicator; and a, b, b_1, and b_2 are the critical values, see Table 3.

Table 3. Threshold values and standardized scoring functions used for soil quality indicators.

Factor	Unit	Scoring Survey	a *	b *	b_1 *	O *	b_2 *	Reference Source
BD	g·cm^{-3}	SSF_2	1.3	1.8				[25]
MWD	mm	SSF_1	0.4	2.0				[26]
EC	mS·cm^{-1}	SSF_2	0.2	4				[25]
pH		SSF_3	3	11	5.5	7	8.5	[25]
SOC*	g·kg^{-1}	SSF_1	3.48	23.2				
TN	g·kg^{-1}	SSF_1	0.5	20				[18]
TP	g·kg^{-1}	SSF_1	0.2	1				
TK	g·kg^{-1}	SSF_1	5	25				

* a = lower threshold at which or below the score is 0.1; b = upper threshold at which or above score is 1.0; b_1 = lower baseline, at which score is 0.5 with bell-shaped relationship; O = optimum level, at which score is 1.0 with bell-shaped relationship; b_2 = upper baseline at which score is 0.5 with bell-shaped relationship. The threshold for soil organic carbon was obtained by conversion based on organic matter = organic carbon × 1.742 [25].

The weights of the indicators can be determined by dividing the common factor of principal component analysis by the total eigenvalues. Finally, after scoring and weighting the selected indicators, the SQI was calculated using the soil quality index formula:

$$\mathrm{SQI}=\sum_{i=1}^{n} W_i \times S_i \tag{7}$$

where W_i is the indicator weight value, S_i is the indicator score, and n is the number of variables integrated in the indicator.

2.4. Data Processing

Experimental data processing was performed using Excel 2019. SPSS Statistics 22.0 (SPSS Inc., Chicago, IL, USA) was used for normal distribution tests, factor analysis, and Pearson correlation analysis. After testing, the data were normally distributed; MANOVA was used to test the differences in soil quality indicators between tillage years and tillage management practices. Origin 2021 was used to plot the principal component analysis to clarify the relationship between soil quality indicators and yield.

3. Results

3.1. Multivariate Analysis of Variance (MANOVA)

The eight selected soil quality indicators were significantly affected by period of tillage, tillage management practices, and their interaction effects (Table 4). We analyzed the data for different soil tillage practices in each period and investigated the effect of soil tillage methods on soil quality indicators in 0–60 cm and seed cotton yield.

Table 4. Multivariate analysis of variance (MANOVA) results to assess the effect of period, treatment, and their interactions for eight measured soil properties and seed cotton yield.

Factors *	DF	Wilk's λ	*p*-Value
Period of tillage	18	0.004	<0.001
Tillage management	27	0.001	<0.001
Period of tillage × Tillage management	54	0.017	0.002

* The results of the data normality tests are detailed in Table S3.

3.2. Soil Properties

TP and TK in the 0–60 cm soil depth range during 2020–2022 were not significantly different between groups among treatments, but there were significant differences between groups of soil indicators including BD, MWD, pH, EC, SOC, and TN (Figure 2). The mean values of BD, MWD, pH, and EC under different tillage management at three years were in the order of CTM > DTM20 > DTM40 > DTM60, while the opposite pattern was seen for SOC and TN. Deep vertical rotary tillage management consistently affected the quality indicators of 0–60 cm soils during the three years (2020–2022). DTM significantly reduced BD and pH compared to CTM, but there was no significant difference in pH between DTM20 and CTM. MWD and EC were significantly lower than those of CTM by 3.79%, 6.40%, and 29.40%, 38.21% for DTM40 and DTM60, respectively; the difference between CTM and DTM20 was not significant. SOC of DTM20, DTM40, and DTM60 were significantly higher than CTM by 2.32%, 4.51%, and 7.47%, respectively; and TN of DTM40 and DTM60 were significantly higher than CTM by 1.51% and 2.97% (the difference between CTM and DTM20 was not significant). There were no significant differences in TP and TK between deep vertical rotary tillage and conventional tillage ($p > 0.05$). Neither deep vertical rotary tillage management nor an increase in the number of years of tillage significantly affected the content of TP and TK in the soil.

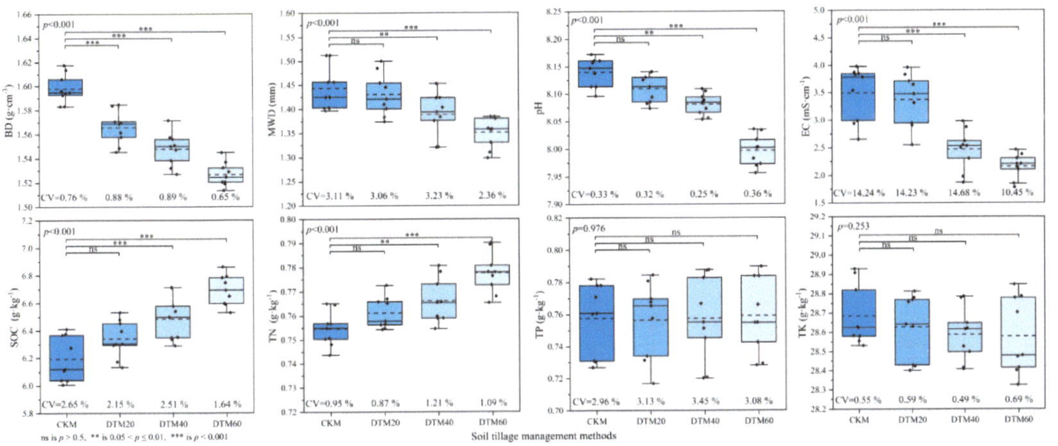

Figure 2. Boxplot and coefficient of variation of eight soil properties of the 0–60 cm layer in 2020, 2021, and 2022. CV is a variable coefficient of each soil property in 2020, 2021, and 2022. p is the level of significance of soil properties between treatments in 2020, 2021, and 2022. BD, soil bulk density; MWD, soil mean weight diameter; EC electrical conductivity; SOC, soil organic carbon; TN, total nitrogen; TK, total potassium; TP, total phosphorus. CTM, conventional tillage; DTM20, 20 cm; DTM40, 40 cm; DTM60, 60 cm, the same as below. Where ns is $p > 0.05$, ** is $0.05 < p \leq 0.01$, *** is $p < 0.001$.

MWD showed a significant positive correlation with BD (0.51), EC showed a significant positive correlation with BD (0.80), and pH showed a significant positive correlation with BD, MWD, and EC (0.54, 0.57, and 0.74, respectively). SOC showed a significant negative correlation with BD, MWD, EC, and pH (−0.83, −0.37, −0.90, and −0.70, respectively). TN showed a significant negative correlation with BD, MWD, EC, and pH (−0.66, −0.49, −0.38, and −0.41, respectively). TN showed a significant positive correlation with SOC (0.46). MWD and pH were significantly positively correlated with TK (0.44 and 0.74, respectively). TP showed no correlation with any of the indicators (Figure 3).

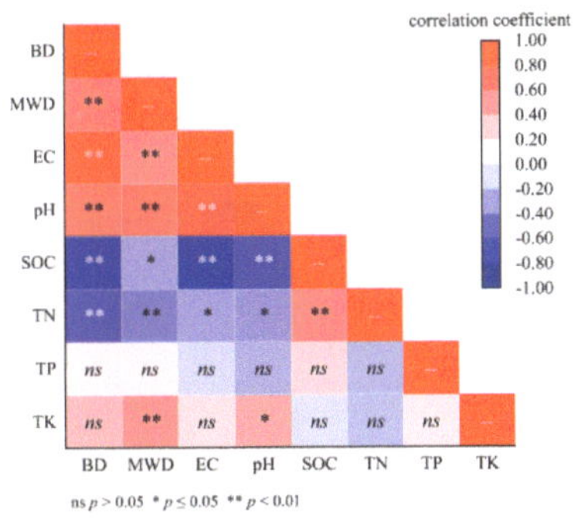

Figure 3. Correlation coefficients among soil properties in the 0–60 cm layer. The correlation coefficients and levels between soil properties are shown in the figure. Where ns is $p > 0.05$, * is $p \leq 0.05$, ** is $p < 0.001$.

3.3. Soil Quality

The common factor variances of the eight soil property indicators were obtained by factor analysis, and all were >0.5 (Table 5). Three components with eigenvalues greater than one were selected, and these components explained a total of 82.53% of the variance in the dataset. The eigenvalues of PC1, PC2, and PC3 were categorized as 3.51, 1.72, and 1.38, explaining 43.88%, 21.45%, and 17.20% of the variance in the dataset, respectively. Both

PC1 and PC2 consisted of positive principal component coefficients, or loadings, on BD, MWD, TK, and loadings on TN. SOC and TP were loadings in PC1 and positive loadings in PC2. Indicators of pH and EC were positive loads in PC1 and loads in PC2. MWD, pH, SOC, TN, and TK were positive loadings in PC3, and BD, EC, and TP were loadings. In the TDS dataset, SOC had the largest common factor variance and weight, 0.91 and 0.14, respectively, and MWD had the smallest, 0.68 and 0.10.

Table 5. Soil properties of factor pattern, common factor variance, and weighting considered in TDS.

Soil Properties	Packet	PCA PC1	PC2	PC3	Communality	Weighting
BD	1	0.886	0.083	−0.288	0.875	0.1325
MWD	1	0.687	0.297	0.350	0.683	0.1035
pH	1	0.826	−0.262	0.279	0.828	0.1254
EC	1	0.875	−0.312	−0.179	0.895	0.1356
SOC	1	−0.882	0.267	0.250	0.912	0.1381
TN	1	−0.674	−0.494	0.212	0.743	0.1125
TP	2	−0.074	0.851	−0.366	0.864	0.1309
TK	3	0.413	0.428	0.669	0.802	0.1215
Principal component eigenvalue		3.510	1.716	1.376		
Of Variance (%)		43.879	21.447	17.199		
Cumulative (%)		43.879	65.327	82.525		

The results showed that deep vertical rotary tillage management could improve the soil quality of saline farmland (Table 6). The scores of eight soil property indices and soil quality indices (SQI) were calculated for each treatment at 0–60 cm of soil for each year (2020–2022) using Equations (4)–(6) and Table 2. Scores under BD, pH, EC, SOC, and TN soil property indexes for each treatment were in the order of CTM < DTM20 < DTM40 < DTM60, with the highest scores obtained under CTM treatment and the lowest scores obtained under DT60 in MWD. The scores were all relatively similar under different tillage management methods in TP. The scores under all treatments were 1.000 because the content of TK was greater than 25 g·kg^{-1} at soil depths of 0–60 cm (Table 2). The magnitude of the soil quality indices (SQI) were in the order of CTM < DTM20 < DTM40 < DTM60, with DTM40 and DTM60 both significantly greater than CTM and DTM20, and DTM60 significantly greater than DTM40 ($p < 0.05$).

Table 6. Scores for the soil properties considered in TDS.

Indicators	Scoring Curve	Weight	TDS CTM	DTM20	DTM40	DTM60
BD	SSF2	0.1325	0.404	0.469	0.504	0.547
MWD	SSF1	0.1035	0.652	0.644	0.618	0.594
PH	SSF2	0.1254	0.353	0.364	0.384	0.409
EC	SSF3	0.1356	0.272	0.305	0.549	0.675
SOC	SSF1	0.1381	0.138	0.145	0.152	0.163
TN	SSF1	0.1125	0.170	0.174	0.177	0.185
TP	SSF1	0.1309	0.697	0.695	0.697	0.691
TK	SSF1	0.1215	1.000	1.000	1.000	1.000
SQI			0.453 c	0.468 c	0.507 b	0.532 a *

* Different letters indicate significant differences at $p < 0.05$.

The results showed that deep vertical rotary tillage management can significantly increase seed cotton yield (Figure 4a). The average seed cotton yield in 2020–2022 was highest under DT60 treatment. There were significant increases of 24.93%, 35.86%, and 43.98% ($p < 0.05$) under DTM20, DTM40, and DTM60, respectively, relative to that of CTM.

The smallest coefficient of variation was 8.53% for CTM treatment, and the highest was 12.65% for DTM60. The Pearson coefficient of SQI and seed cotton yield was 0.93, indicating significant correlation ($p < 0.001$). The R^2 value of 0.8658 indicated a good correlation and fit (Figure 4b), indicating this evaluation method is meaningful. Improvements in soil quality can contribute to increased seed cotton yields.

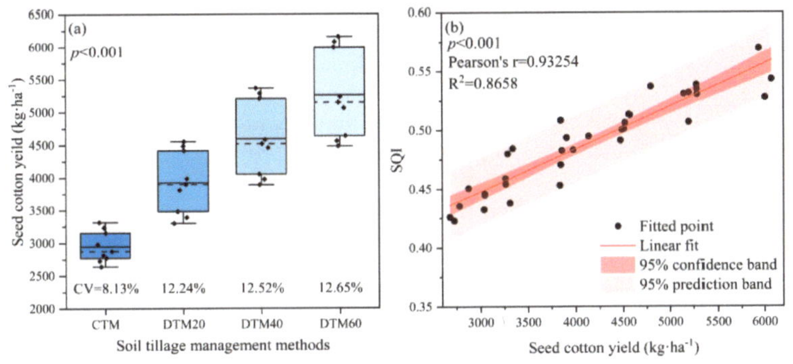

Figure 4. Interrelationship of soil quality index and crop yield. p is the level of significance of seed cotton yield between treatments in 2020, 2021, and 2022. Figure (**a**) shows seed cotton yield under different tillage management practices in 2020, 2021 and 2022. (**b**) shows the linear fit of seed cotton yield and SQI index.

3.4. Relationship between Soil Quality Indicators and Yield of Crops

The relationship between all soil quality indicators and seed cotton yield under different tillage management practices was assessed by principal component analysis (Figure 5). In principal component analysis, PC1, PC2, and PC3 explained 54.60%, 17.80%, and 11.20% of the variance in the data, respectively. BD, pH, and EC were significantly negatively correlated with seed cotton yield, SOC, and TN, indicating that a reduction in BD, MWD, pH, and EC promotes an increase in seed cotton yield, SOC, and TN. None of the relationships between TK and seed cotton yield were significant, and the direction of their vectors was almost perpendicular to the direction of the yield vectors. TP showed a positive correlation with seed cotton yield. Therefore, BD, MWD, pH, and EC were limiting sensitive indicators of seed cotton yield, and SOC and TN were positive sensitive indicators. The data distribution of each treatment shifted leftward toward the vector direction of seed cotton yield with increased depth of tillage. Overall, the results showed that different tillage management practices had significant effects on soil quality indicators and seed cotton yield.

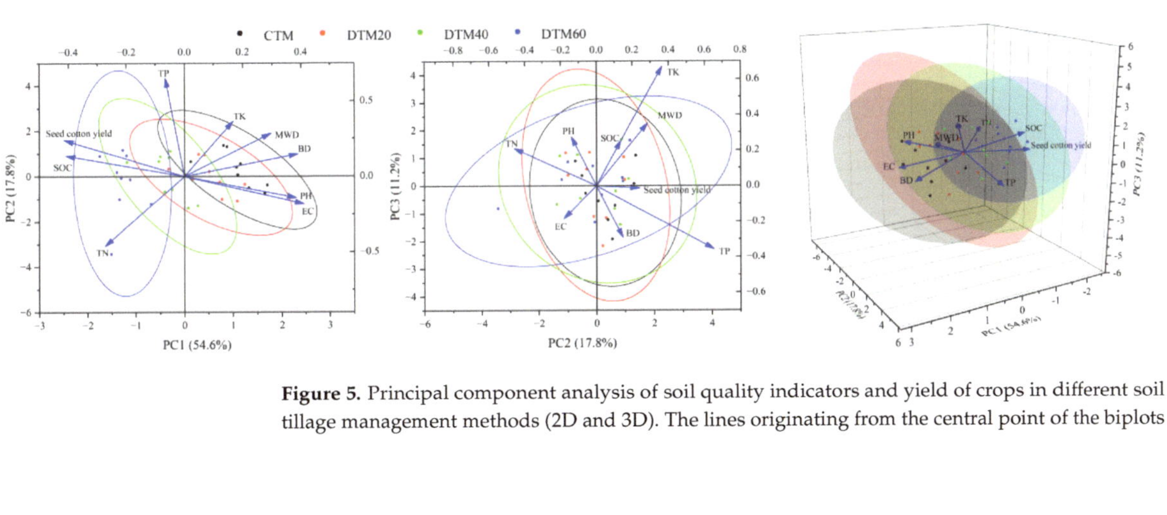

Figure 5. Principal component analysis of soil quality indicators and yield of crops in different soil tillage management methods (2D and 3D). The lines originating from the central point of the biplots

show negative or positive correlations of different variables, and their closeness indicates strength of correlation with a particular treatment.

4. Discussion

4.1. Effect of Different Soil Tillage Management Methods on Soil Quality Indicators

High soil salinity severely limits land productivity and crop production in the southern Xinjiang region [27]. Modified soil tillage management can improve soil properties, and comparing different tillage management helps us to explore ways to improve soil quality in this region. The mean values of BD, MWD, pH, EC, SOC, and TN of 0–60 cm soil were all significantly different under different tillage strategies during 2020–2022, and use of deep vertical rotary tillage sustainably affected the soil structure and environment (Figure 3). Increased depth of deep vertical rotary tillage resulted in greater reductions in BD and MWD, while decreased soil pH and EC resulted in increased levels of SOC and TN. A reduction in soil salinity content by tillage to increase SOC and TN was seen previously [28]. Soil total nitrogen and organic carbon content showed a significant correlation, probably because most of the nitrogen was bound in the organic matter matrix (Figure 3) [29], and the application of organic fertilizer added carbon to the soil. Soil TP content under different tillage management methods did not differ significantly from 2020 to 2022, but TP content decreased in 2020 (Table S1). This was because deep vertical rotary tillage can increase soil permeability and phosphorus leaching, but chemical fertilizer application and straw return to the field can provide carbon, nitrogen, and phosphorus to replenish the nutrients [30]. Different tillage management methods had less effect on TK content, probably because tillage in potassium-rich farmland is always in equilibrium with potassium replenishment and depletion. The results of this study showed that deep vertical rotary tillage decreased the average diameter of soil aggregates (Figure 2). Previous studies have shown that a reduction in soil macroaggregates decreases aggregate stability [31] and also leads to a decrease in carbon and nitrogen contents in the soil [32]. However, in this study, carbon and nitrogen content did not decrease. This may be due to the long-term use of straw returning. The fragmentation of the agglomerates can increase the contact between crushed straw and soil, and this promotes the decomposition of the natural organic matter and enables the soil to obtain more carbon [33–35]. In addition, the crushing of large agglomerates can facilitate the washing out of saline and alkaline ions from the agglomerates under the action of spring irrigation.

4.2. Effect of Different Soil Tillage Management Methods on Soil Quality

Soil quality evaluation using sensitivity indicators helps to determine the functional capacity of tillage management in an agroecosystem [36]. There have been few long-term experimental studies using deep vertical rotary tillage. Our results suggest that the effects of deep vertical rotary tillage last longer than three years, as seen by the maintenance of soil structure in the third year. Alternating conventional tillage with deep vertical rotary tillage during this period is a more economical and efficient way to improve saline farmland [37]. Improvement of soil quality can be achieved by enhancing or maintaining soil-related properties, with SOC significantly contributing to soil quality and improving the structural stability and carbon and nitrogen cycling [38]. The results of this study showed that SOC had the highest weight in the evaluation of soil quality, and tillage management practices that scored high in SOC also had the highest SQI. The second highest weight was EC and the lowest was MWD, which is consistent with previous studies [23,39]. Sadiq et al. [40] concluded that deep tillage is important and can reduce soil salinity and promote cotton yield formation. A loose soil structure creates favorable conditions for salt leaching and cotton root growth. The score function for MWD was SSF1, and tillage to a depth that is too deep reduces the average diameter of soil aggregates, explaining the lowest score of the DTM60 under this indicator. The observed scores for total soil potassium under all tillage management practices were all 1.00, indicating that the soil is rich in potassium, sufficient for the soil to reach its full potential. Raiesi and Kabiri [39] found that MWD is less sensitive to tillage than other physical properties, consistent with the results of this study (Table 6).

DTM40 and DTM60 were able to significantly improve soil quality compared to CTM and DTM20, and seed cotton yield showed a significant linear relationship with SQI, indicating the soil quality evaluation was meaningful. Crop yield is directly related to climatic and hydrological factors and field management level, and the soil quality index in this study showed a good fit with seed cotton yield (Figure 5), indicating that improving soil quality is beneficial to crop yield [41].

4.3. Relationship between Soil Quality Indicators and Seed Cotton Yield under Different Soil Tillage Management Methods

Soil tillage alters soil organic matter mineralization and nutrient cycling rates [42], which affect soil quality and cotton production. Deep tillage can effectively increase seed cotton yield, and here, seed cotton yield showed a significant linear correlation with SQI. The coefficient of variation (CV) of CTM was the smallest among seed cotton yield, with higher yields and CV values for deep vertical rotary tillage treatments. In the principal component analysis of soil quality indicators and seed cotton yield, BD, MWD, pH, and EC were identified as limiting sensitive indicators for seed cotton yield, and SOC and TN were identified as positive sensitive indicators, consistent with previous results [43]. Therefore, assessing SQI allows determination of the effect of different tillage methods on soil quality. Although soil quality varies in different locations and is associated with different land use and tillage management methods [14], our results clearly suggest that deep tillage management can help to improve saline farmland in the southern Xinjiang region. In the short term, efforts to improve saline soil should focus on desalination, followed by the application of organic fertilizer to improve soil fertility. However, future work should investigate whether the DTM60 tillage management method can result in long-term improvements to saline farmland for crop yield increase. Additionally, future work should investigate the yield increase threshold of this tillage management method. With long-term use, irrigation and fertilization can increase the accumulation of soil salts. Modified tillage management can only temporarily alleviate the limitations of soil salinity on crop yields; the best strategy to solve this problem is to remove the soil salts. Therefore, in future research, we will continue to investigate the effect of deep vertical rotary tillage management and subsurface pipe drainage technology on the improvement of saline farmland. Of course, soil quality is also affected by rainfall and temperature and additional agricultural practices, so future work should investigate the general applicability of the SQI evaluation method.

5. Conclusions

Different tillage management methods can significantly affect soil quality indicators and crop yield. The results of this study showed that deep vertical rotary tillage optimizes soil properties and improves the soil quality index (SQI) compared to CTM. Deep tillage reduces BD, MWD, EC, and pH, and promotes the accumulation of SOC and TN in the soil. The average diameter of aggregates in the soil also becomes smaller with an increase in the tillage depth, and a reduction in MWD does not reduce crop yield but promotes soil salinity leaching. The significant linear relationship between seed cotton yield and soil quality ($p < 0.001$) indicated that improving soil quality can increase crop yield. By principal component analysis, we identified BD, MWD, pH, and EC as limiting sensitive indicators for seed cotton yield, and SOC and TN as positive sensitive indicators. Utilizing these sensitivity indicators and evaluating soil quality can help expand our understanding of how to improve land productivity and crop yield potential of saline farmland in southern Xinjiang.

Supplementary Materials: The following Supporting Information can be downloaded at: https://www.mdpi.com/article/10.3390/agriculture13101864/s1, Table S1: Soil quality indicators for 0–60 cm, 2020–2022.; Table S2: Seed cotton yield in 2020–2022; Table S3: Data normality distribution test.

Author Contributions: Z.L., H.L. and T.W. conceived this study, and H.Y. collected and analyzed the data. All authors contributed critically to the drafts and gave final approval for publication. All authors have read and agreed to the published version of the manuscript.

Funding: This research was supported by the National Natural Science Foundation of China (No. 52069026, U1803244).

Institutional Review Board Statement: Not applicable.

Data Availability Statement: Not applicable.

Acknowledgments: Thank you to all participants for their strong support.

Conflicts of Interest: The authors declare no conflict of interest.

Abbreviations

BD, soil bulk density; MWD, soil mean weight diameter; EC electrical conductivity; SOC, soil organic carbon; TN, total nitrogen; TK, total potassium; TP, total phosphorus. CTM, conventional tillage; DTM20, 20 cm; DTM40, 40 cm; DTM60, 60 cm.

References

1. Announcement of National Bureau of Statistics on Cotton Production in 2022. Available online: http://www.stats.gov.cn/xxgk/sjfb/zxfb2020/202212/t20221227_1891259.html (accessed on 19 August 2023).
2. Li, X.W.; Jin, M.G.; Zhou, N.Q.; Huang, J.O.; Jiang, S.M.; Telesphore, H. Evaluation of evapotranspiration and deep percolation under mulched drip irrigation in an oasis of Tarim basin, China. *J. Hydrol.* **2016**, *538*, 677–688. [CrossRef]
3. Lipiec, J.; Horn, R.; Pietrusiewicz, J.; Siczek, A. Effects of soil compaction on root elongation and anatomy of different cereal plant species. *Soil Tillage Res.* **2012**, *121*, 74–81. [CrossRef]
4. Daly, E.J.; Kim, K.; Hernandez-Ramirez, G.; Klimchuk, K. The response of soil physical quality parameters to a perennial grain crop. *Agric., Ecosyst. Environ.* **2023**, *343*, 108265. [CrossRef]
5. Matosic, S.; Birkás, M.; Vukadinovic, V.; Kisic, I.; Bogunovic, I. Tillage, Manure and Gypsum Use in Reclamation of Saline-Sodic Soils. *Acs Agric. Conspec. Scintificus* **2018**, *83*, 131–138.
6. Ding, Z.L.; Kheir, A.M.S.; Ali, O.A.M.; Hafez, E.M.; ElShamey, E.A.; Zhou, Z.X.; Wang, B.Z.; Lin, X.E.; Ge, Y.; Fahmy, A.E.; et al. A vermicompost and deep tillage system to improve saline-sodic soil quality and wheat productivity. *J. Environ. Manag.* **2021**, *277*, 111388. [CrossRef]
7. Li, L.; Liu, H.G.; Gong, P.; Lin, E.; Bai, Z.T.; Li, P.F.; Wang, C.X.; Li, J. Multi-objective optimization of winter irrigation for cotton fields in salinized freeze-thaw areas. *Eur. J. Agron.* **2023**, *143*, 126715. [CrossRef]
8. Bai, W.; Sun, Z.X.; Zheng, J.M.; Liu, Y.; Hou, Z.Y.; Feng, L.S.; Qian, C.; Yang, N.; Chen, F.; Zhe, Z. The combination of subsoil and the incorporation of corn stover affect physicochemical properties of soil and corn yield in semi-arid China. *Toxicol. Environ. Chem.* **2016**, *98*, 561–570. [CrossRef]
9. Zhai, L.; Xu, P.; Zhang, Z.; Li, S.; Xie, R.; Zhai, L.; Wei, B. Effects of deep vertical rotary tillage on dry matter accumulation and grain yield of summer maize in the Huang-Huai-Hai Plain of China. *Soil Tillage Res.* **2017**, *170*, 167–174. [CrossRef]
10. Sun, X.F.; Ding, Z.S.; Wang, X.B.; Hou, H.P.; Zhou, B.Y.; Yue, Y.; Ma, W.; Ge, J.Z.; Wang, Z.M.; Zhao, M. Subsoiling practices change root distribution and increase post-anthesis dry matter accumulation and yield in summer maize. *PLoS ONE* **2017**, *12*, e0174952. [CrossRef]
11. Qiang, X.; Sun, J.; Ning, H. Impact of Subsoiling on Cultivated Horizon Construction and Grain Yield of Winter Wheat in the North China Plain. *Agriculture* **2022**, *12*, 236. [CrossRef]
12. Shan, F.; Li, D.; Zhu, J.; Kang, S.; Wang, J. Effects of Vertical Smashing Rotary Tillage on Root Growth Characteristics and Yield of Broccoli. *Agriculture* **2022**, *12*, 928. [CrossRef]
13. Yang, J.J.; Tan, W.J.; Han, J.R.; Li, F.M.; Zhang, F. Distribution pattern of rainwater in soil under vertical deep rotary tillage in dryland farmland. *Agric. Water Manag.* **2022**, *273*, 107891. [CrossRef]
14. Raiesi, F. A minimum data set and soil quality index to quantify the effect of land use conversion on soil quality and degradation in native rangelands of upland arid and semiarid regions. *Ecol. Indic.* **2017**, *75*, 307–320. [CrossRef]
15. Andrews, S.S.; Karlen, D.L.; Mitchell, J.P. A comparison of soil quality indexing methods for vegetable production systems in Northern California. *Agric. Ecosyst. Environ.* **2002**, *90*, 25–45. [CrossRef]
16. Zuber, S.M.; Behnke, G.D.; Nafziger, E.D.; Villamil, M.B. Multivariate assessment of soil quality indicators for crop rotation and tillage in Illinois. *Soil Tillage Res.* **2017**, *174*, 147–155. [CrossRef]
17. Meng, Q.F.; Li, D.W.; Zhang, J.; Zhou, L.R.; Ma, X.F.; Wang, H.Y.; Wang, G.C. Soil properties and corn (*Zea mays* L.) production under manure application combined with deep tillage management in solonetzic soils of Songnen Plain, Northeast China. *J. Integr. Agric.* **2016**, *15*, 879–890. [CrossRef]
18. Soil Science Database. Available online: http://vdb3.soil.csdb.cn/extend/jsp/introduction (accessed on 19 August 2023).

19. Millan, H.; Gonzalez-Posada, M.; Aguilar, M.; Dominguez, J.; Cespedes, L. On the fractal scaling of soil data. Particle-size distributions. *Geoderma* **2003**, *117*, 117–128. [CrossRef]
20. Bao, S.D. *Soil and Agricultural Chemistry Analysis*, 3rd ed.; Chinese Agriculture Publishing House: Beijing, China, 2018; pp. 40–110.
21. Ibrahimi, K.; Mowrer, J.; Amami, R.; Belaid, A. Burn Effects on Soil Aggregate Stability and Water Repellency of Two Soil Types from East and North Tunisia. *Commun. Soil Sci. Plant Anal.* **2019**, *50*, 827–837. [CrossRef]
22. Fernandes, J.C.; Gamero, C.A.; Rodrigues, J.G.L.; Miras-Avalos, J.M. Determination of the quality index of a Paleudult under sunflower culture and different management systems. *Soil Tillage Res.* **2011**, *112*, 167–174. [CrossRef]
23. Nakajima, T.; Lal, R.; Jiang, S.G. Soil quality index of a crosby silt loam in central Ohio. *Soil Tillage Res.* **2015**, *146*, 323–328. [CrossRef]
24. Andrews, S.S.; Carroll, C.R. Designing a soil quality assessment tool for sustainable agroecosystem management. *Ecol. Appl.* **2001**, *11*, 1573–1585. [CrossRef]
25. Amami, R.; Ibrahimi, K.; Sher, F.; Milham, P.J.; Khriji, D.; Annabi, H.A.; Abrougui, K.; Chehaibi, S. Effects of conservation and standard tillage on soil physico-chemical properties and overall quality in a semi-arid agrosystem. *Soil Res.* **2022**, *60*, 485–496. [CrossRef]
26. LeBissonnais, Y. Aggregate stability and assessment of soil crustability and erodibility: I. Theory and methodology. *Eur. J. Soil Sci.* **1996**, *47*, 425–437. [CrossRef]
27. Pisinaras, V.; Tsihrintzis, V.A.; Petalas, C.; Ouzounis, K. Soil salinization in the agricultural lands of Rhodope District, northeastern Greece. *Environ. Monit. Assess.* **2010**, *166*, 79–94. [CrossRef]
28. Li, L.; Guan, J.X.; Chen, S.Y.; Zhang, X.Y. Intermittent Deep Tillage on Improving Soil Physical Properties and Crop Performance in an Intensive Cropping System. *Agronomy* **2022**, *12*, 688. [CrossRef]
29. Yan, Y.; Tian, J.; Fan, M.S.; Zhang, F.S.; Li, X.L.; Christie, P.; Chen, H.Q.; Lee, J.; Kuzyakov, Y.; Six, J. Soil organic carbon and total nitrogen in intensively managed arable soils. *Agric. Ecosyst. Environ.* **2012**, *150*, 102–110. [CrossRef]
30. Pinitpaitoon, S.; Bell, R.W.; Suwanarit, A. The significance of available nutrient fluxes in N and P budgets for maize cropping on a Rhodic Kandiustox: A study with compost, NP fertilizer and stubble removal. *Nutr. Cycl. Agroecosyst.* **2011**, *89*, 199–217. [CrossRef]
31. Six, J.; Bossuyt, H.; Degryze, S.; Denef, K. A history of research on the link between (micro)aggregates, soil biota, and soil organic matter dynamics. *Soil Tillage Res.* **2004**, *79*, 7–31. [CrossRef]
32. Huang, S.; Sun, Y.N.; Rui, W.Y.; Liu, W.R.; Zhang, W.J. Long-Term Effect of No-Tillage on Soil Organic Carbon Fractions in a Continuous Maize Cropping System of Northeast China. *Pedosphere* **2010**, *20*, 285–292. [CrossRef]
33. Aziz, I.; Mahmood, T.; Islam, K.R. Effect of long term no-till and conventional tillage practices on soil quality. *Soil Tillage Res.* **2013**, *131*, 28–35. [CrossRef]
34. Marinari, S.; Radicetti, E.; Petroselli, V.; Allam, M.; Mancinelli, R. Microbial Indices to Assess Soil Health under Different Tillage and Fertilization in Potato (*Solanum tuberosum* L.) Crop. *Agriculture* **2022**, *12*, 415. [CrossRef]
35. Gonzalez-Rosado, M.; Parras-Alcantara, L.; Aguilera-Huertas, J.; Lozano-Garcia, B. No-Tillage Does Not Always Stop the Soil Degradation in Relation to Aggregation and Soil Carbon Storage in Mediterranean Olive Orchards. *Agriculture* **2022**, *12*, 407. [CrossRef]
36. Singh, A.K.; Bordoloi, L.J.; Kumar, M.; Hazarika, S.; Parmar, B. Land use impact on soil quality in eastern Himalayan region of India. *Environ. Monit. Assess.* **2014**, *186*, 2013–2024. [CrossRef]
37. Bai, Z.T.; Liu, H.G.; Wang, T.A.; Gong, P.; Li, H.Q.; Li, L.; Xue, B.; Cao, M.H.; Feng, J.P.; Xu, Y.B. Effect of Smashing Ridge Tillage Depth on Soil Water, Salinity, and Yield in Saline Cotton Fields in South Xinjiang, China. *Water* **2021**, *13*, 3592. [CrossRef]
38. Bhaduri, D.; Purakayastha, T.J. Long-term tillage, water and nutrient management in rice-wheat cropping system: Assessment and response of soil quality. *Soil Tillage Res.* **2014**, *144*, 83–95. [CrossRef]
39. Raiesi, F.; Kabiri, V. Identification of soil quality indicators for assessing the effect of different tillage practices through a soil quality index in a semi-arid environment. *Ecol. Indic.* **2016**, *71*, 198–207. [CrossRef]
40. Sadiq, M.; Hassan, G.; Mehdi, S.M.; Hussain, N.; Jamil, M. Amelioration of saline-sodic soils with tillage implements and sulfuric acid application. *Pedosphere* **2007**, *17*, 182–190. [CrossRef]
41. Yang, S.C.; Wang, C.B.; Huo, L.; Jiang, W.L.; Wen, M.J. Effects of different tillage practices on soil organic carbon of cultivated land in Gansu Yellow River irrigation district. *Trans. Chin. Soc. Agric. Eng.* **2019**, *35*, 114–121. (In Chinese) [CrossRef]
42. Al-Kaisi, M.M.; Yin, X.H.; Licht, M.A. Soil carbon and nitrogen changes as influenced by tillage and cropping systems in some Iowa soils. *Agric. Ecosyst. Environ.* **2005**, *105*, 635–647. [CrossRef]
43. Zhao, H.X.; Wu, L.; Zhu, S.W.; Sun, H.C.; Xu, C.L.; Fu, J.D.; Ning, T.Y. Sensitivities of Physical and Chemical Attributes of Soil Quality to Different Tillage Management. *Agronomy* **2022**, *12*, 1153. [CrossRef]

agriculture

MDPI

Article

Influence of Nitrogen Fertilizer Application on Soil Acidification Characteristics of Tea Plantations in Karst Areas of Southwest China

Yanling Liu [1,2], Meng Zhang [1,2], Yu Li [1,2,*], Yarong Zhang [1,2], Xingcheng Huang [1,2], Yehua Yang [1,2], Huaqing Zhu [1,2], Han Xiong [1,2] and Taiming Jiang [1,2]

1 Institute of Soil and Fertilizer, Guizhou Academy of Agricultural Sciences, Guiyang 550006, China; lyl890615@163.com (Y.L.)

2 Scientific Observing and Experimental Station of Arable Land Conservation and Agricultural Environment, Ministry of Agriculture, Guiyang 550006, China

* Correspondence: liyu83110@163.com; Tel.: +86-0851-83761735

Abstract: Nitrogen (N) fertilizer application is one of the causes of soil acidification at tea plantations. However, the effect of N fertilizer application on the soil acidification characteristics of tea plantations with different acidities remains unclear. In this study, field experiments were conducted to investigate the effects of different nitrogen fertilizer application rates on the pH, pH buffer capacity (pHBC), exchangeable total acidity (ETA), exchangeable base cations (EBCs), and cation exchange capacity (CEC) in the topsoil of non-acidified (NA), mildly acidified (MA), and heavily acidified (HA) tea plantations. The results showed that the exchangeable Al^{3+} (E-Al) and CEC were HA > MA > NA in all tea plantations, whereas the EBCs and base saturation percentage (BSP) were HA < MA < NA. In the tea plantations with pH > 4.0, the pH, EBCs, and BSP showed decreasing trends with increasing N fertilizer application, whereas E-Al showed an increasing trend. In the tea plantations with pH < 4.0, the soil pH showed a small increasing trend with the increase in N fertilizer application, whereas the soil exchangeable H^+ (E-H), E-Al, and CEC showed decreasing trends. Meanwhile, in the pH range of 4–6, the soil acid–base buffer curve rose sharply, and an excessive application of N fertilizer (N900) significantly reduced the pHBC. In addition, a stepwise regression analysis showed that the BSP, EBCs, and exchangeable Mg^{2+} (E-Mg) had significant direct effects on the soil pH, whereas the CEC and N application had significant direct effects on the soil pHBC. In conclusion, a decrease in the BSP and an increase in E-Al were the main mechanisms of acidification at tea plantations, whereas a decrease in the BSP caused by the application of N fertilizer was the main cause of exacerbated soil acidification in non-acidified tea plantations.

Keywords: soil acidification; tea plantation; N fertilization; pH buffer capacity; exchangeable function

Citation: Liu, Y.; Zhang, M.; Li, Y.; Zhang, Y.; Huang, X.; Yang, Y.; Zhu, H.; Xiong, H.; Jiang, T. Influence of Nitrogen Fertilizer Application on Soil Acidification Characteristics of Tea Plantations in Karst Areas of Southwest China. *Agriculture* **2023**, *13*, 849. https://doi.org/10.3390/agriculture13040849

Academic Editors: Anetta Siwik-Ziomek and Anna Figas

Received: 15 March 2023
Revised: 4 April 2023
Accepted: 5 April 2023
Published: 11 April 2023

1. Introduction

Soil acidification is a serious aspect of soil degradation worldwide and has been reported in various ecosystems and regions [1,2]. Soil acidification induces soil nutrient imbalances [3] and losses of natural flora and fauna species [4], reduces agricultural production [5] and reduced belowground processes [6,7], and increases greenhouse gas emissions [8]. Soil acidification is ascribed to a combination of high-N fertilization [9], plant uptake and the removal of base cations from the soil [10], and acid deposition [11]. The overuse of N fertilizer is the dominant factor that has resulted in soil acidification in conventional agricultural systems aimed at maximizing profits [12,13]. It is thought that accelerated soil acidification due to N fertilization is directly caused by the production of protons via nitrification after ammonium nitrogen fertilization occurs [14]. In addition, under heavy rainfall, nitrate nitrogen ions (NO_3^-) leach out of the soil and carry away a large number of base ions, leaving more H^+, which is the indirect cause of the soil acidification caused by nitrogen fertilizer application [15].

The tea plant (*Camellia sinensis*) is an important cash crop that is cultivated in many tropical and subtropical countries. Owing to its high economic value, tea cultivation has been rapidly expanding in China [16]. The optimal soil pH for tea growth is 4.5–5.5 [17]. Tea growth is inhibited when the soil pH is lower than 4.0, and both the quality and quantity of the tea that is produced are negatively affected [18]. Nitrogen (N) fertilizer is applied to improve the yield and quality of tea because N is required for the production of amino acids, which are key quality indicators of tea [19]. High rates of N fertilization, as high as 444 kg ha^{-1}, further accelerate soil acidification [20]. In China, 46.0% of soil samples had a pH < 4.5, indicating that the soil acidification trend of tea plantations is severe [21].

Most studies have shown that the higher the amount of nitrogen fertilizer that is applied, the more serious the soil acidification [22]. However, the soil acidification rate may be reduced by less nitrate leaching because the nitrification rate is typically inhibited by a low soil pH [23]. Additionally, rather than the total production of protons, soil acidification strongly depends on the soil buffering capacity and the depletion of the soil base cation pool. The soil acid–base buffer system mainly depends on the soil pH [24]. Therefore, a basic soil pH has an important effect on the degree of soil acidification caused by nitrogen application.

The transformation of soil nutrients and microbial activities are significantly affected by soil pH. The nutrient transformation and microbial activities are different in soils with different pH values [25]. Most previous studies have been carried out under the same pH conditions; therefore, the responses of the soil acidification characteristics of tea plantations with different pH values to N fertilizer application is not completely clear. In addition, the effect of fertilization on soil properties can be more accurately evaluated using the multiyear positioning test. At present, there are few studies on the effect of the long-term application of N fertilizer on the soil properties at tea plantations. In this study, we conducted three field experiments that considered a range of N additions in Guizhou, the province with the largest tea planting area in China, and 86.9% of the soil samples had soil pH values < 4.5 [21]. We tested the soil pH, exchangeable total acidity (ETA), exchangeable base cations (EBCs), and soil acid–base buffering capacity (pHBC). Our objectives were (1) to reveal the contribution of N fertilization to soil acidification at tea plantations at different pH levels, (2) to evaluate the main factors controlling the soil pH at tea plantations, and (3) to provide a reference for rational N application and acidification improvement at tea plantations.

2. Materials and Methods

2.1. Description of the Study Site

The experimental sites were located in the main region for green tea cultivation in Guizhou Province, Southwest China, and all field experiments were conducted from 2016 to 2020. The Guiding test site (GD) was located in Baoguan Township, Guiding County, Qiannan Prefecture, Guizhou Province (107°8′53.8″ E, 26°13′44.6″ N, altitude 1244 m). The experimental site has a subtropical monsoon climate with a frost-free period of 280 days, an average annual temperature of 13.2 °C, and an average annual precipitation of 1200 mm. The soil at the site was a yellow soil and was classified as an Acrisol in the World Reference Base for Soil Resources (WRB). Before the experiment, tea plants of the 'Niaowang' variety grew in the studied field for 10 years, and the planting density was approximately 60,000 plants ha^{-1}. A N-P-K ternary compound fertilizer (N-P-K: 15/6.5/12.4, 400 kg ha^{-1}) and urea (120 kg ha^{-1}) were applied annually before the experiment.

The Meitan test site (MT) was located in Xinglong Town, Meitan County, Zunyi City, Guizhou Province (107°33′4.4″ E, 27°45′33.3″ N, altitude 831 m). The experimental site has a subtropical monsoon climate with a frost-free period of 284 days, an average annual temperature of 15.3 °C, and an average annual precipitation of 1100 mm. The soil at the site was a yellow soil and was classified as an Acrisol in the World Reference Base for Soil Resources (WRB). Before the experiment, tea plants of the 'Fuding' variety grew in the studied field for 35 years, and the planting density was approximately 60,000 plants ha^{-1}. An organic–inorganic compound fertilizer (N-P-K 11/2.2/3.3, 3000 kg ha^{-1}) was applied annually before the experiment.

The Xixiu test site (XX) was located in Jichang Township, Xixiu District, Anshun City, Guizhou Province (106°4′33.1″ E, 26°5′20.4″ N, altitude 1233 m). This experimental site has a subtropical monsoon climate with a frost-free period of 250 days, an average annual temperature of 13.9 °C, and an average annual precipitation of 1200 mm. The soil at the site was yellow soil and was classified as an Acrisol in the World Reference Base for Soil Resources (WRB). Before the experiment, tea plants of the 'Fuding' variety grew in the studied field for 34 years, and the planting density was approximately 60,000 plants ha^{-1}. A total of 750 kg ha^{-1} of a N-P-K ternary compound fertilizer, 450–675 kg ha^{-1} of urea, and 1500 kg ha^{-1} of organic fertilizer (rapeseed cake) were applied annually before the experiment.

The surface (0–20 cm) soil properties that existed at each site before the experiment are shown in Table 1. According to the impact of the pH on the growth of tea plants, the pH was divided into three levels: pH < 4 was heavily acidified (HA), 4 < pH < 4.5 was mildly acidified (MA), and pH > 4.5 was non-acidified (NA). The pH value of the soil at the Guiding test site (GD) was NA, the soil at the Meitan test site (MT) was MA, while the soil at the Xixiu test site (XX) was HA.

Table 1. Soil properties that existed at each site before the experiment.

Test Site	pH	SOM (g kg^{-1})	CEC (cmol kg^{-1})	TN (g kg^{-1})	AN (mg kg^{-1})	AP (mg kg^{-1})	AK (mg kg^{-1})
GD	5.01	19.0	9.30	1.44	110.2	14.60	187.0
MT	4.14	28.7	14.2	1.76	147.9	41.30	118.1
XX	3.74	68.4	24.8	3.29	181.6	54.20	181.3

Note: SOM—soil organic matter; CEC—cation exchange capacity; TN—total nitrogen; AN—alkali-hydrolyzed nitrogen; AP—available phosphorus; AK—available potassium.

2.2. Experimental Design

The experiment consisted of five treatments, and each treatment was repeated three times according to a randomized complete block design (RCBD). The area of each plot was 22.5 m^2 (1.5 m × 15.0 m). The treatments included N0 (P 43.7 kg ha^{-1} and K 83.9 kg ha^{-1}), N150 (N 150 kg ha^{-1}, P 43.7 kg ha^{-1}, and K 83.9 kg ha^{-1}), N300 (N 300 kg ha^{-1}, P 43.7 kg ha^{-1}, and K 83.9 kg ha^{-1}), N600 (N 600 kg ha^{-1}, P 43.7 kg ha^{-1}, and K 83.9 kg ha^{-1}), and N900 (N 900 kg ha^{-1}, P 43.7 kg ha^{-1}, and K 83.9 kg ha^{-1}). The fertilizers used in the test included urea (46.0% N), superphosphate (7.0% P), and potassium sulfate (41.9% K). Nitrogen fertilizer was applied in three stages: base (30%), spring (40%), and summer (30%). The base fertilizer was applied from October to November every year, the spring fertilizer was applied in early February of the following year, and the summer fertilizer was applied in May–June every year. Phosphorus and potassium fertilizers were applied as a base fertilizer in October–November every year. All fertilizers were applied in the band furrows (at a depth of 15–20 cm) about 20–30 cm from the roots of the tea plants and then covered with soil after their application.

2.3. Sampling and Measurement

Soil samples from depths of 0–20 cm were collected between rows of tea trees from 10 randomly selected spots in the main experimental area before fertilization. Soil samples were collected before the experiment in October 2016, and the soil samples for this study were collected in October 2020. The soil samples were composited, and visible impurities and roots were removed. Then, the samples were naturally air-dried, ground, and passed through 2 mm and 0.15 mm sieves to determine their chemical properties. The chemical properties of the soil were determined according to the method described by Bao [26]. The soil pH was measured with a 1:2.5 extraction mixture (soil/water, *w/v*) using a pH meter

(FE20K, Mettler Toledo, Zurich, Switzerland). The organic matter (OM) was determined by oxidation with potassium dichromate and titration with ferrous ammonium sulfate. The total N (TN) was determined using the Kjeldahl method. The available nitrogen (AN) was measured using the alkaline hydrolysis diffusion method. The available phosphorus (AP) was extracted using a 0.03 mol L^{-1} NH_4F–0.025 mol L^{-1} HCl solution and analyzed using an ultraviolet-visible spectrophotometer (T6 New Century, Beijing, China) via a molybdenum blue colorimetric analysis. The available potassium (AK) contents were extracted using 1 mol L^{-1} NH_4AC (pH 7.0) and measured using a flame photometer (AP1200, Shanghai, China). The exchangeable total acids (E-Al and E-H) were determined using 1 mol L^{-1} potassium chloride solution drenching and NaOH-neutralization titration. The CEC was determined using a 1 mol L^{-1} ammonium acetate exchange and a distillation method. EBCs were extracted using a 1 mol L^{-1} ammonium acetate (pH 7) solution, the Ca and Mg in the extracts were determined using atomic absorption spectrophotometry, and the K and Na were determined using flame photometry.

Three of the treatments (N0, N300, and N900) were selected for the soil acid–base buffer titration curve. A 0.5 g soil sample was weighed into each of the 15 beakers (numbered 1–15). Then, 0, 0.25, 0.5, 1.0, 2.0, 4.0, 6.0, and 9.0 mL of a 0.1 mol L^{-1} HCl solution was added to beakers 1–8, and 0.25, 0.5, 1.0, 2.0, 4.0, 6.0, and 9.0 mL of a 0.1 mol L^{-1} NaOH solution was added to beakers 9–15, and finally deionized water was added to fix the volume to 25.0 mL. The solutions were shaken well, and the pH values were measured after 30 min of standing. The pHBC was determined by the linear fitting of the data between two inflection points [27]. The calculation formula was as follows:

$$\text{pHBC} = 1/|a|$$

where pHBC indicates the acid–base buffer capacity at the end of the test and a is the slope of the linear fitting equation.

2.4. Statistical Analysis

The experimental data were calculated using Excel 2010. Variance, correlation, and stepwise regression analyses were performed using SPASS 20.0. Differences between treatments were analyzed using a one-way ANOVA combined with Duncan's multiple range test ($p < 0.05$).

3. Results

3.1. Effect of N Fertilizer Application Rates on Soil pH values of Tea Plantations

At the NA and MA plantation, the pH of each treatment was N0 < N150 < N300 < N600 < N900, whereas at the HA plantation the pH of the N0 treatment was significantly lower than those of the N600 and N900 treatments, and there was no significant difference between the N application treatments. Compared with the N0 treatment, the pH decreased by 11.3–45.0% at the NA plantation and by 1.4–12.7% at the MA plantation, whereas the pH increased by 3.1–41.7% at the HA plantation for the N fertilizer application treatments (Figure 1). The results of the linear fit of the soil pH and N application rates (Table 2) showed that the coefficient of determination of the NA and MA equations reached a highly significant level ($p < 0.01$), and the slope of the NA equations was 3.3 times higher than that of the MA equation, whereas the coefficient of determination of the HA equation did not reach significance ($p > 0.05$). This indicates that the lower the degree of acidification at tea plantations, the greater the effect of nitrogen fertilizer application on the pH. The results of fitting the quadratic equation for one variable to the soil pH and N application rates (Table 2) showed that the coefficients of determination of the equations reached a highly significant level ($p < 0.01$) at all experimental sites, with the pH values corresponding to the inflection points of the equations for the NA, MA, and HA plantations, which were 3.72, 3.82 and 4.04, respectively. This indicates that the pH decreased continuously with increasing nitrogen application when the soil pH was > 4.0, whereas the pH showed an increasing trend with the increasing nitrogen application when the soil pH was < 4.0.

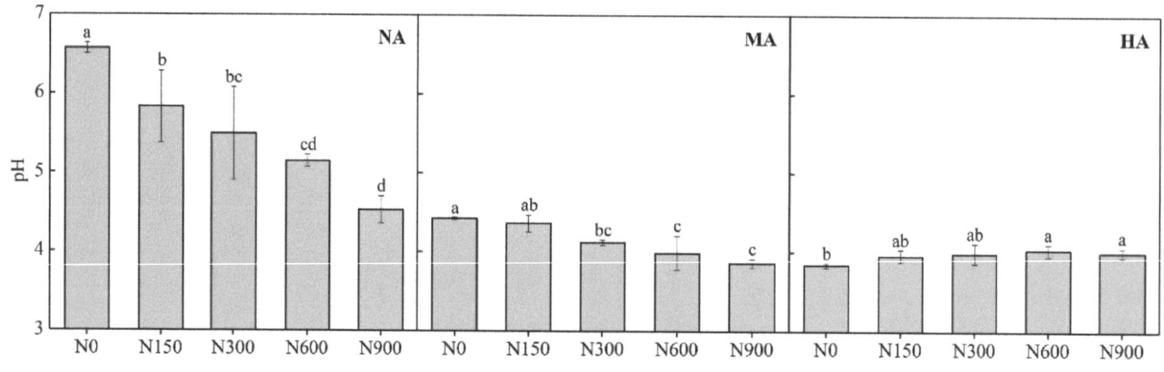

Figure 1. Effect of N application rate on soil pH at tea plantations with different degrees of acidification. Note: the different lowercase letters indicate significant differences at $p < 0.05$ for different N application rates at the same degree of acidification.

Table 2. Response equations of soil pH to N application rate at tea plantations with different degrees of acidification.

Soil Acidification Degree	Response Equation of Soil pH to Nitrogen Application Rate	R^2	pH Value Corresponding to the Inflection Point of the Equation
NA	$y = -0.0020x + 6.3069$	0.9350 **	—
MA	$y = -0.0006x + 4.3967$	0.9398 **	—
HA	$y = 0.0001x + 3.9074$	0.5472	—
NA	$y = 1 \times 10^{-6}x^2 - 0.0033x + 6.4443$	0.9626 **	3.72
MA	$y = 4 \times 10^{-7}x^2 - 0.001x + 4.4404$	0.9684 **	3.82
HA	$y = -5 \times 10^{-7}x^2 + 0.0006x + 3.8566$	0.9601 **	4.04

Note: ** represents significance at 0.01 probability level.

3.2. Effect of N Fertilizer Application Rates on Soil pHBC Values of Tea Plantations

Figure 2 shows the soil acid–base titration curves for different N fertilizer application rates. The results show that all curves were "S" shaped (Figure 2). In the pH range of 4–6, the soil acid–base buffer curves rose sharply, indicating that the soil acid–base buffer capacity was weak in this pH range. When the soil pH was <4 or >6, the soil acid–base buffer curves became flat, indicating that the soil acid–base buffer capacity was sharply enhanced. The soil pHBC was calculated via a linear fitting of the soil acid–base buffer curve in the pH range of 4–6 (Table 3). The pHBC values of the NA, MA, and HA soils were 1.09–1.38 cmol kg^{-1}, 1.21–1.52 cmol kg^{-1}, and 1.52–3.54 cmol kg^{-1}, respectively. Compared with the N0 treatment, the pHBC values of the N300 and N900 treatments decreased by 20.9% and 20.2% in the NA soils, respectively. The pHBC was reduced by 57.2% and 47.9% in HA soils in the N900 treatment compared with the N0 and N300 treatments, respectively, whereas the pHBC was not different in the MA soils. This indicates that heavy soil acidification increased the pHBC, whereas an excessive application of N fertilizer reduced the pHBC.

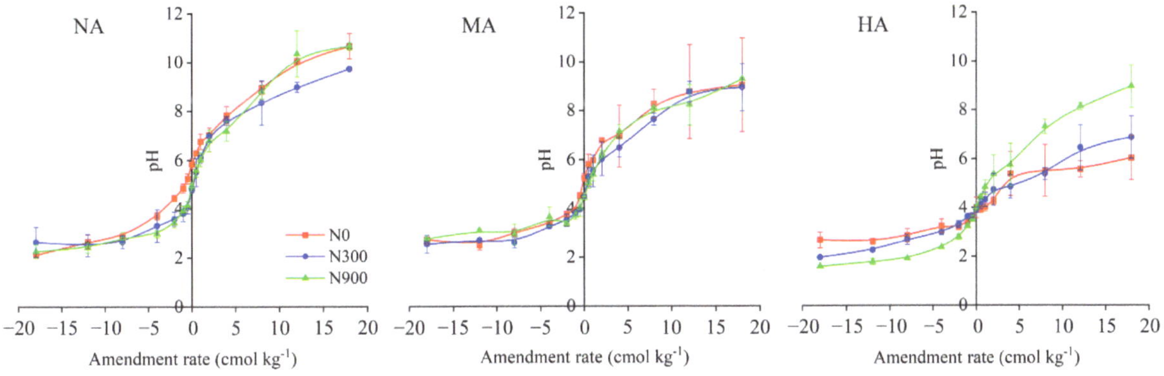

Figure 2. Titration curves with different N application rates at tea plantations with different degrees of acidification. Note: on the *X* axis, negative values indicate the amount of acid, and positive values indicate the amount of alkali.

Table 3. Soil pH buffering capacity under different N application rates at tea plantations with different degrees of acidification.

Soil Acidification Degree	Treatments	Linear Fitting Equation			pHBC
		a	b	R^2	(cmol kg^{-1})
NA	N0	0.7240	5.7981	0.9628 **	1.38 ± 0.055 a
	N300	0.9348	4.9919	0.9415 **	1.09 ± 0.195 b
	N900	0.9106	4.9938	0.9684 **	1.10 ± 0.089 b
MA	N0	0.8321	5.1636	0.9562 **	1.21 ± 0.071 a
	N300	0.7086	4.6700	0.9370 **	1.52 ± 0.407 a
	N900	0.7576	4.6164	0.9722 **	1.32 ± 0.057 a
HA	N0	0.277	3.7981	0.9336 **	3.54 ± 0.206 a
	N300	0.3646	3.9676	0.9830 **	2.91 ± 0.794 a
	N900	0.6876	4.0471	0.9838 **	1.52 ± 0.385 b

Note: the different lowercase letters indicate significant differences at $p < 0.05$ for different N application rates at the same degree of acidification. ** represents significance at 0.01 probability level.

3.3. *Effect of N Fertilizer Application Rates on Exchange Performances of Tea Plantations*

3.3.1. Exchangeable Total Acidity

Both E-Al and ETA showed HA > MA > NA at all the tea plantations (Table 4). In NA soils, the E-H, E-Al, and ETA contents increased continuously with increasing N application rates, whereas the contents for the N900 treatment were 4.9, 3.0, and 3.2 times higher than those of the N0 treatment, respectively. In the MA soils, the E-Al and ETA contents kept increasing with increasing N application rates, whereas in the N900 treatment, the contents significantly increased by 33.1% and 29.3%, respectively, compared to the N0 treatment. In the HA soils, the E-H, E-Al, and ETA contents all tended to decrease with increasing N application rates, whereas for the N900 treatment, the contents significantly decreased by 27.0%, 19.0%, and 19.2%, respectively, compared to the N0 treatment.

3.3.2. Exchangeable Base Cations

The CEC showed HA > MA > NA at tea plantations with different acidities, whereas the E-Ca, E-Mg, E-K, E-Na, TEB, and BSP showed NA > MA > HA (Tables 5 and 6). The application of N fertilizer significantly reduced the E-Ca, E-Mg, E-K, E-Na, TEB, and BSP in NA soils, whereas the N900 treatment significantly reduced these values by 65.4%, 67.3%, 60.8%, 37.9%, 65.0%, and 63.7%, respectively, compared with the N0 treatment. The E-Ca, E-Mg, TEB, and BSP trended to increase and then decrease with increasing N application rates in the MA soils. Compared to the N150 treatment, the E-Ca, E-Mg, TEB, and BSP of the N900 treatment were significantly reduced by 41.9%, 42.9%, 41.1%, and 42.7%, respectively. However, the application of N fertilizer reduced the CEC, and the CEC of the N900 treatment was significantly reduced by 14.9% compared to that of the N0 treatment.

Table 4. Effects of N application rates on soil exchangeable acids at tea plantations with different degrees of acidification.

Soil Acidification Degree	Treatments	E-H (cmol kg^{-1})	E-Al (cmol kg^{-1})	ETA (cmol kg^{-1})
NA	N0	0.054 ± 0.024 ^{b}	0.87 ± 0.149 ^{b}	0.92 ± 0.173 ^{b}
	N150	0.241 ± 0.114 ^{a}	2.15 ± 0.329 ^{ab}	2.39 ± 0.321 ^{ab}
	N300	0.157 ± 0.058 ^{ab}	3.11 ± 1.683 ^{a}	3.27 ± 1.626 ^{a}
	N600	0.238 ± 0.140 ^{a}	3.47 ± 0.664 ^{a}	3.71 ± 0.660 ^{a}
	N900	0.265 ± 0.073 ^{a}	2.64 ± 0.417 ^{a}	2.90 ± 0.345 ^{a}
MA	N0	0.258 ± 0.173 ^{a}	6.13 ± 0.173 ^{b}	6.38 ± 0.000 ^{b}
	N150	0.072 ± 0.026 ^{a}	5.89 ± 1.453 ^{b}	5.96 ± 1.478 ^{b}
	N300	0.059 ± 0.006 ^{a}	6.78 ± 0.074 ^{ab}	6.84 ± 0.080 ^{ab}
	N600	0.090 ± 0.000 ^{a}	6.63 ± 0.400 ^{ab}	6.72 ± 0.400 ^{ab}
	N900	0.092 ± 0.002 ^{a}	8.16 ± 0.158 ^{a}	8.25 ± 0.160 ^{a}
HA	N0	0.178 ± 0.014 ^{a}	13.29 ± 0.249 ^{a}	13.47 ± 0.235 ^{a}
	N150	0.124 ± 0.019 ^{bc}	11.29 ± 0.977 ^{b}	11.41 ± 0.990 ^{b}
	N300	0.148 ± 0.023 ^{ab}	11.40 ± 1.137 ^{b}	11.54 ± 1.158 ^{b}
	N600	0.099 ± 0.016 ^{c}	10.15 ± 0.769 ^{b}	10.25 ± 0.767 ^{b}
	N900	0.130 ± 0.023 ^{bc}	10.76 ± 0.359 ^{b}	10.89 ± 0.376 ^{b}

Note: the different lowercase letters indicate significant differences at $p < 0.05$ for different N application rates at the same degree of acidification.

Table 5. Effects of N application rates on EBCs at tea plantations with different degrees of acidification.

Soil Acidification Degree	Treatments	E-Ca (cmol kg^{-1})	E-Mg (cmol kg^{-1})	E-K (cmol kg^{-1})	E-Na (cmol kg^{-1})
NA	N0	9.33 ± 1.23 ^{a}	1.53 ± 0.404 ^{a}	0.904 ± 0.271 ^{a}	0.116 ± 0.022 ^{a}
	N150	5.03 ± 2.54 ^{b}	0.83 ± 0.503 ^{b}	0.558 ± 0.172 ^{b}	0.104 ± 0.023 ^{ab}
	N300	5.00 ± 2.49 ^{b}	0.53 ± 0.252 ^{b}	0.546 ± 0.199 ^{b}	0.085 ± 0.018 ^{ab}
	N600	3.83 ± 1.63 ^{b}	0.43 ± 0.115 ^{b}	0.388 ± 0.036 ^{b}	0.072 ± 0.007 ^{b}
	N900	3.23 ± 0.68 ^{b}	0.50 ± 0.200 ^{b}	0.354 ± 0.072 ^{b}	0.072 ± 0.014 ^{b}
MA	N0	3.45 ± 0.21 ^{ab}	0.45 ± 0.071 ^{ab}	0.384 ± 0.181 ^{a}	0.098 ± 0.005 ^{a}
	N150	4.65 ± 0.92 ^{a}	0.70 ± 0.141 ^{a}	0.384 ± 0.109 ^{a}	0.087 ± 0.031 ^{a}
	N300	3.15 ± 0.35 ^{ab}	0.50 ± 0.000 ^{ab}	0.185 ± 0.009 ^{a}	0.058 ± 0.010 ^{a}
	N600	2.85 ± 0.78 ^{ab}	0.35 ± 0.071 ^{b}	0.467 ± 0.118 ^{a}	0.080 ± 0.000 ^{a}
	N900	2.70 ± 0.85 ^{b}	0.40 ± 0.141 ^{b}	0.269 ± 0.036 ^{a}	0.065 ± 0.010 ^{a}
HA	N0	2.53 ± 0.32 ^{a}	0.23 ± 0.058 ^{a}	0.350 ± 0.036 ^{a}	0.072 ± 0.004 ^{a}
	N150	3.00 ± 0.75 ^{a}	0.30 ± 0.000 ^{a}	0.388 ± 0.118 ^{a}	0.087 ± 0.011 ^{a}
	N300	2.67 ± 0.42 ^{a}	0.30 ± 0.000 ^{a}	0.354 ± 0.018 ^{a}	0.065 ± 0.004 ^{a}
	N600	2.57 ± 0.72 ^{a}	0.23 ± 0.058 ^{a}	0.222 ± 0.027 ^{a}	0.065 ± 0.004 ^{a}
	N900	2.53 ± 0.47 ^{a}	0.27 ± 0.058 ^{a}	0.234 ± 0.009 ^{a}	0.101 ± 0.023 ^{a}

Note: the different lowercase letters indicate significant differences at $p < 0.05$ for different N application rates at the same degree of acidification.

Table 6. Effects of N application rates on BSP at tea plantations with different degrees of acidification.

Soil Acidification Degree	Treatments	TEB (cmol kg^{-1})	CEC (cmol kg^{-1})	BSP (%)
NA	N0	11.89 ± 1.41 ^{a}	14.9 ± 0.70 ^{a}	80.2 ± 12.19 ^{a}
	N150	6.53 ± 3.20 ^{b}	14.4 ± 1.27 ^{a}	44.7 ± 20.68 ^{b}
	N300	6.16 ± 2.98 ^{b}	16.1 ± 2.18 ^{a}	37.5 ± 14.41 ^{b}
	N600	4.73 ± 1.82 ^{b}	15.6 ± 1.67 ^{a}	29.8 ± 9.29 ^{b}
	N900	4.16 ± 1.02 ^{b}	14.3 ± 0.68 ^{a}	29.1 ± 7.27 ^{b}
MA	N0	4.38 ± 0.33 ^{ab}	18.7 ± 0.40 ^{a}	23.4 ± 2.24 ^{b}
	N150	5.82 ± 1.20 ^{a}	16.2 ± 1.19 ^{a}	35.8 ± 4.77 ^{a}
	N300	3.89 ± 0.33 ^{ab}	16.1 ± 1.27 ^{a}	24.4 ± 4.02 ^{ab}
	N600	3.75 ± 0.73 ^{ab}	17.3 ± 2.83 ^{a}	21.6 ± 0.69 ^{b}
	N900	3.43 ± 0.96 ^{b}	17.0 ± 1.45 ^{a}	20.5 ± 7.39 ^{b}
HA	N0	3.62 ± 0.39 ^{a}	30.2 ± 3.04 ^{ab}	12.2 ± 1.05 ^{a}
	N150	3.76 ± 0.89 ^{a}	32.3 ± 1.80 ^{a}	11.7 ± 3.36 ^{a}
	N300	3.38 ± 0.41 ^{a}	30.0 ± 2.43 ^{a}	11.4 ± 2.06 ^{a}
	N600	3.08 ± 0.76 ^{a}	27.7 ± 1.14 ^{bc}	11.1 ± 2.27 ^{a}
	N900	3.11 ± 0.53 ^{a}	25.7 ± 0.92 ^{c}	12.1 ± 1.76 ^{a}

Note: the different lowercase letters indicate significant differences at $p < 0.05$ for different N application rates at the same degree of acidification.

3.3.3. Inter-Subject Effect Test

The results of a two-way ANOVA showed that the N application rate had a significant effect on the pH, BSP, pHBC, TEB, E-Ca, E-Mg, E-K, and E-Na, whereas there was no effect on the E-H, E-Al, ETA, and CEC (Table 7). The degree of soil acidification had a significant effect on all indicators of the soil exchange properties and the pHBC. However, the interaction of the N application rate and acidification degree of the soil had no effect on the E-Na but had significant effects on all other indicators.

Table 7. F values of inter-subject effect test of N application rate (N) and acidification degree of soil (A) on soil acidification characteristics.

Source	pH	pHBC	ETA	TEB	CEC	BSP	E-H	E-Al	E-Ca	E-Mg	E-K	E-Na
N	14.17 **	7.84 **	1.11	4.72 **	2.03	5.37 **	0.49	1.10	4.08 *	4.70 **	3.93 *	4.60 **
A	197.45 **	39.86 **	484.87 **	19.49 **	283.59 **	51.17 **	4.40 **	480.42 **	17.01 **	20.73 **	12.85 **	7.52 **
N × A	10.00 **	6.80 **	7.26 **	3.72 **	2.63 **	5.52 **	3.92 **	6.68 **	3.36 **	4.45 **	2.40 **	1.57 **

Note: * represents significance at 0.05 probability level. ** represents significance at 0.01 probability level.

3.4. Relationship between Soil Exchangeable Function and N Application Rate, pH, and pHBC

A correlation analysis was performed, including all three soils, and the results showed that the N application rate was significantly or highly significantly negatively correlated with the pH, E-Ca, E-Mg, E-K, E-Na, TEB, and BSP (Figure 3). The pH was negatively correlated with the E-Al, ETA, CEC, and pHBC, whereas it was positively correlated with the E-Ca, E-Mg, E-K, E-Na, TEB, and BSP. In addition, the pHBC was positively correlated with the E-Al, ETA, and CEC but negatively correlated with the BSP.

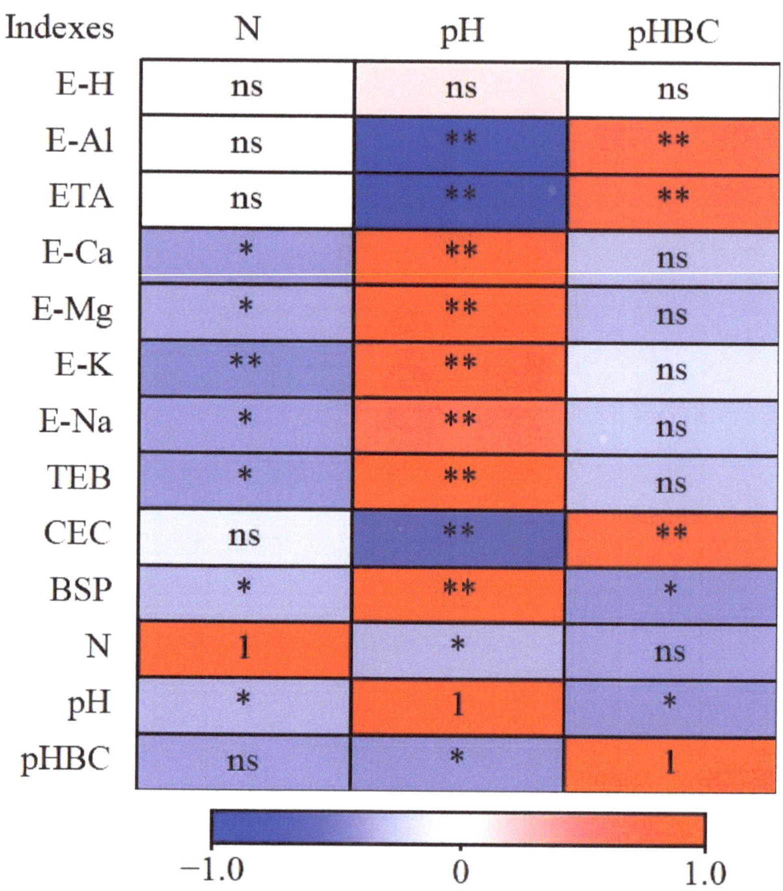

Figure 3. Correlation coefficient and path coefficient between soil exchangeable function and N application rate, pH, and pH buffer capacity. Note: ns represents no difference. * represents significance at 0.05 probability level. ** represents significance at 0.01 probability level.

We performed a stepwise regression analysis of the pH and pHBC with the soil exchange properties and N application rate and obtained the following regression equations:

$$\text{pH} = 3.238 + 0.048\text{BSP} - 1.977\text{E-Mg} + 0.212\text{TEB}\ (R = 0.977^{**}) \quad (1)$$

$$\text{pHBC} = -0.194 + 0.108\text{CEC} - 0.001\text{N}\ (R = 0.851^{**}) \quad (2)$$

In Equation (1), the direct path coefficients of the BSP, E-Mg, and TEB were 1.161, 0.683, and −0.906, respectively, and their partial regression coefficients reached extremely significant levels ($p < 0.01$). In Equation (2), the direct path coefficients of the CEC and N were 0.770 and −0.263, respectively, and their partial regression coefficients reached a significant level ($p < 0.05$). This indicated that the BSP, E-Mg, and TEB had significant direct effects on the pH, with the BSP having the greatest effect. The CEC and N had significant direct effects on the pHBC, with the CEC having the greatest effect.

4. Discussion

4.1. Characteristics of Soil pHBC at Tea Plantations

The soil pH buffer capacity (pHBC) is an indicator of soil resistance to acidification or alkalization. A higher pHBC value indicates a smaller change in soil pH for the same acid–base input [28]. In this study, the soil acid–base buffer curve in the pH range of 4–6 rose sharply, indicating that the soil had poor buffering performance against acid–base addition in this pH range [29]. Meanwhile, the results of this study showed that the pHBC of the HA tea plantation was significantly higher than those of the NA and MA tea plantations, which may have been due to the fact that the soil buffering substances at the HA tea plantation were mainly an iron–aluminum buffering system (pH < 4) that had a strong soil acid–base buffering capacity [30]. However, the CEC is an important factor that affects the soil acid–base buffering capacity. Many studies have shown that the soil pHBC has a significant positive correlation with the CEC [31–33]. In this study, a stepwise regression analysis showed that the CEC had a significant direct effect on the pHBC, which was consistent with the above results. In addition, the results of this study also showed that the pHBC was significantly positively correlated with the E-Al, whereas it was significantly negatively correlated with both the TEB and BSP, suggesting that soil acidification makes Al^{3+} play a greater role than EBCs in the acid–base buffering performances of tea plantation soils [9]. It is noteworthy that the stepwise regression analysis showed a significant direct negative effect of the N application rate on the pHBC, which indicates that the excessive application of N fertilizer is an important factor in the decrease in the soil pHBC at tea plantations.

4.2. Relationship between Soil pH and Exchangeable Base Cations

In this study, the pH was significantly negatively correlated with the E-Al, and the E-Al accounted for more than 90% of the ETA, whereas there was no significant correlation with the E-H, which suggests that E-Al plays a determinant role in driving the acidification at tea plantations [22]. Generally, a tea tree is an aluminum-loving crop, and Al can be returned to the soil by fallen leaves and trimmings to improve the E-Al content [34]. The leaching of exchangeable base cations (EBCs) is another important reason for soil acidification [35]. In this study, EBCs showed E-Ca > E-Mg > E-K > E-Na, whereas the pH was highly significantly and positively correlated with the E-Ca, E-Mg, E-K, and E-Na, and the correlation coefficients with the E-Ca and E-Mg were higher, which indicated that the E-Ca and E-Mg had a greater effect on the soil pH. In addition, the proportion of E-Al in the CEC gradually increased, whereas the proportion of EBCs in the CEC gradually decreased (Figure 4), suggesting that replacing EBCs with E-Al as the major cation is the main mechanism of soil acidification at tea plantations [13]. However, the correlation between the pH and E-Al and EBCs was not linear but was a highly significant power function correlation (Figure 4). The results showed that as E-Al increased, the pH decreased to approximately 4.0 and then did not continue to decrease, whereas as the EBCs increased, the pH increased to approximately 6.0 and then did not continue to increase. This may have been because when the pH was <4 or >6, the acid–base buffering capacity of the soil increased sharply and the pH hardly changed. This was more consistent with the non-linear relationship between the pH and BSP that was considered in earlier studies [36,37] but was inconsistent with the linear relationship between the pH and BSP suggested in the study by Hao et al. [38].

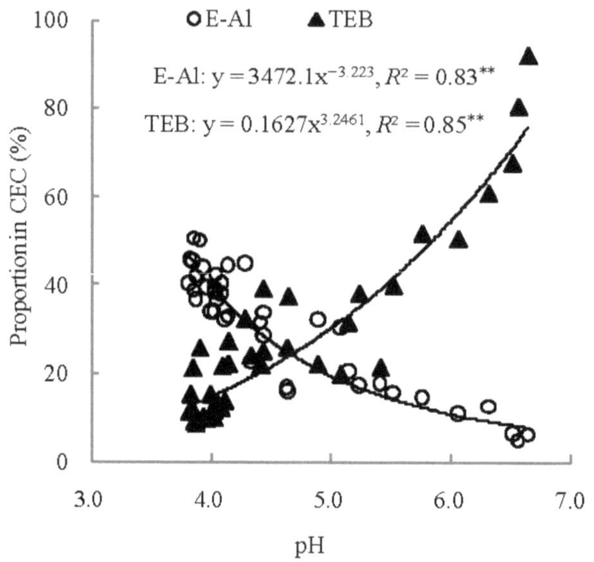

Figure 4. The ratio of TEB and E-Al in CEC changes with soil pH. Note: ** represents significance at 0.01 probability level.

4.3. *Effect of N Application on Soil Acidification at Tea Plantations*

Excessive application of N fertilizer was found to be the main anthropogenic factor exacerbating soil acidification at tea plantations [39]. In this study, the effect of the N application rate on the pH decreased with an increase in the degree of acidification. This may have been because the pHBC increased with an increasing acidification degree at the tea plantation, and the pH change was small. When the soil pH was >4, the E-Al content increased with an increasing N application rate, whereas the opposite was true for the TEB and BSP, which led to a constant decrease in the soil pH. At the same time, the higher the soil pH before the experiment, the more obvious the decrease in pH due to N application. At tea plantations with pH < 4, N application reduced the contents of E-H and E-Al, which did not lead to a further decrease in pH and even led to a small increase. This may have been due to the fact that the soil nitrogen nitrification was affected by the soil pH, and the soil nitrification was inhibited, thereby reducing the production of ETA in the HA soils. However, enhanced soil nitrification increased the contents of ETA and NO_3^- in the NA and MA soils, which were eventually lost with salt-based cations [40]. Meanwhile, the correlation analysis showed that N application was significantly negatively correlated with the pH and EBCs but was not significantly correlated with the ETA, which indicated that the loss of soil EBCs due to N application was the main cause of soil acidification at tea plantations. It is worth noting that the excessive application of N fertilizer also reduced the pHBC, which may be one of the reasons for exacerbated soil acidification at tea plantations.

In practical agricultural production, reasonable N fertilizer management measures to increase EBCs (especially E-Ca and E-Mg) and reduce E-Al contents can prevent and improve soil acidification at tea plantations. For heavily acidified tea plantations with pH < 4, N fertilizer application is no longer the main factor causing soil acidification. It is recommended to reduce the E-Al content and increase the EBC content to improve the soil pH by increasing limestone, organic fertilizer, biochar, and fertilizer, which are rich in base ions [41–44]. For tea plantations with pH > 4, the unreasonable application of N fertilizer is an important factor that exacerbates soil acidification. Therefore, strictly controlling the N fertilizer rate, reducing nitrification by adding nitrification inhibitors, and increasing the CEC content by using organic fertilizers instead of chemical fertilizers are recommended as important measures to prevent the further acidification of tea plantation soils [45–47].

5. Conclusions

N fertilizer is an important factor affecting soil acidification at tea plantations. When the tea plantations had pH values > 4.0, the E-Al contents increased with increasing N application rates, whereas the EBC contents decreased, which in turn led to decreases in soil pH. When the tea plantations had pH values < 4.0, the application of N fertilizer reduced the ETA content, which in turn prevented the soil pH from continuing to decrease with the increase in the N application rates. The acid–base buffering capacity of the soils at tea plantations was weak at pH values of 4.0–6.0, while the excessive application of N fertilizer reduced the soil pHBC. The loss of EBCs owing to N application is the main mechanism of soil acidification at tea plantations. In agricultural production, the amount of nitrogen fertilizer should be strictly controlled for tea plantations that are not seriously acidified, and measures such as applying nitrogen fertilizer synergists and organic fertilizers should be taken to prevent further acidification of the soil. For severely acidified tea plantations, alkaline biomass materials should be appropriately applied to improve the soil acidification.

Author Contributions: Conceptualization, Y.L. (Yanling Liu), M.Z., Y.L. (Yu Li) and T.J.; Data curation, Y.L. (Yanling Liu), X.H., Y.Z., H.X., Y.Y. and H.Z.; Formal analysis, M.Z. and Y.L. (Yanling Liu).; Supervision, Y.L. (Yu Li); Writing—original draft, Y.L. (Yanling Liu); Writing—review and editing, Y.L. (Yanling Liu), M.Z. and Y.L (Yu Li). All authors have read and agreed to the published version of the manuscript.

Funding: This research was funded by the Guizhou Provincial Key Technology R&D Program (No. [2022]YB110 and No. [2020]1Y119), the National Major Agricultural Science and Technology Projects (No. NK2022180303), and the National Natural Science Foundation of China (No. 32060302).

Data Availability Statement: Raw data can be provided to researchers on request by corresponding with the first author or the corresponding author.

Conflicts of Interest: The authors declare no conflict of interest.

References

1. Kirk, G.J.D.; Bellamy, P.H.; Lark, R.M. Changes in soil pH across England and Wales in response to decreased acid deposition. *Glob. Change Biol.* **2010**, *16*, 3111–3119. [CrossRef]
2. Wu, Z.; Sun, X.; Sun, Y.; Yan, J.; Zhao, Y.; Chen, J. Soil acidification and factors controlling topsoil pH shift of cropland in central China from 2008 to 2018. *Geoderma* **2022**, *408*, 115586. [CrossRef]
3. De Vries, W.; Hettelingh, J.P.; Posch, M. *Critical Loads and Dynamic Risk Assessments: Nitrogen, Acidity and Metals in Terrestrial and Aquatic Ecosystems*; Springer: Dordrecht, The Netherlands, 2015; Volume 25.
4. Tian, Q.; Liu, N.; Bai, W.; Li, L.; Chen, J.; Reich, P.B.; Yu, Q.; Guo, D.; Smith, M.D.; Knapp, A.K.; et al. A novel soil manganese mechanism drives plant species loss with increased nitrogen deposition in a temperate steppe. *Ecology* **2016**, *97*, 65–74. [CrossRef] [PubMed]
5. Dai, Z.; Zhang, X.; Tang, C.; Muhammad, N.; Wu, J.; Brookes, P.C.; Xu, J. Potential role of biochars in decreasing soil acidificationb-A critical review. *Sci. Total Environ.* **2017**, *581–582*, 601–611. [CrossRef] [PubMed]
6. Zhang, X.; Liu, W.; Zhang, G.; Jiang, L.; Han, X. Mechanisms of soil acidification reducing bacterial diversity. *Soil Biol. Biochem.* **2015**, *81*, 275–281. [CrossRef]
7. Meng, C.; Tian, D.; Zeng, H.; Li, Z.; Yi, C.; Niu, S. Global soil acidification impacts on belowground processes. *Environ. Res. Lett.* **2019**, *14*, 074003. [CrossRef]
8. Zheng, Q.; Ding, J.; Lin, W.; Yao, Z.; Li, Q.; Xu, C.; Zhuang, S.; Kou, X.; Li, Y. The influence of soil acidification on N_2O emissions derived from fungal and bacterial denitrification using dual isotopocule mapping and acetylene inhibition. *Environ. Pollut.* **2022**, *303*, 119076. [CrossRef]
9. Tian, D.; Niu, S. A global analysis of soil acidification caused by nitrogen addition. *Environ. Res. Lett.* **2015**, *2*, 024019. [CrossRef]
10. Tarkalson, D.D.; Payero, J.O.; Hergert, G.W.; Cassman, K.G. Acidification of soil in a dry land winter wheat-sorghum/corn-fallow rotation in the Semiarid U.S. Great Plains. *Plant Soil* **2016**, *283*, 367–379. [CrossRef]
11. Cecchini, G.; Andreetta, A.; Marchetto, A.; Carnicelli, S. Atmospheric deposition control of soil acidification in central Italy. *Catena* **2019**, *182*, 104102. [CrossRef]
12. Schroder, J.L.; Zhang, H.; Girma, K.; Raun, W.R.; Penn, C.J.; Payton, M.E. Soil acidification from long-term use of nitrogen fertilizers on winter wheat. *Soil Sci. Soc. Am. J.* **2011**, *75*, 957. [CrossRef]
13. Cai, Z.; Wang, B.; Xu, M.; Zhang, H.; He, X.; Zhang, L.; Gao, S. Intensified soil acidification from chemical N fertilization and prevention by manure in an 18-year field experiment in the red soil of southern China. *J. Soils Sediments* **2014**, *15*, 260–270. [CrossRef]

14. Zhou, J.; Xia, F.; Liu, X.; He, Y.; Xu, J.; Brookes, P.C. Effects of nitrogen fertilizer on the acidification of two typical acid soils in South China. *J. Soils Sediments* **2013**, *14*, 415–422. [CrossRef]
15. Lucas, R.W.; Klaminder, J.; Futter, M.N.; Bishop, K.H.; Egnell, G.; Laudon, H.; Högberg, P. A meta-analysis of the effects of nitrogen additions on base cations: Implications for plants, soils, and streams. *For. Ecol. Manag.* **2011**, *262*, 95–104. [CrossRef]
16. Su, S.; Zhou, X.; Wan, C.; Li, Y.; Kong, W. Land use changes to cash crop plantations: Crop types, multilevel determinants and policy implications. *Land Use Policy* **2016**, *50*, 379–389. [CrossRef]
17. Ruan, J.; Gerendás, J.; Härdter, R.; Sattelmacher, B. Effect of nitrogen form and root-zone pH on growth and nitrogen uptake of tea (*Camellia sinensis*) plants. *Ann. Bot.* **2007**, *99*, 301–310. [CrossRef]
18. Fung, K.F.; Carr, H.P.; Zhang, J.; Wong, M.H. Growth and nutrient uptake of tea under different aluminum concentrations. *J. Sci. Food Agric.* **2008**, *88*, 1582–1591. [CrossRef]
19. Qiao, C.; Xu, B.; Han, Y.; Wang, J.; Wang, X.; Liu, L.; Liu, W.; Wan, S.; Tan, H.; Liu, Y.; et al. Synthetic nitrogen fertilizers alter the soil chemistry, production and quality of tea: A meta-analysis. *Agron. Sustain. Dev.* **2018**, *38*, 10. [CrossRef]
20. Ni, K.; Liao, W.Y.; Yi, X.Y.; Niu, S.G.; Ma, L.F.; Shi, Y.Z.; Zhang, Q.F.; Liu, M.Y.; Ruan, J.Y. Fertilization status and reduction potential in tea plantations of China. *J. Plant Nutr. Fertil.* **2019**, *25*, 421–432, (In Chinese with English Abstract).
21. Yan, P.; Wu, L.; Wang, D.; Fu, J.; Shen, C.; Li, X.; Zhang, L.; Zhang, L.; Fan, L.; Wenyan, H. Soil acidification in Chinese tea plantations. *Sci. Total Environ.* **2020**, *715*, 136963. [CrossRef]
22. Yang, X.; Ni, K.; Shi, Y.; Yi, X.; Zhang, Q.; Fang, L.; Ma, L.; Ruan, J. Effects of long-term nitrogen application on soil acidification and solution chemistry of a tea plantation in China. *Agric. Ecosyst. Environ.* **2018**, *252*, 74–82. [CrossRef]
23. Kemmitt, S.J.; Wright, D.; Jones, D.L. Soil acidification used as a management strategy to reduce nitrate losses from agricultural land. *Soil Biol. Biochem.* **2005**, *37*, 867–875. [CrossRef]
24. Bowman, W.D.; Cleveland, C.C.; Halada, Ĺ.; Hreško, J.; Baron, J.S. Negative impact of nitrogen deposition on soil buffering capacity. *Nat. Geosci.* **2008**, *1*, 767–770. [CrossRef]
25. Wang, X.; Feng, J.; Ao, G.; Qin, W.; Han, M.; Shen, Y.; Liu, M.; Chen, Y.; Zhu, B. Globally nitrogen addition alters soil microbial community structure, but has minor effects on soil microbial diversity and richness. *Soil Biol. Biochem.* **2023**, *179*, 108982. [CrossRef]
26. Bao, S.D. *Soil and Agricultural Chemistry Analysis*; China Agricultural Press: Beijing, China, 2000.
27. Liu, L.; Cheng, Y.; Li, Z.; Xie, D. Influence of fitting method on the accuracy of measuring soil acid buffering capacity by acid-base titration. *Acta Pedol. Sin.* **2022**, *59*, 509–516. [CrossRef]
28. Xu, R.; Zhao, A.; Yuan, J.; Jiang, J. pH buffering capacity of acid soils from tropical and subtropical regions of China as influenced by incorporation of crop straw biochars. *J. Soils Sediments* **2012**, *12*, 494–502. [CrossRef]
29. Su, Y.; Wang, Y.; Zhang, Y.; Luo, Y.; Sun, L.; Song, L.; Liao, W. Soil pH buffer capacity of tea plantation with different planting years. *Chin. J. Appl. Ecol.* **2014**, *25*, 2914–2918. [CrossRef]
30. Ulrich, B. Natural and anthropogenic components of soil acidification. *Z. Pflanz. Bodenkd.* **1986**, *149*, 702–717. [CrossRef]
31. Nelson, P.N.; Su, N. Soil pH buffering capacity: A descriptive function and its application to some acidic tropical soils. *Aust. J. Soil Res.* **2010**, *48*, 201. [CrossRef]
32. Wei, H.; Liu, Y.; Xiang, H.; Zhang, J.; Li, S.; Yang, J. Soil pH responses to simulated acid rain leaching in three agricultural soils. *Sustainability* **2019**, *12*, 280. [CrossRef]
33. Liu, L.; Xie, D.; Li, Z.; Liu, F. Cations exchange and its effect on acid buffering capacity of acid purple soil. *Acta Pedol. Sin.* **2020**, *57*, 887–897. [CrossRef]
34. Wang, L.; Butterly, C.R.; Tian, W.; Herath, H.M.S.K.; Xi, Y.; Zhang, J.; Xiao, X. Effects of fertilization practices on aluminum fractions and species in a wheat soil. *J. Soils Sediments* **2016**, *16*, 1933–1943. [CrossRef]
35. Dong, Y.; Yang, J.; Zhao, X.; Yang, S.; Mulder, J.; Dörsch, P.; Peng, X.; Zhang, G. Soil acidification and loss of base cations in a subtropical agricultural watershed. *Sci. Total Environ.* **2022**, *827*, 154338. [CrossRef] [PubMed]
36. Ranney, R.W.; Ciolkosz, E.J.; Petersen, G.W.; Matelski, R.P.; Johnson, L.J.; Cunningham, R.L. The pH base-saturation relationship in B and C horizons of Pennsylvania soils. *Soil Sci.* **1974**, *118*, 247–253. [CrossRef]
37. Bowman, J.R.; Lannan, J.E. Evaluation of soil pH-percent base saturation relationships for use in estimating the lime requirements of earthen aquaculture ponds. *J. World Aquac. Soc.* **1995**, *26*, 172–182. [CrossRef]
38. Hao, T.; Zhu, Q.; Zeng, M.; Shen, J.; Shi, X.; Liu, X.; Zhang, F.; de Vries, W. Impacts of nitrogen fertilizer type and application rate on soil acidification rate under a wheat-maize double cropping system. *J. Environ. Manag.* **2020**, *270*, 110888. [CrossRef]
39. Li, S.; Li, H.; Yang, C.; Wang, Y.; Xue, H.; Niu, Y. Rates of soil acidification in tea plantations and possible causes. *Agric. Ecosyst. Environ.* **2016**, *233*, 60–66. [CrossRef]
40. Cheng, Y.; Wang, J.; Mary, B.; Zhang, J.; Cai, Z.; Chang, S.X. Soil pH has contrasting effects on gross and net nitrogen mineralizations in adjacent forest and grassland soils in central Alberta, Canada. *Soil Biol. Biochem.* **2013**, *57*, 848–857. [CrossRef]
41. Rheinheimer, D.S.; Tiecher, T.; Gonzattoa, R.; Zafar, M.; Brunetto, G. Residual effect of surface-applied lime on soil acidity properties in a long-term experiment under no-till in a Southern Brazilian sandy Ultisol. *Geoderma* **2018**, *313*, 7–16. [CrossRef]
42. Crusciol, C.A.C.; Foltran, R.; Rossato, O.B.; McCray, J.M.; Rossetto, R. Effects of surface application of calcium-magnesium silicate and gypsum on soil fertility and sugarcane yield. *Rev. Bras. Ciência Solo* **2014**, *38*, 1843–1854. [CrossRef]
43. Zhu, Q.C.; Liu, X.J.; Hao, T.X.; Zeng, M.F.; Shen, J.B.; Zhang, F.S. Modeling soil acidification in typical Chinese cropping systems. *Sci. Total Environ.* **2018**, *613–614*, 1339–1348. [CrossRef] [PubMed]

44. Yang, W.; Li, C.; Wang, S.; Zhou, B.; Mao, Y.; Rensing, C.; Xing, S. Influence of biochar and biochar-based fertilizer on yield, quality of tea and microbial community in an acid tea orchard soil. *Appl. Soil Ecol.* **2021**, *166*, 104005. [CrossRef]
45. Qiao, C.; Mia, S.; Wang, Y.; Hou, J.; Xu, B. Assessing the effects of nitrification inhibitor DMPP on acidification and inorganic N leaching loss from tea (*Camellia sinensis* L.) cultivated soils with increasing urea-N rates. *Sustainability* **2021**, *13*, 994. [CrossRef]
46. Wu, X.; Zhang, B.; Yu, Y.; Yan, M.; Ni, K.; Cui, J.; Wang, S.; Cheng, Y. Effects of nitrification inhibitors on nitrification rate of urea in four typical tea soils. *J. Agro-Environ. Sci.* **2017**, *36*, 2063–2070. [CrossRef]
47. Shi, R.; Liu, Z.; Li, Y.; Jiang, T.; Xu, M.; Li, J.; Xu, R. Mechanisms for increasing soil resistance to acidification by long-term manure application. *Soil Tillage Res.* **2019**, *185*, 77–84. [CrossRef]

agriculture

MDPI

Article

Effects of Tillage and Sowing Methods on Soil Physical Properties and Corn Plant Characters

Ying Wang [1,2], Sen Yang [1,2], Jian Sun [1], Ziguang Liu [2], Xinmiao He [2] and Jinyou Qiao [1,*]

1 College of Engineering, Northeast Agricultural University, Harbin 150030, China
2 Key Laboratory of Combining Farming and Animal Husbandry, Ministry of Agriculture and Rural Affairs, Harbin 150030, China
* Correspondence: S200701052@neau.edu.cn

Abstract: In the northeast plains of China, the intensive utilization of agricultural soils has been a persistent issue, and finding ways to utilize soil resources efficiently and sustainably through a scientifically-driven management system has become a crucial challenge for agricultural production. Conservation tillage is a crucial technology for sustainable agriculture. Currently, plow and rotary tillage are the dominant methods used in Mollisols, but there is limited information on the effects of different conservation tillage practices in this region. The objective of this study was to investigate the short-term impact of tillage and sowing methods on soil physical properties and corn plant growth and to examine the relationship between soil physical properties and plant characteristics during various stages of growth. This study consisted of four tillage and sowing methods: plow tillage and precision seeder sowing (PTS), rotary tillage and precision seeder sowing (RTS), no-tillage and no-tillage seeder sowing (NTS), and no-tillage and precise sowing in stubble field (STS) (all four treatments involved total straw return). The results indicated that the soil penetration resistance (SPR) in the 10–40 cm soil layer under the PTS treatment was significantly lower (by 11.9% to 18%) compared to the other treatments ($p < 0.05$). On average, the soil moisture content in the NTS treatment was 2.7% and 1.4% higher than that of the PTS and RTS treatments. Additionally, soil temperature was 5.6% to 8.6% lower under the STS treatment compared to the other treatments during late corn growth. The RTS treatment also significantly reduced the bulk density of surface soil. High SPR impeded early crop growth but did not impact mid-crop development, while low soil temperature was one of the main factors affecting late corn growth and development as temperatures decreased. Based on the comparisons, we found that the short-term implementation of conservation tillage did not result in a significant decrease in corn yield. We believe that the short-term implementation of NTS tillage sowing practices in Mollisol regions is a feasible option.

Keywords: tillage practice; crop residue management; Mollisols

Citation: Wang, Y.; Yang, S.; Sun, J.; Liu, Z.; He, X.; Qiao, J. Effects of Tillage and Sowing Methods on Soil Physical Properties and Corn Plant Characters. *Agriculture* **2023**, *13*, 600. https://doi.org/10.3390/agriculture13030600

Academic Editors: Anetta Siwik-Ziomek and Anna Figas

Received: 16 January 2023
Revised: 17 February 2023
Accepted: 25 February 2023
Published: 1 March 2023

1. Introduction

Northeast China boasts a significant amount of straw production and low average temperatures, making it a critical and challenging region for the comprehensive mineralization of straw in the country [1]. To effectively utilize straw and minimize the negative impact of straw returning to the fields during actual production, combining tillage measures can significantly improve the utilization efficiency of straw. However, extended and extensive tillage leads to a rapid depletion of soil nutrients [2]. This is especially true in the black soil region of Northeast China, where long-term and intensive tillage causes the soil environment to deteriorate and the plow layer to become shallow [3]. The objective of soil tillage is to create an environment suitable for crop growth by storing water and maintaining moisture and to enhance crop production. Selecting the appropriate tillage methods to improve soil structure and balance the relationship between various components in the soil environment will create favorable conditions for crop growth and increase production and efficiency through mechanical action [4]. Currently, plowing and rotary tillage are the main tillage methods used in Northeast China. Deep-plowing combined with straw or organic fertilizer can significantly improve soil physical structure, optimize soil three-phase ratio, and increase corn yield and harvest index [5]. A study conducted by Misbah in Ethiopia found that short-term deep tillage significantly reduced soil bulk density and penetration resistance, significantly increased soil infiltration rates, and effectively countered soil degradation [6]. Rotary tillage can decrease soil bulk density in the 0–20 cm soil layer, increase the field's water-holding capacity and soil porosity, and significantly enhance seedling plant height, stem diameter, leaf area, and dry weight per plant [7,8]. In his study of the Mediterranean region, Pietro believes that rototilling exposes organic matter to oxidation processes, crushing it and destroying soil aggregates [9]. Conservation tillage has significant ecological benefits, improving soil quality, enhancing water and fertilizer retention capacity, and controlling soil erosion. Currently, conservation tillage technologies centered on straw mulching and reduced/no-tillage are widely promoted in Northeast China. No-tillage has a positive impact on soil physical properties, but the extent of change varies with time and soil texture, and the implementation of no-tillage should be combined with straw mulching and crop rotation [10,11]. In comparison to traditional tillage, the yield-increasing effect of no-tillage with stubble increased as the planting years expanded [12]. Kahlon found that long-term use of conservation tillage and straw mulch enhances soil structural stability and carbon sequestration and that no-tillage has a higher infiltration capacity and better hydraulic conductivity than plow tillage and a lower stockpile density and infiltration resistance than tillage in the east-central region of the United States [13]. The impact of soil's physical properties on plant growth is significant, and no-tillage practices can have negative effects. No-tillage increases soil penetration resistance (SPR) and reduces soil temperature, which can hinder the accumulation of dry matter in maize during the seedling stage and negatively impact yield formation [14]. On the other hand, deep plowing can effectively reduce soil penetration resistance and promote the early above-ground growth of crops [15]. Research has indicated that SPR can not only affect soil physical properties but also have a profound impact on crop growth and yield [16,17]. Currently, research has been focused on the effects of various tillage and planting methods on soil physical properties and crop yield. However, it remains unclear how soil physical properties impact crop growth and development under different tillage and sowing methods. Conservation tillage is a key technology for sustainable agricultural development, but various conservation tillage practices can have different impacts on the soil environment. When compared to no-tillage with stubble cutting, no-tillage and precise sowing in stubble fields are more effective in protecting the soil from wind and rain erosion. Nonetheless, studies are mostly limited to the performance and adaptability of no-tillage precision seeders, and there has been limited research on the impact of no-tillage and precise sowing in stubble fields on soil physical properties and crop growth. In the Mollisols region of Northeast China, field positioning tests were conducted to analyze the effects of different tillage and planting methods on soil physical properties and crop yield under the condition of total straw return. Four methods were selected for this analysis: plow tillage and sowing, rotary tillage and sowing, no-tillage and sowing, and no-tillage and precise sowing in stubble fields. This study also explores the impact of soil physical properties on plant growth during different stages, using redundancy analysis. These findings will provide a foundation for enhancing soil quality in the Northeast Mollisol region and selecting the most appropriate tillage and planting methods based on local conditions.

2. Materials and Methods

2.1. Experimental Site

Experiments in the field were carried out at the Xiangyang Experimental Base (located at latitude 126.92° E and longitude 45.77° N) of Northeast Agricultural University in Heilongjiang Province, China. The region experiences a continental monsoon climate, with an average temperature of 3.7 °C per year and an average rainfall of 400–600 mm. The area enjoys a frost-free period that lasts 135–145 days annually. The temperature and rainfall patterns during the experiment are depicted in Figure 1. The soil used in this study was silty clay loam Mollisols, and its physical and chemical characteristics are presented in Table 1 [18].

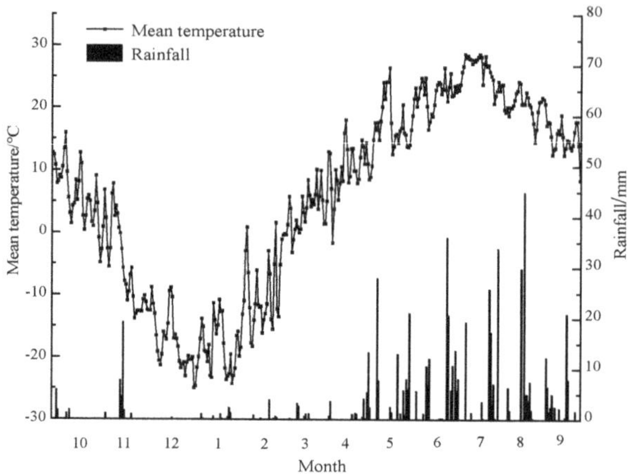

Figure 1. Mean temperature and rainfall during the experimental period.

Table 1. Physical and chemical properties of test soil.

Parameter	Value
Soil organic matter/g·kg^{-1}	24.5
Available phosphorus/mg·kg^{-1}	20.5
Available potassium/mg·kg^{-1}	100
Ammonium nitrogen/mg·kg^{-1}	32.5
Nitrate nitrogen/mg·kg^{-1}	14.3
Moisture content/%	36.3
pH	6.03

2.2. Experimental Design

The test was carried out from October 2020 to September 2021, and the data was collected on 20 June 2021 (50 days after corn sowing), 10 August 2021 (100 days after corn sowing), and 30 September 2021 (150 days after corn sowing). The experiment consisted of four treatments: plow tillage and precision seeder sowing (PTS); rotary tillage and precision seeder sowing (RTS); no-tillage and no-tillage seeder sowing (NTS); and no-tillage and precise sowing in stubble fields (STS). All treatments were carried out under conditions of total straw return. Table 2 lists the details of the four tillage and sowing methods. The treatments were designed using a completely randomized block method with three repetitions each. Each treatment was 8 m in length and 2.6 m in width. The corn was planted in ridges with a 65 cm row spacing and an average plant spacing of 23 cm. All other planting and management practices were in line with the local conventional farming methods.

Table 2. Test plan.

Cultivation and Sowing Mode of Corn Stubble	Concrete Measure
Plow tillage and sowing (PTS)	After the corn is harvested, all the straws are crushed with a length of ≤10 cm [Model:6B-1204,Manufacturer: John Deere(U.S.)], and evenly scattered on the ground. Then, a moldboard plow shall be used to turn it 25 cm deep[Model:Case PUMA 210, Manufacturer: Case IH(Shanghai, China)]. After turning it, it shall be harrowed twice, with a depth of 15–18 cm. Then, it shall be ridged and suppressed, and the next spring shall be seeded with a precision seeder [Model:BY484-2,Manufacturer: John Deere(U.S.)].
Rotary tillage and sowing (RTS)	After the corn is harvested, all the straws are crushed with a length of ≤10 cm[Model:6B-1204,Manufacturer: John Deere(U.S.)], evenly scattered on the ground, and then deeply loosened for 30 cm [Model:1SB-570, Manufacturer: WRNZ(Harbin,China)]. Use rotary cultivator to mix straw in 0–15 cm soil layer [Model:TM140, Manufacturer: New Holland(U.S.)], and then ridge and suppress. The next spring, precision seeders are used for sowing [Model:BY484-2, Manufacturer: John Deere(U.S.)].
No-tillage and sowing (NTS)	After the corn is harvested, all the corn stalks will be crushed with a length of ≤10 cm, and the ground will be covered evenly. The next spring, no-tillage seeder will be used for direct sowing [Model:2BMF-4 No-tillage seeder, Manufacturer: Jilin Kangda(Jilin,China)].
No-tillage and precise sowing in stubble field (STS)	During the harvest of corn, 40–60 cm stubble shall be left, and the next spring, do not conduct any treatment on the corn stubble and soil before sowing, no-tillage precision planter shall be used to directly sow the original stubble field [Model:2BMFJ-BLY2 stubble precision seeder, Manufacturer: Northeast Agricultural University].

2.3. Methods for Determination of Indicators and Analysis

The soil physical properties and corn plant characters data were measured at 50 days, 100 days, and 150 days after corn sowing, and the yield was measured after harvest under different tillage and sowing methods.

2.3.1. Methods for Determination of Soil Physical Properties

Soil penetration resistance (SPR) and soil temperature were measured directly in the field, while soil bulk density and soil moisture content were calculated by taking in situ soil and measuring the relevant indexes in the laboratory, and there was no rainfall in the test field for 5 consecutive days before the data testing. SPR data were measured using a PV6.08 penetration resistance meter (test accuracy: 0.1 MPa) manufactured by Eijkelkamp. (There was no rainfall for five consecutive days before the SPR measurements were taken, with the focus on the ultimate effect of the SPR under different treatments. Therefore, the analysis of the SPR data was not adjusted for covariates using soil water content data). One profile was selected for each plot, with three replications, and the measurement depth was 0–80 cm. The center line of the seedling strip was used as the central measurement point of the section, and a pair of measurement points were set up every 11 cm on both sides, with a section width of 66 cm. The SPR data were collected automatically at 1 cm depth intervals [19].

Determination of soil temperature at 5 cm, 15 cm, 25 cm, and 35 cm depth using TPJ-21 soil temperature recorder (test accuracy: 0.5 °C), 4 layers, and 3 repetitions [20]. The undisturbed soil at a depth of 5 cm, 15 cm, 25 cm, and 35 cm was taken using a ring knife (volume V = 100 cm^3), 4 layers, and 3 repetitions. Using JE1002 electronic balance (testing accuracy: 0.01 g) to weigh the aluminum box, the soil sample, and the total weight of the aluminum box. Drying the soil sample using DG-101-1S electric constant temperature drying oven (testing accuracy: ±1 °C), and the moisture content at different soil depths was determined according to national standard NY/T 52-1987 [21], based on which the soil bulk density was calculated [20].

2.3.2. Methods for Determination of Plant Growth Characters and Yield Components

Ten plants were randomly selected from each plot, and multiple measurements were taken to study their growth. The height of the corn plant was measured from the top of the spike to the base of the stem using steel tape with a precision of 0.1 cm. The diameter of the stem was determined by measuring the maximum width of the second section using a vernier caliper with a precision of 0.01 mm [22]. The spike height from the corn base to the bottom was measured using steel tape with a precision of 0.1 cm 150 days after sowing [23]. The full length of the cob was measured with steel tape having an accuracy of 0.1 cm, and the spike length was calculated. The length of the cob's convex tip, without the corn kernels, was measured using a vernier caliper with a precision of 0.01 mm. The sample plants were further examined by removing the corn bush leaves and whiskers, counting the number of rows per ear and grains per row, and determining the weight of 100 harvested corn seeds [20].

2.3.3. Method for Determination of Corn Yield

After corn ripening, we randomly select two adjacent ridges in each test plot, 3.85 m long (5 m^2), as the production test area. After manual threshing, the grain quality was measured by electronic balance. The dry matter quality of corn grains was measured by the whole grain-drying method according to national standard GB/T5009.3-2016 [24]. The corn grain moisture content C_0 (%) in the test plot is calculated as follows:

$$C_0 = \frac{m_0 - m_1}{m_0 - m} \times 100\% \tag{1}$$

where
m_0 = the mass of the aluminum box and the sample before baking, g,
m_1 = the mass of the aluminum box and the sample after baking, g, and
m = the mass of the aluminum box, g.

The corn yield M (g) under the condition of standard moisture content in the test plot is as follows:

$$M = \frac{(m_0 - m)(1 - C_0)}{(1 - C)} \tag{2}$$

where
C = the standard moisture content of corn; here, C = 14%.

2.4. Methods of Data Analyses

The test data was plotted with Origin 2018. Statistical analysis was performed by SPSS 23.0. Correlation analysis was made by R software, and redundancy analysis was made by Canoco 5.0 [25].

3. Results

3.1. Effects of Tillage and Sowing Methods on Soil Physical Properties

3.1.1. Effects of Tillage and Sowing Methods on SPR

In order to visually compare the differences of SPR in different periods under each treatment method, the tested 0–80 cm was divided into 8 soil layers equally with 10 cm as a soil layer, and the SPR of different soil layers in different periods under different tillage and sowing methods were analyzed for significance, and the results are shown in Table 3.

Table 3. Significance analysis of soil penetration resistance.

Test Period	Treatment	Soil Depth/cm							
		0–10	10–20	20–30	30–40	40–50	50–60	60–70	70–80
50 days after corn sowing	PTS	1.01 b	1.25 c	1.34 c	1.41 c	1.53 b	1.61 b	1.62 b	1.74 bc
	RTS	1.1 ab	1.53 a	1.66 a	1.57 b	1.58 b	1.61 b	1.63 b	1.65 c
	NTS	1.15 a	1.49 a	1.63 a	1.76 a	1.84 a	1.89 a	1.97 a	2.32 a
	STS	1.01 b	1.43 b	1.55 b	1.56 b	1.55 b	1.51 c	1.62 b	1.79 b
100 days after corn sowing	PTS	0.85 ab	1.13 b	1.42 a	1.45 b	1.39 c	1.39 c	1.49 c	1.6 c
	RTS	0.93 a	1.28 a	1.43 a	1.46 b	1.44 c	1.53 b	1.57 c	1.7 b
	NTS	0.82 b	1.17 b	1.31 b	1.48 b	1.55 b	1.64 a	2.05 a	2.22 a
	STS	0.83 b	1.34 a	1.46 a	1.66 a	1.66 a	1.6 ab	1.68 b	1.72 b
150 days after corn sowing	PTS	0.86 b	1.28 b	1.26 c	1.35 c	1.38 b	1.42 c	1.48 d	1.44 c
	RTS	0.83 b	1.09 d	1.26 c	1.32 c	1.41 b	1.53 b	1.71 b	1.73 b
	NTS	1.06 a	1.51 a	1.55 a	1.46 b	1.59 a	1.71 a	1.92 a	1.92 a
	STS	0.76 b	1.2 c	1.46 b	1.63 a	1.56 a	1.56 b	1.61 c	1.66 b

Note: PTS—plow tillage and sowing; RTS—rotary tillage and sowing; NTS—no-tillage and sowing; STS—no-tillage and precise sowing in stubble field. Different small letters indicate that there are significant differences between different tillage modes in the same test period ($p < 0.05$). The same below.

The results indicate that the tillage sowing methods had a significant impact on soil penetration resistance (SPR) at various depths and times. Overall, SPR tended to increase with increasing soil depth. At 50 days after sowing, the PTS treatment showed an 11.9% to 18% lower SPR compared to other treatments in the 10–40 cm soil layer ($p < 0.05$). Conversely, the NTS treatment displayed a 21.6% to 23.6% higher SPR compared to other treatments in the 30–80 cm soil layer ($p < 0.05$). At 100 days after sowing, the differences among treatments in the 0–40 cm soil layer were minimal. However, in the 50–80 cm soil layer, the NTS treatment showed a higher SPR compared to other treatments by 18.2% to 31.9% ($p < 0.05$). At 150 days after sowing, the NTS treatment had a higher SPR than other treatments in all soil layers except for the 30–40 cm layer. In general, the average SPR variation among soil layers was in descending order for the STS treatment (12.8%), followed by NTS (12.7%), RTS (9.9%), and PTS (9.9%) treatments. Between 50 to 100 days after sowing, the SPR for each treatment generally showed a decrease. The largest decrease was observed in the 10–20 cm soil layer under the NTS treatment. From 100 to 150 days after sowing, the changes in SPR were primarily concentrated in the 10–30 cm soil layer. The SPR for the PTS, RTS, and STS treatments decreased, while it increased for the NTS treatment.

3.1.2. Effects of Tillage and Sowing Methods on Soil Moisture Content and Temperature

According to Figure 2, the impact of tillage methods on soil moisture content is primarily evident in the soil depth range of 5 to 15 cm. At 50 days after sowing (Figure 2a), the tillage method had a pronounced effect on the moisture content of the 5 cm soil layer. The PTS treatment resulted in 11.2% and 10.7% lower moisture content compared to the RTS and STS treatments, respectively. Furthermore, the PTS treatment was significantly lower than the NTS treatment at 12.7% ($p < 0.05$). The highest mean variation in moisture content with increasing soil depth was observed in the PTS treatment, reaching 7.5%. At 150 days after sowing (Figure 2c), the tillage methods had a notable impact on the soil layer between 5 to 15 cm. In the 5 cm soil layer, the moisture content of the PTS treatment was significantly lower compared to the other treatments, with a range of 6.5% to 9.8% ($p < 0.05$). Meanwhile, in the 15 cm soil layer, the moisture content of the NTS treatment was significantly higher compared to the RTS treatment by 4.8% ($p < 0.05$). Among the three tests, the mean soil moisture content values in each soil layer followed a descending order of NTS, STS, RTS, and PTS. The NTS treatment increased the soil moisture content by 2.7% and 1.4% in comparison to the PTS and RTS treatments, respectively. On the other hand, the STS treatment increased the soil moisture content by 1.8% and 0.5% in comparison to the PTS and RTS treatments, respectively.

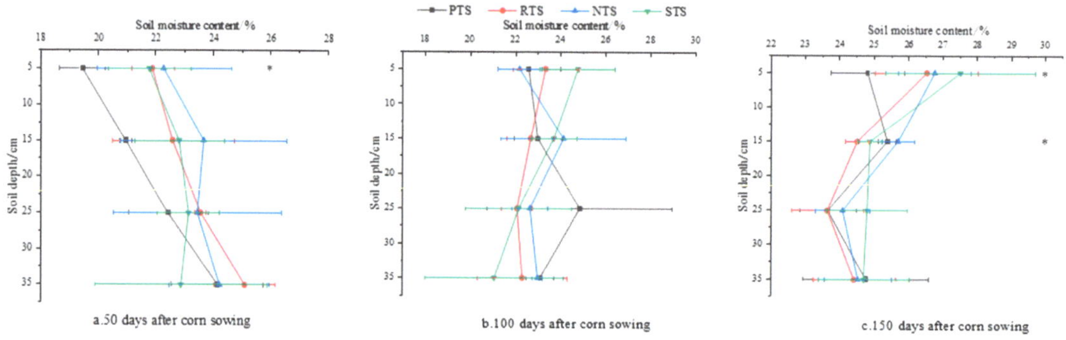

Figure 2. Effect of tillage and sowing methods on soil moisture content. Note: * Indicates a significant difference at 0.05 level. PTS—plow tillage and sowing; RTS—rotary tillage and sowing; NTS—no-tillage and sowing; STS—no-tillage and precise sowing in stubble field. The same below.

The impact of the tillage method on surfer soil temperature was significant (Figure 3). At 50 days after sowing (Figure 3a), the soil temperature in the 5 cm layer was 17.4% and 15.2% higher in the STS treatment compared to the RTS and NTS treatments, respectively. Additionally, the STS treatment was found to be 23.7% higher than the PTS treatment ($p < 0.05$). At 100 days after sowing (Figure 3b), the RTS treatment showed an increase in soil temperature by 5.2% and 4.1% compared to the PTS and NTS treatments, respectively. Moreover, the RTS treatment was 7.8% higher than the STS treatment ($p < 0.05$). In the 5–35 cm soil layer, the soil temperature was found to be higher in the RTS treatment compared to all other treatments. At 150 days after sowing (Figure 3c), the soil temperature in the 5–35 cm layer of the STS treatment was lower than the other three treatments, ranging from a decrease of 5.6% to 8.6%. The soil temperature showed a declining trend with increasing depth at 50 and 100 days after sowing. However, as the temperature drops, at 150 days after sowing, the overall trend of soil temperature showed an increase with the increase of soil depth, suggesting the transmission of heat from the bottom to the top. In the 5 cm soil layer, the soil temperature of the NTS, STS, RTS, and PTS treatments decreased by 5.9%, 8.7%, 9.1%, and 9.9%, respectively, with the order of soil temperature being RTS > PTS > NTS > STS.

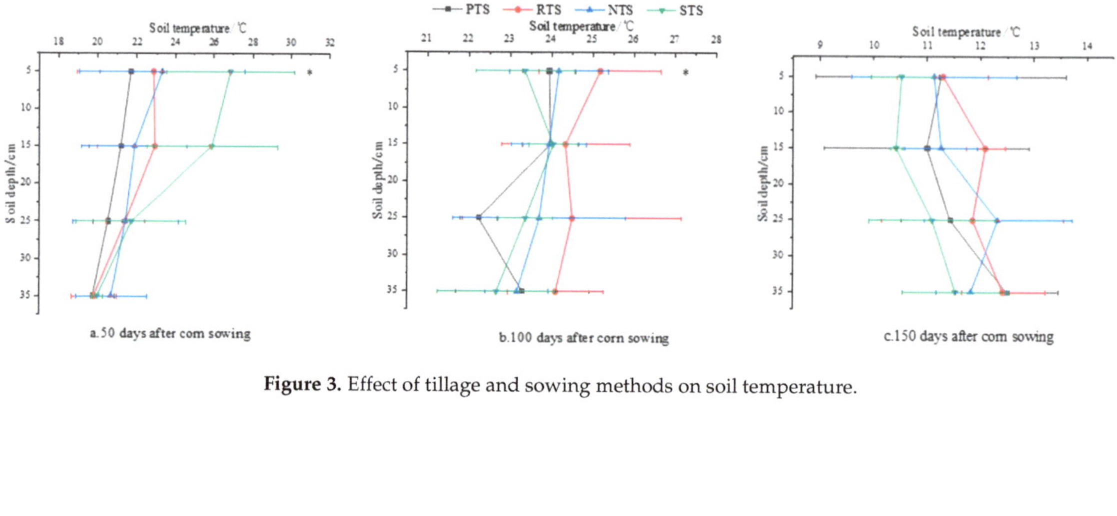

Figure 3. Effect of tillage and sowing methods on soil temperature.

3.1.3. Effects of Tillage and Sowing Methods on Soil Bulk Density

The response law of soil bulk density under four tillage and sowing methods is shown in Figure 4.

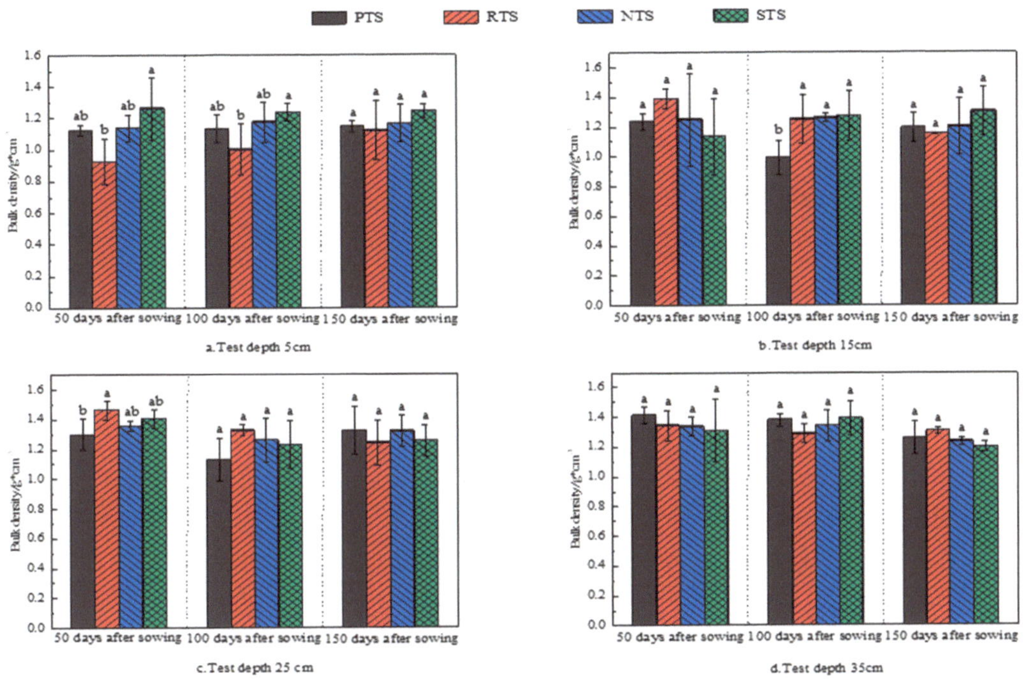

Figure 4. Effect of tillage and sowing methods on soil bulk density. Note: PTS—plow tillage and sowing; RTS—rotary tillage and sowing; NTS—no-tillage and sowing; STS—no-tillage and precise sowing in stubble field. Different lowercase letters indicate significant differences between the same test time, the same soil depth, and different tillage methods ($p < 0.05$).

Overall, soil bulk density increases with depth. However, in the 5-cm soil layer (Figure 4a), the soil bulk density of RTS decreases significantly by 26.5% and 18.9% compared to STS after 50 and 100 days of sowing, respectively ($p < 0.05$). As time goes by, the soil bulk density of RTS increases, and after 150 days of sowing, there is no significant difference compared to other treatments. On average, the soil bulk density of PTS is 2% and 8.9% lower than NTS and STS, respectively. Meanwhile, the soil bulk density of RTS is, on average, 12.3% and 18.4% lower than NTS and STS, respectively. In the 15 cm soil layer (Figure 4b), the PTS treatment resulted in a significantly lower soil bulk density of 20.9% to 22.1% compared to the other three tillage sowing methods after 100 days of sowing ($p < 0.05$). In the 25 cm soil layer (Figure 4c), the soil bulk density of the RTS treatment was significantly higher than that of the PTS treatment by 12.8% after 50 days of sowing ($p < 0.05$). The difference in soil bulk density between treatments is more significant in the 5 cm soil at 50 and 100 days after corn sowing but is not significant in the 35 cm soil 50–150 days after corn sowing. At the same depth, the difference in soil bulk density between treatments gradually decreases with time. RTS exhibits a smaller average soil bulk density in 0–30 cm soil depth 150 days after corn sowing. From 50 to 100 days after sowing, there was an overall decreasing trend in soil bulk density for PTS, RTS, and NTS, while STS showed an increasing trend. In the 25 cm soil layer, all four treatments showed a decrease in soil bulk density, with the greatest decrease observed in the PTS treatment and the smallest decrease in the NTS treatment. From 100 to 150 days after sowing, RTS, NTS, and STS showed a decreasing trend in soil bulk density, while the PTS treatment showed an increasing trend. The change in soil bulk density was not consistent among the different treatments across different soil layers.

3.2. Effects of Tillage and Sowing Methods on Plant Characters and Yield

The growth of corn plants under different tillage and sowing methods is shown in Figure 5.

Under different tillage and sowing methods, there is a significant difference in corn plant height 50 days after sowing (Figure 5a). The plant height of NTS is 22.3–27.5% lower compared to the other treatments, and the difference is significant ($p < 0.05$). However, there is no significant difference in plant height between the treatments 100 and 150 days after sowing. NTS shows the fastest growth rate in plant height from 50 to 100 days after sowing, but its subsequent growth rate is slower than the other treatments. Although there is no significant difference in plant height at 150 days after sowing, NTS still has a shorter average plant height compared to the other treatments. The stem diameter of corn also shows significant differences across the three growth periods (Figure 5b). At 50 days after sowing, the stem diameter of NTS is 33.2–35.3% lower compared to the other treatments, with the difference being significant ($p < 0.05$). NTS has the fastest growth rate in stem diameter from 50 to 100 days after sowing, which follows the same trend as the corn plant height of NTS. From 100 to 150 days, RTS has the fastest growth rate in stem diameter, while STS has the slowest growth rate. At 150 days after sowing, the stem diameter of STS is the lowest, being 9.9% and 10.5% lower than RTS and NTS, respectively, with the difference being significant ($p < 0.05$). The spike height of corn was measured at 150 days after sowing (Figure 5c). The spike height of PTS is 6.3% higher than that of STS, and the difference is significant ($p < 0.05$). There is no difference in spike height between NTS and RTS. According to Equations (1) and (2), the yield of corn in the experiment plot under standard water content was calculated, and a significant analysis of the yield was performed, as shown in Table 4.

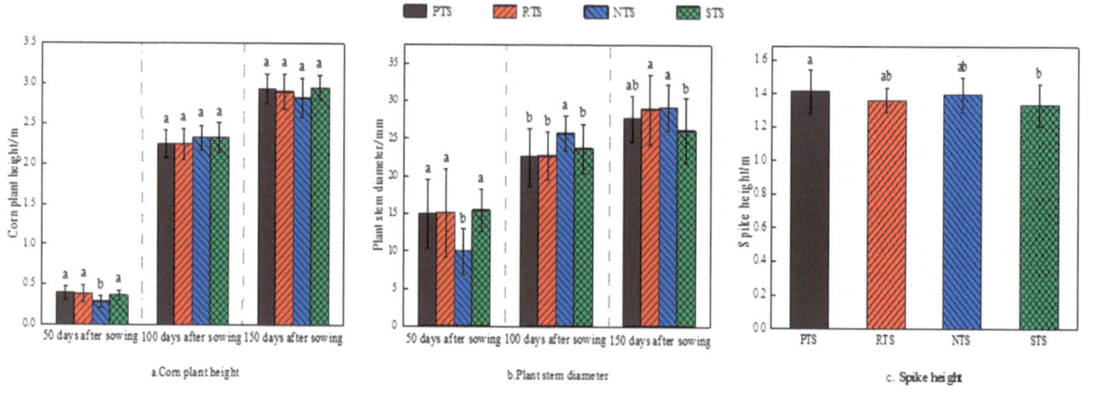

Figure 5. Changes in crop growth under different tillage and sowing methods.Note: PTS—plow tillage and sowing; RTS—rotary tillage and sowing; NTS—no-tillage and sowing; STS—no-tillage and precise sowing in stubble field. Different small letters indicate that there are significant differences between different tillage and sowing methods at the same test time ($p < 0.05$). The same below.

The results show that the convex tip length, number of rows per ear, and 100-grain weight of NTS are slightly higher than the other treatments, but the difference does not reach a significant level between treatments. The corn yield of different tillage and sowing methods follows RTS > PTS > NTS > STS, but there is no significant difference between treatments.

Table 4. Effects of tillage and sowing methods on corn yield.

Tillage Methods	Yield Components					Yield (kg·hm^{-2})
	Spike Length (cm)	Convex Tip Length (cm)	Number of Rows per Ear	Number of Grains per Row (Individual)	100 Grain Weight (g)	
PTS	22.59 ± 2.00 a	0.8 ± 1.0 a	18.87 ± 2.01 a	39.53 ± 5.39 a	38.42 ± 1.03 a	2725.44 ± 272.53 a
RTS	22.72 ± 2.55 a	0.85 ± 0.69 a	18.53 ± 1.74 a	40.13 ± 4.99 a	38.73 ± 4.96 a	2805.76 ± 294.29 a
NTS	22.52 ± 2.23 a	1.16 ± 1.06 a	19.07 ± 1.80 a	40.27 ± 3.21 a	40.12 ± 3.57 a	2686.92 ± 177.43 a
STS	23.68 ± 2.31 a	0.73 ± 1.02 a	18.53 ± 1.89 a	40.10 ± 4.60 a	36.9 ± 2.09 a	2358.5 ± 325.12 a

PTS—plow tillage and sowing; RTS—rotary tillage and sowing; NTS—no-tillage and sowing; STS—no-tillage and precise sowing in stubble field.

3.3. Correlation Analysis of Soil Physical Properties, Plant Characters, and Yield under Different Tillage and Sowing Methods

3.3.1. Correlation Analysis

By Pearson correlation analysis, the correlation relationship among soil physical properties, corn plant characters, and yield in 0–40 cm soil at 150 days after corn sowing is shown in Figure 6.

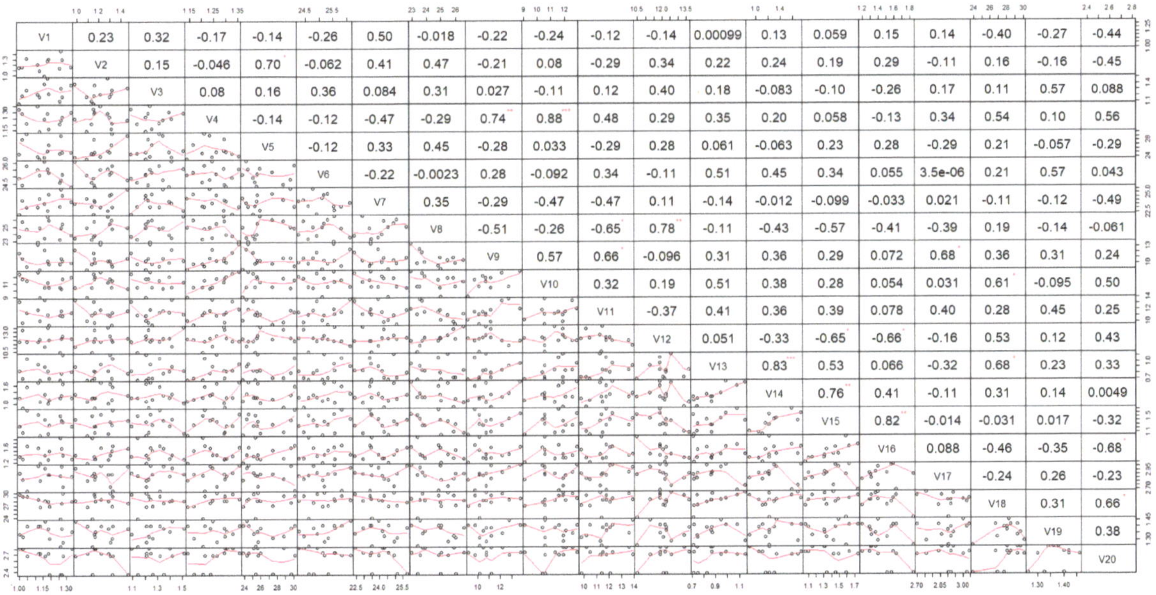

Figure 6. Pearson correlation among soil physical properties, plant characters, and yield. Note: V1, V2, V3, and V4 are, respectively, mean soil bulk density in 5 cm, 15 cm, 25 cm, and 35 cm soil; V5, V6, V7, and V8 are mean moisture content in 5 cm, 15 cm, 25 cm, and 35 cm soil; V9, V10, V11, V12 are, respectively, mean temperature in 5 cm, 15 cm, 25 cm, and 35 cm soil; V13, V14, V15, and V16 are mean SPR in 5 cm, 15 cm, 25 cm, and 35 cm soil; V17 are mean corn plant height; V18 is mean plant stem diameter; V19 is mean spike height; V20 is mean corn yield. *, ** and *** indicate significant correlations at 0.10, 0.05, and 0.01 probability, respectively.

Plant stem diameter and yield of corn are strongly correlated and positively correlated. SPR in 35 cm soil is negatively correlated with corn yield. The soil temperature in 5 cm soil is strongly correlated with corn plant height, and the soil temperature in 15 cm soil and SPR in 5 cm soil are positively correlated with corn stem diameter. SPR of the adjacent soil layer is correlated.

3.3.2. Redundancy Analysis

By redundancy analysis of soil physical property indexes, corn plant height, and stem diameter in 0–40 cm soil layer in three periods, the two-dimensional ranking of the effects of soil physical properties on plant characters, and the explanatory degree of impact factors are obtained, as shown in Figure 7 and Table 5.

At 50 days after corn sowing, the explanatory contribution of each impact factor to corn plant growth is SPR (51.8%), moisture content (23.6%), and temperature (7.4%) in descending order. Among the factors, SPR in 35 cm soil has the highest explanatory degree to plant growth (42.2%) and is negatively correlated with corn plant height and stem diameter (Figure 7a). At 100 days after corn sowing, the explanatory contribution of each impact factor to corn plant growth is soil bulk density (47.8%), temperature (23%), moisture content (4.3%), and SPR (1.6%) in descending order. Among the factors, soil bulk density in 25 cm soil has the highest explanatory degree to plant growth (38.9%) and is positively correlated with corn plant height and stem diameter (Figure 7b). At 150 days after corn sowing, the explanatory contribution of each impact factor to corn plant growth is the temperature (43.6%), penetration resistance (35.7%), bulk density (15.3%), and moisture content (5.4%), in descending order. Among the factors, SPR in 5 cm soil has the highest explanatory degree to plant growth (33.2%) and is positively correlated with corn stem diameter and negatively correlated with corn plant height (Figure 7c).

On the whole, SPR has a high degree of explanation to plant growth in the early and later stages of corn growth, reaching 51.8% and 35.7%, respectively. The soil moisture content has a great influence on the early growth of the plant, and the interpretability is 23.6% 50 days after corn sowing. The importance of soil temperature in the whole plant growth process gradually increased, from 7.4% in the early stage of corn growth to 43.6% in the later stage. Soil bulk density has a great influence on the mid-term growth of corn, and its interpretability is as high as 47.8% 100 days after corn sowing.

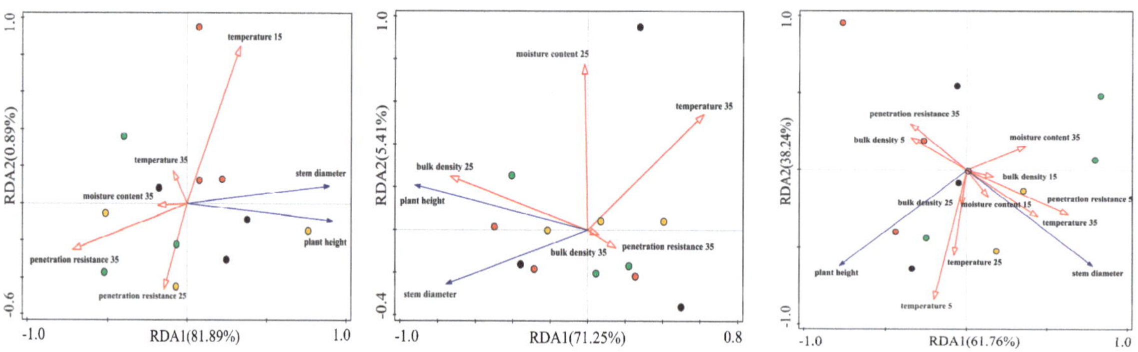

(**a**) 50 days after corn sowing (**b**) 100 days after corn sowing (**c**) 150 days after corn sowing

Figure 7. RDA two-dimensional ranking diagram of plant characters and soil physical properties. Note: SPR 5, 15, 25, and 35 are, respectively, mean soil penetration resistance in 5 cm,15 cm, 25 cm, and 35 cm; Moisture content 5, 15, 25, and 35 are, respectively, mean soil moisture content in 5 cm, 15 cm, 25 cm, and 35 cm; Temperature 5, 15, 25, and 35 are, respectively, mean soil temperature in 5 cm, 15 cm, 25 cm, and 35 cm; Bulk density 5, 15, 25, and 35 are, respectively, mean soil bulk density in 5 cm, 15 cm, 25 cm, and 35 cm. The same below.

Table 5. Explanatory degree and significance test of impact factors.

50 Days after Corn Sowing				100 Days after Corn Sowing				150 Days after Corn Sowing			
Influence Factors	**Explanation Rate (%)**	*F*	*p*	**Influence Factors**	**Explanation Rate (%)**	*F*	*p*	**Influence Factors**	**Explanation Rate (%)**	*F*	*p*
SPR 35	42.2	7.3	0.014	Bulk density 25	38.9	6.4	0.018	SPR 5	33.2	8	0.008
SPR 25	9.6	3.1	0.102	Bulk density 35	8.9	2.4	0.128	SPR 35	2.5	1.7	0.28
Moisture content 35	23.6	6.2	0.032	Temperature 35	23	5.4	0.02	Temperature 5	29.3	4.1	0.03
Temperature 35	6.7	2.6	0.148	Moisture content 25	4.3	1.2	0.278	Temperature 35	13.8	4.7	0.038
Temperature 15	0.7	0.2	0.7	SPR 35	1.6	0.4	0.672	Temperature 25	0.5	308	0.15
								Bulk density 5	10	5.1	0.022
								Bulk density 15	4.1	5	0.028
								Bulk density 25	1.2	5.2	0.11
								Moisture content 15	3.8	2.3	0.134
								Moisture content 35	1.6	2.9	0.18

4. Discussion

SPR is a crucial factor impacting soil physical properties. Research indicates that SPR increases with soil depth, a result of both cover pressure and internal soil friction [26,27]. During the early stages of corn growth, the SPR of the 10–40 cm soil layer under the PTS treatment was significantly lower compared to the other treatments. This is because deep tillage creates larger soil pores, which in turn reduces soil bulk density and penetration resistance [28]. The NTS treatment resulted in higher soil penetration resistance and sustained high SPR values throughout the corn reproductive period, which is consistent with previous studies [29]. The variation among soil layers under PTS and RTS treatment was comparatively small, which could be attributed to tillage improving soil pore conditions [2] and having a more pronounced effect on subsoil improvement, thus reducing SPR differences among soil layers. Soil moisture and temperature play a crucial role in the growth and development of crops [30]. Studies show that tillage practices have a greater impact on the surface soil moisture content. Conservation tillage, in particular, can improve soil moisture levels due to the presence of straw cover and stubble that delay runoff, increase water infiltration, and reduce surface water evaporation [31]. In the early stage of corn growth, the soil moisture content under PTS increases the most with the soil depth, indicating that deep plowing can increase the water infiltration rate [32]. The effect of tillage methods on soil temperature was mainly observed in the topsoil layer. The impact of temperature on plant growth increased with plant growth and reached its peak 150 days after sowing. At this stage, the temperature of the 5–35 cm soil layer under the STS treatment was lower than the other treatments. As the temperature drops, the soil temperature showed an increasing trend with increasing depth 150 days after sowing, which indicated heat conduction from bottom to top, with NTS having the best thermal conductivity, followed by STS, RTS, and PTS. The results are consistent with previous studies [33]. The temperature of the 35 cm soil layer was higher under RTS treatment, which might be due to the loosened and porous soil structure that has a warming effect on the deeper layers [34].

The rotary tillage (RTS) method resulted In a reduction of surface soil bulk density due to the shallow depth of tillage and the porous soil structure created by it. This had a greater impact on the topsoil. At a depth of 5 cm, the average values of bulk density under the PTS and RTS treatments were lower than those of the NTS and STS treatments, and the differences between treatments became smaller as the soil depth increased. This suggests that tillage can indeed reduce soil bulk density, but primarily in the top 10 cm of soil [10,35]. Over time, the difference between treatments becomes smaller, indicating that the effects of subsoiling and tillage on soil bulk density only persist for one to two seasons in fertile soil [26]. The early growth of corn was impacted by tillage and sowing methods, with the differences between treatments becoming less pronounced as the corn grew. The NTS treatment hampered the initial growth of the crop, but from 50 to 100 days, its growth rate was significantly higher than the other treatments. Wang et al. showed that under no-tillage conditions, straw mulching could improve the seedling emergence rate, but it was not conducive to the accumulation of dry matter at the jointing stage, which may be improved with the further decomposition of straw [36]. The most rapid growth period of corn was from 50 to 100 days after planting; the growth rate slowed down in 100 to 150 days; during this period, the growth rate of NTS and STS treatments was lower than that of PTS and RTS treatments. Wang et al. showed that plow tillage treatment significantly increased the number of ears, number of grains per row, and 100-grain weight of corn, which was inconsistent with the results of this study, that may be related to the soil type and the gene shape of the corn variety itself [37]. According to the findings of Pittelkow, a combination of no-tillage, straw mulching, and rotation can mitigate the adverse effects of no-tillage on crop yield and, under certain conditions, even yield the same or greater output compared to traditional tillage methods [11]. There was no difference in corn yield among the four tillage and sowing methods, which is inconsistent with the research results of Xu, which may be related to soil texture, weather, and climate [38]. Crop growth is sensitive to SPR and soil moisture in the early stage of corn growth, which is consistent with the results given by Xiong et al. [39]. Yu showed that corn growth was sensitive to waterlogging and negatively correlated with it [40]. Colombi et al. discovered that SPR could affect seedling emergence rate, nutrient uptake, root morphology, and crop yield [41]. As the plants grow, the effect of soil moisture content on corn decreases while the impact of temperature increases, with the explanation degree reaching 43.6% 150 days after corn sowing. The reason is that the temperature decreases with corn plant growth, and the soil temperature are different under different tillage and sowing methods, which affects the growth and development of the plants in the later period. Hatfield et al. found that corn is a temperature-sensitive crop, and even a 1 °C difference can greatly impact its growth and development [42]. As shown in Figures 3c and 7c, the low-soil temperature may be the reason for the slow growth of stem diameter and plant height in the late stages under NTS and STS treatment. In summary, corn at different growth stages has a different sensitivity to soil physical properties. Under NTS, SPR and moisture content were higher, which affected the early growth of corn, but with the change of the sensitivity of corn growth to indexes, the early deficiency will be made up in the middle growth of corn. With the temperature decreasing in the later period, temperature becomes one of the main factors affecting corn growth, which is caused by the differences in the cropping system and climate in northeast China.

5. Conclusions

The results of this study indicate that the short-term application of different tillage and sowing methods has a significant impact on soil physical properties and plant growth. High soil penetration resistance impeded early crop growth, but the PTS treatment method reduced this resistance in the 0–40 cm soil layer. The effect of tillage on soil moisture content was concentrated in the 5–15 cm soil layer, and both the NTS and STS treatments increased soil moisture content. The impact of tillage and sowing methods on soil bulk density was most pronounced in the 5 cm soil layer, with RTS treatment significantly reducing surface soil bulk density. As the temperature decreased, its influence on plant growth increased, and the STS treatment resulted in low soil temperature, which might inhibit the later development of the crop. Among the measured plant characters, stem diameter was found to have a strong correlation with yield. The results showed that the short-term implementation of conservation tillage did not cause a significant decrease in corn yield and that the short-term implementation of NTS treatment is considered feasible.

Author Contributions: Y.W. was responsible for the experiments, performed the statistical analysis, and wrote this manuscript. S.Y. was responsible for the experiments and revised this manuscript. J.Q., J.S., Z.L. and X.H. supervised the experiments and revised this manuscript. All authors have read and agreed to the published version of the manuscript.

Funding: This study was supported by National Key R&D Projects during the Fourteenth Five Year Plan "Integration of key technologies for high-performance seeding and demonstration application of intelligent seeder" (2021YFD2000405), Open project of Key Laboratory of Combining Farming and Animal Husbandry, Ministry of Agriculture and Rural Affairs.

Institutional Review Board Statement: Not applicable.

Data Availability Statement: The data presented in this study are available on request from the corresponding author.

Conflicts of Interest: The authors declare no conflict of interest.

References

1. Cong, H.B.; Meng, H.B.; Yu, J.D. Analysis of long-term mechanism for development of straw industry in northeast China under guidance of "Green-concept". *Trans. CSAE* **2021**, *37*, 314–321. [CrossRef]
2. Yan, L.; La, Y.P.; Dong, T.H. Soil physical properties and vertical distribution of root systems affected by tillage methods in black soil slope farmlands in Northeast China. *Trans. CSAE* **2021**, *37*, 125–132. [CrossRef]
3. Zhang, X.Y.; Liu, X.B. Key issues of Mollisols research and soil erosion control strategies in China. *Bull. Soil Water Conserv.* **2020**, *40*, 340–344. [CrossRef]
4. Chen, N.N.; Li, J.; Lv, W.; Wang, S.L. Effects of different rotational tillage patterns on soil physical properties and yield of winter wheat-spring maize rotation field in Weibei highland. *Chin. J. Eco-Agric.* **2015**, *23*, 1102–1111. [CrossRef]
5. Li, N.; Long, J.H.; Han, X.Z.; Zhang, F.Q.; Lei, W.Y.; Sheng, M.; Han, Z.Y. Effects of short-term plowing and organic amendments on soil physical properties and maize yield in dark brown soil in Northeast China. *Trans. CSAE* **2021**, *37*, 99–107. [CrossRef]
6. Misban, A.H.; Habtamu, M.; Petra, S.; Prossie, N.; Seifu, A.T.; Simon, L.; Jennie, B.; Tammo, S.S. Deep Tillage Improves Degraded Soils in the (Sub) Humid Ethiopian Highlands. *Land* **2019**, *8*, 159. [CrossRef]
7. Dong, J.X.; Song, W.J.; Cong, P. Improving Farmland Soil Physical Properties by Rotary Tillage Combined with High Amount of Granulated Straw. *Sci. Agric. Sin.* **2021**, *54*, 2789–2803. [CrossRef]
8. Zhu, M.; Shi, Y.X.; Sun, Z.Y.; Wang, X.L.; Zhang, D.; Kong, F.L.; Yuan, J.C. Effect of straw return and rotary tillage on soil physical properties and mechanical sowing quality of maize in Central Sichuan. *Chin. J. Eco–Agric.* **2017**, *25*, 1025–1033. [CrossRef]
9. Catania, P.; Badalucco, L.; Laudicina, V.A.; Vallone, M. Effects of tilling methods on soil penetration resistance, organic carbon and water stable aggregates in a vineyard of semiarid Mediterranean environment. *Environ. Earth Sci.* **2018**, *77*, 348. [CrossRef]
10. Blanco-Canqui, H.; Ruis, S.J. No-tillage and soil physical environment. *Geoderma* **2018**, *326*, 164–200. [CrossRef]
11. Pittelkow, C.M.; Liang, X.; Linquist, B.A.; Groenigen, K.J.; Lee, J.; Lundy, M.E.; Gestel, N.; Six, J.; Venterea, R.T.; Kessel, C. Productivity limits and potentials of the principles of conservation agriculture. *Nature* **2015**, *517*, 365. [CrossRef] [PubMed]
12. Lu, B.L.; Bao, X.G.; Che, Z.X.; Zhang, J.D.; Wu, K.S.; Yang, X.J. Effects of long-term no-tillage with crop stubbles on yield and stability of spring wheat in Hexi Oasis Irrigated Areas. *Trans. CSAE* **2022**, *38*, 117–126. [CrossRef]
13. Kahlon, M.S.; Lal, R.; Ann-Varughese, M. Twenty two years of tillage and mulching impacts on soil physical characteristics and carbon sequestration in Central Ohio. *Soil Tillage Res.* **2013**, *126*, 151–158. [CrossRef]
14. Gu, S.Y.; Zhu, Y.W.; Guo, X.J.; Cai, Y.T.; Wu, S.; Zhang, Z.H.; Zhang, H.H. Effects of Different Tillage Ways on Seedling Growth of Maize and Soil Physical Properties in Mollisol Region. *Acta Agric. Boreali-Sin.* **2018**, *33*, 226–231. [CrossRef]
15. Zhao, H.B.; He, J.; Li, H.W.; Wang, Q.J.; Li, W.Y.; Liu, W.Z. Effect of Straw Returning Manners on Seedbed Soil Physical Properties and Winter Wheat Growth. *Trans. Chin. Soc. Agric. Mach.* **2018**, *49*, 60–67. [CrossRef]
16. Orzech, K.; Wanic, M.; Zaluski, D. The Effects of Soil Compaction and Different Tillage Systems on the Bulk Density and Moisture Content of Soil and the Yields of Winter Oilseed Rape and Cereals. *Agriculture* **2021**, *11*, 666. [CrossRef]
17. Linhares, A.; Gonalves, W.G.; Costa, K. Are the yield of sunflower and Paiaguas palisadegrass biomass influenced by soil physical quality? *Soil Tillage Res.* **2021**, *208*, 104873. [CrossRef]
18. Yan, L.; Jiang, X.X.; Ji, X.N.; Zhou, L.; Li, S.Y.; Chen, C.; Li, P.; Zhu, Y.C.; Dong, T.H.; Meng, Q.F. Distribution of water-stable aggregates under soil tillage practices in a black soil hillslope cropland in Northeast China. *J. Soils Sediments* **2020**, *20*, 24–31. [CrossRef]
19. Qiao, J.Y.; Zhang, D.; Zhang, H.B.; Zhang, B.; Chen, H.T.; Shen, L.P.; Zheng, D.M.; Sun, J. Effects of soil compaction by various tractors on soil penetration resistance and soybean yields. *Trans. CSAE* **2019**, *35*, 26–33. [CrossRef]
20. Wang, Y.J. Soil Physical Properties and Effect Evaluation of Mechanized Tillage Planting Methods in Black Sloping Farmland. Ph.D. Thesis, Northeast Agricultural University, Harbin, China, 2021.
21. *NY/T52-1987*; National Standards of the People's Republic of China. Method for the Determination of soil Water Content. National Bureau of Standards: Beijing, China, 1987.

22. Wang, Q.J.; Jiao, F.; Liu, F.; Chang, B.C.; Jiang, H.; Jiang, Y.; Mi, G.; Zhou, X. Effect of straw pulverization and concentrated deep-buried into field on growth and yield of maize. *Trans. CSAE* **2018**, *34*, 153–159. [CrossRef]
23. You, S.T.; Deng, C.; Li, H.M.; Lv, M.; Li, Z.M.; Liu, H.F. QTL Mapping of Plant Height and Ear Height in Maize. *J. Henan Agric. Sci.* **2019**, *48*, 20–25. [CrossRef]
24. *GB5009.3—2016*; State health and Family Planning Commission of the People's Republic of China. Determination of moisture in Food. China Standards Press: Beijing, China, 2016.
25. Canoco 5. Canoco for Visualization of Multivariate Data—WUR. Available online: https://www.wur.nl/en/show/Canoco-for-visualization-of-multivariate-data.htm#:~:text=Canoco%20is%20a%20software%20package,of%20the%20two%2C%20constrained%20ordination (accessed on 1 February 2022).
26. Bogunovic, I.; Pereira, P.; Kisic, I.; Sajko, K.; Sraka, M. Tillage management impacts on soil compaction, erosion and crop yield in Stagnosols (Croatia). *CATENA* **2018**, *160*, 376–384. [CrossRef]
27. Gao, W.; Whalley, W.R.; Tian, Z.; Liu, J.; Ren, T. A simple model to predict soil penetrometer resistance as a function of density, drying and depth in the field. *Soil Tillage Res.* **2016**, *155*, 190–198. [CrossRef]
28. Wang, Y.K.; Zhang, Z.B.; Jiang, F.H.; Guo, Z.C.; Peng, X.H. Evaluating soil physical quality indicators of a Vertisol as affected by different tillage practices under wheat-maize system in the North China Plain. *Soil Tillage Res.* **2021**, *209*, 104970. [CrossRef]
29. Burgos Hernández, T.D.; Slater, B.K.; Corbalá, R.T.; Shaffer, J.M. Assessment of long-term tillage practices on physical properties of two Ohio soils. *Soil Tillage Res.* **2019**, *186*, 270–279. [CrossRef]
30. Yang, Y. Effcets of Different Tillage Measures on Soil Physical and Chemical Characters and Crop Yield in Cropland. Ph.D. Thesis, Northwest A&F University, Xianyang, China, 2017.
31. Li, Y.J.; Wu, J.Z.; Huang, M.; Yao, Y.Q.; Zhang, C.J.; Cai, D.X.; Jin, K. Effects of different tillage systems on photosynthesis characteristics of flag leaf and water use efficiency in winter wheat. *Trans. CSAE* **2006**, *22*, 44–48.
32. Mohanty, M.; Bandyopadhyay, K.K.; Painuli, D.K.; Ghosh, P.K.; Misra, A.K.; Hati, K.M. Water transmission characteristics of a Vertisol and water use efficiency of rainfed soybean (*Glycine max* (L.) Merr.) under subsoiling and manuring. *Soil Tillage Res.* **2007**, *93*, 420–428. [CrossRef]
33. Lu, Y.L.; Zhang, M.; Liu, X.N.; Ren, T.S.; Wang, Y.J. Effects of soil water content and bulk density on thermal conductivity of plough layer soil in arid land and its prediction. *Trans. CSAE* **2018**, *34*, 146–151. [CrossRef]
34. Yang, J.Y.; Gu, S.Y.; Li, Y.H.; He, W.Y.; Tang, Y.; Zhai, C.; Du, L. Effects of Deep Ploughing-rotary Tillage Combined with Organic Fertilizer on Black Soil Physical Properties. *Chin. J. Soil Sci.* **2021**, *52*, 1290–1298. [CrossRef]
35. Awe, G.O.; Reichert, J.M.; Fontanela, E. Sugarcane production in the subtropics: Seasonal changes in soil properties and crop yield in no-tillage, inverting and minimum tillage. *Soil Tillage Res.* **2020**, *196*, 104447. [CrossRef]
36. Wang, L.X.; Li, Z.P.; Liu, S.P.; Du, J.P. Effects of Different Tillage and Mulching Methods on the Growth and Photosynthetic Characteristics of Maize. *J. Maize Sci.* **2016**, *24*, 142–146. [CrossRef]
37. Wang, Y.H.; Xi, L.L.; Zhou, X.; Shen, D.F. Effects of deep tillage treatments on summer maize yield and soil physicochemical properties. *Shanxi Agric. Univ. Nat. Sci. Ed.* **2018**, *38*, 9–14. [CrossRef]
38. Xu, Y.Y.; Wang, J.H.; Liu, Y.T.; Gao, P.; Wang, Y.X.; Yang, H.Y.; Yu, K.C.; Ge, X.L.; Chi, L.; Fan, S.J. Effects of Different Returning Methods of Straw on Soil Physical Property, Yield of Corn. *Maize Sci.* **2018**, *26*, 78–84. [CrossRef]
39. Xiong, P.; Zhang, Z.; Hallett, P.D.; Peng, X. Variable responses of maize root architecture in elite cultivars due to soil compaction and moisture. *Plant Soil* **2020**, *455*, 79–91. [CrossRef]
40. Yu, W.D.; Feng, L.P.; Sheng, S.X.; Shi, L. Effect of waterlogging at jointing and tasseling stages on growth and yield of summer maize. *Trans. CSAE* **2014**, *30*, 127–136. [CrossRef]
41. Colombi, T.; Torres, L.C.; Walter, A.; Keller, T. Feedbacks between soil penetration resistance, root architecture and water uptake limit water accessibility and crop growth—A vicious circle. *Sci. Total Environ.* **2018**, *626*, 1026–1035. [CrossRef] [PubMed]
42. Hatfield, J.L.; Allmaras, R.R.; Rehm, G.W.; Lowery, B. Ridge tillage for corn and soybean production: Environmental quality impacts. *Soil Tillage Res.* **1998**, *48*, 145–154. [CrossRef]

agriculture

MDPI

Review

Taxonomy, Ecology, and Cellulolytic Properties of the Genus *Bacillus* and Related Genera

Jakub Dobrzyński [1,*], Barbara Wróbel [1] and Ewa Beata Górska [2]

[1] Institute of Technology and Life Sciences—National Research Institute, Falenty, 3 Hrabska Avenue, 05-090 Raszyn, Poland; b.wrobel@itp.edu.pl

[2] Department of Biochemistry and Microbiology, Institute of Biology, Warsaw University of Life Sciences—SGGW, 02-776 Warsaw, Poland; e.b.gorska@wp.pl

* Correspondence: j.dobrzynski@itp.edu.pl

Abstract: Bacteria of the genus *Bacillus* and related genera (e.g., *Paenibacillus*, *Alicyclobacillus* or *Brevibacillus*) belong to the phylum Firmicutes. Taxonomically, it is a diverse group of bacteria that, to date, has not been well described phylogenetically. The group consists of aerobic and relatively anaerobic bacteria, capable of spore-forming. *Bacillus* spp. and related genera are widely distributed in the environment, with a particular role in soil. Their abundance in the agricultural environment depends mainly on fertilization, but can also depend on soil cultivated methods, meaning whether the plants are grown in monoculture or rotation systems. The highest abundance of the phylum Firmicutes is usually recorded in soil fertilized with manure. Due to the great abundance of cellulose in the environment, one of the most important physiological groups among these spore-forming bacteria are cellulolytic bacteria. Three key cellulases produced by *Bacillus* spp. and related genera are required for complete cellulose degradation and include endoglucanases, exoglucanases, and β-glucosidases. Due to probable independent evolution, cellulases are encoded by hundreds of genes, which results in a large structural diversity of these enzymes. The microbial degradation of cellulose depends on its type and environmental conditions such as pH, temperature, and various substances including metal ions. In addition, *Bacillus* spp. are among a few bacteria capable of producing multi-enzymatic protein complexes called cellulosomes. In conclusion, the taxonomy of *Bacillus* spp. and related bacteria needs to be reorganized based on, among other things, additional genetic markers. Also, the ecology of soil bacteria of the genus *Bacillus* requires additions, especially in the identification of physical and chemical parameters affecting the occurrence of the group of bacteria. Finally, it is worth adding that despite many spore-forming strains well-studied for cellulolytic activity, still few are used in industry, for instance for biodegradation or bioconversion of lignocellulosic waste into biogas or biofuel. Therefore, research aimed at optimizing the cellulolytic properties of spore-forming bacteria is needed for more efficient commercialization.

Keywords: taxonomy; Firmicutes; spore-forming bacteria; cellulases

Citation: Dobrzyński, J.; Wróbel, B.; Górska, E.B. Taxonomy, Ecology, and Cellulolytic Properties of the Genus *Bacillus* and Related Genera. *Agriculture* **2023**, *13*, 1979. https://doi.org/10.3390/agriculture13101979

Academic Editors: Ana P. G. C. Marques and Ramakrishna Wusirika

Received: 9 August 2023
Revised: 2 October 2023
Accepted: 5 October 2023
Published: 12 October 2023

1. Introduction

Bacillus spp. and related genera, including *Paenibacillus*, *Alicyclobacillus*, or *Brevibacillus*, are mostly Gram-positive and have the ability to produce spores and display metabolic capabilities under aerobic as well as relatively anaerobic conditions (Figure 1). Due to their characteristics, bacteria of this group have high resistance to environmental stresses such as drought, water stress, UV radiation, or low nutrient content in the environment [1,2]. *Bacillus* spp. and related genera commonly populate the Earth and occur in a variety of environments of both natural and anthropogenic origin [2–4].

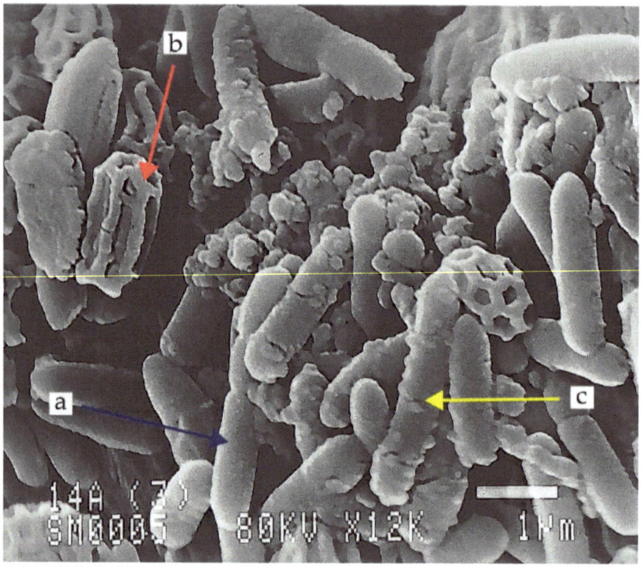

Figure 1. Scanning electron microscope (SEM) image of *Paenibacillus polymyxa* EG14 cultivated on medium with 0.5% cellobiose: vegetative cell (**a**), endospores (**b**), cellulosomes (**c**) (own photo).

In recent decades, rapid population growth has resulted in a significant intensification of agriculture, which has contributed to environmental pollution affecting both community structure and physiology of most microbial groups in the soil [5–7]. Because of the high cellulose abundance, organisms that have cellulolytic activity gained importance. Cellulases are synthesized by almost all groups of systematic organisms including microorganisms such as bacteria, fungi, protists, plants, and nematodes [8]. The bacteria capable of producing cellulolytic enzymes include both aerobic bacteria, e.g., *Butyrivibrio* spp. and *Cellulomonas* spp., as well as anaerobic bacteria, e.g., *Clostridium* spp. or *Ruminococcus* spp. bacteria [9]. However, due to their resistance to unfavorable environmental conditions, aerobic and relatively anaerobic, spore-forming bacteria of the phylum Firmicutes (i.e., bacterial strains of the genus *Bacillus* and related genera) are the most interesting [10]. So far, hundreds of cellulolytic spore-forming strains belonging to the phylum Firmicutes were isolated, including the genus *Bacillus* (e.g., *B. subtilis*) [11]; the genus *Alicyclobacillus* (e.g., *A. cellulosilyticus* [12] and *A. acidocaldarius* [13]); the genus *Geobacillus* (e.g., *Geobacillus* sp. HTA426) [14], or the genus *Lysinibacillus* (e.g., *Lysinibacillus fusiformis*) [15]. Despite the large number of isolated strains, still only a small part of them is commercialized, e.g., in the biodegradation of lignocellulosic waste.

The aim of the review is to summarize current knowledge of the taxonomy, ecology and properties of cellulolytic bacteria and to find gaps, the filling of which may lead to a better understanding of the ecology of *Bacillus* spp. and related genera, improving taxonomy and to a better exploitation of the cellulolytic potential of the bacteria group.

2. Taxonomy of the Genus *Bacillus* and Related Genera

The genus *Bacillus* and related genera (e.g., *Paenibacillus* or *Alicylobacillus*) are a very diverse group of bacteria that belongs to the phylum Firmicutes. The phylum includes several classes such as Bacilli, Clostridia, Mollicutes, and Erysipelotrichia. The group of bacteria which is the subject of the review belongs to the class Bacilli. Currently, it is classified into several families including Bacillaceae, Paenibacillaceae, etc. [16,17]. However, earlier, there was only the genus *Bacillus* which was first described in 1874. One of the first species specified in the genus *Bacillus* is type species-*B. subtilis*. The species is also one of the best-studied organisms belonging to the prokaryota and thus is extensively used as a model microorganism for Gram-positive bacteria. Also, in the past, *B. subtilis* was the model organism in studies conducted to understand spore formation mechanisms [18,19]. However, despite numerous and extensive studies on *Bacillus* and related bacteria, the overall phylogenetic and evolutionary history of these genera remains unclear and relatively unexplored.

Initially, bacteria were identified by phenotypic methods using light microscopy and staining techniques including Gram staining [20]. Other older techniques that remain helpful nowadays include evaluation of bacterial biochemical properties, for example determining the metabolic profile, which can be used to differentiate between bacterial species. An example of such identification methods is the API® 50 CHB/E system, which is based on 50 biochemical tests that test the carbohydrate metabolism of the data from *Bacillus* spp. and related genera [21]. On the other hand, an improved version for bacterial identification using rapid tests is the Biolog OmniLog System. In addition to carbon source metabolism, the method also includes 23 chemical tests that determine, for example, the bacteria's tolerance to salinity or sensitivity to other chemicals [22]. In both cases, the obtained results can be compared with databases and, to some extent, determine the taxonomic affiliation of the studied bacteria. However, it was not until the development of sequencing techniques in the 1990s that major changes in the taxonomy of spore-forming bacteria occurred. Then, other genera began to be separated from the genus *Bacillus*. Most phylogenetic studies are based on 16S rRNA gene sequences [23]. Based on branching in phylogenetic trees, initial phylogenetic studies delineated and identified five clusters of *Bacillus* species [24]. One of these clusters including *B. subtilis* was named *Bacillus sensu stricto* [24], while bacterial species from the other clusters were subsequently reclassified to form the following genera: *Paenibacillus, Lysinibacillus Brevibacillus,* and *Geobacillus* [23,25,26]. In subsequent years, based on phylogenetic and phenotypic results, many other *Bacillus* species were reclassified to form several new genera, for instance *Aneurinibacillus, Alicyclobacillus, Alkalicoccus, Sporosarcina, Gracilibacillus, Virgibacillus, Hydrogenibacillus, Ureibacillus, Solibacillus*. [23,25,27,28]. The bacterial genera listed above belong to different families, as shown in Table 1. For instance, *Alicylobacillus*, along with *Tumebacillus, Effusibacillus, Kyrpidia,* and *Sulfobacillus*, have been assigned to the family Alicylobacillace; a particularly important genus among those listed is *Alicylobacillus* [29,30]. On the other hand, the Paenibacillace family includes 14 genera of spore-forming bacteria of which *Paenibacillus* and *Brevibacillus* are the most interesting in terms of potential industrial use [31].

Table 1. List of families assigned to the order Bacillales [17,23,29,32–39].

Family Name	Proposed by	Type Genus	Other Example of Genus
Alicyclobacillaceae	da Costa and Rainey	*Alicyclobacillus*	*Effusibacillus, Kyrpidia, Tumebacillus*
Bacillaceae	Fischer	*Bacillus*	*Perribacillus, Weizmannia, Neobacillus, Metabacillus, Ferdinandcohnia, Gottfriedia, Heyndrickxia, Lederbergia*
Caryophanaceae	Peshkoff	*Caryophanon*	*Bhargavaea, Chryseomicrobium, Chungangia, Filibacter, Indiicoccus, Jeotgalibacillus, Kurthia, Lysinibacillus, Marinibacillus, Metalysinibacillus, Metaplanococcus, Metasolibacillus, Paenisporosarcina, Planococcus, Psychrobacillus, Rummeliibacillus, Savagea, Solibacillus, Sporosarcina, Ureibacillus*
Desulfuribacillaceae	Sorokin et al.	*Desulfuribacillus*	—
Listeriaceae	Ludwig et al.	*Listeria*	*Brochothrix*

Table 1. *Cont.*

Family Name	Proposed by	Type Genus	Other Example of Genus
Paenibacillaceae	De Vos et al.	*Paenibacillus*	*Ammoniibacillus, Aneurinibacillus group, (Ammoniphilus, Aneurinibacillus, Oxalophagus), Brevibacillus, Chengkuizengella, Cohnella, Fontibacillus, Gorillibacterium, Longirhabdus, Marinicrinis, Paludirhabdus, Saccharibacillus, Thermobacillus, Xylanibacillus*
Pasteuriaceae	Laurent	*Pasteuria*	—
Sporolactobacillaceae	Ludwig et al.	*Sporolactobacillus*	*Caenibacillus, Camelliibacillus Pullulanibacillus, Scopulibacillus, Sinobaca, Tuberibacillus*
Staphylococcaceae	Schleifer and Bell	*Staphylococcus*	*Abyssicoccus, Aliicoccus, Auricoccus Corticicoccus, Gemella, Jeotgalicoccus Macrococcus, Nosocomiicoccus, Salinicoccus*
Thermoactinomycetaceae	Matsuo et al.	*Thermoactinomyces*	*Baia, Croceifilum, Desmospora, Geothermomicrobium, Hazenella, Kroppenstedtia, Laceyella, Lihuaxuella, Marininema, Marinithermofilum, Mechercharimyces, Melghirimyces, Novibacillus, Paludifilum, Planifilum, Polycladomyces, Risungbinella, Salinithrix, Seinonella, Shimazuella, Thermoflavimicrobium*

Although studies using sequences coding 16S rRNA have led to the reclassification of many species to new genera, according to many researchers, analyzes based on this variable gene are not fully sufficient to correctly distinguish taxa at the species level [40–43]. Similarly to other taxa, the previous classification of order Bacillales and other related orders was mostly based on 16S rRNA gene sequences. Moreover, research based on this type of analysis contributes to the formation of various types of anomalies. The occurrence of anomalies among the order is confirmed by the fact that several families and genera forming spores and non-spores were placed in it. Such patterns suggest that one gene marker is not sufficient to determine the phylogenetic structure of the Bacillales order [41]. Phylogenetic analyses have also been carried out using several other gene or protein sequences [44–46]. However, due to the relatively small number of *Bacillus* species studied in these researches, the analysis is insufficient to elucidate species relationships within this large genus. Consequently, *Bacillus* spp. is still a highly heterogeneous genus characterized by extensive polyphyletic branching with other genera of the family Bacillaceae [47,48]. Furthermore, as a result of the diverse branching of current species in the genus *Bacillus*, it was difficult to limit the addition of new species to this genus, even despite the large differences between the new species and the type species. Therefore, more valid methods should be studied and used to delineate the genus *Bacillus* and limit the placement of unrelated species within it [23]. For instance, comparative analysis of whole genomes (based on NCBI available sequences/genomes) makes it possible to study the evolutionary relations of species, and thus provide opportunities to identify molecular markers (molecular synpomorphies) [23,49]. For example, molecular synapomorphies that contain conserved insertions and signature deletions in protein sequences are good means of differentiating species from the two major clades of the genus *Bacillus*, i.e., the "Subtilis clade" and the "Cereus clade". According to ICNP rule 56a, the transfer of a species from the Cereus clade to a new genus may play some part in human health; therefore, transfer to another species is not advisable. As evidenced by a comprehensive genomic analysis of Bacillaceae species, 36 new genetic markers (i.e., conserved signature indels (CSIs)) were detected [23]. Importantly, based on new CSIs, the monophyletic groups found in all reconstructed or new phylogenetic trees were named as follows: Simplex, Firmus, Alcalophilus, Niacini, Fastidiosus, and Jeotgali clades, and collectively included from 5 to 23 *Bacillus* species. In addition, researchers also performed a phylogenomic analysis on various Firmicutes proteins including core and conserved proteins. Moreover, the combined sequences of highly conserved proteins such as GyrB, GyrA, RpoC, RpoB, UvrD, or PolA were also studied, and confirmed by an extended comparative analysis of the genome of the above-mentioned protein sequences [23]. The authors of this study, based on robust evidence from many lines of research (conducted in parallel) confirming the existence of six distinct *Bacillus* clades, propose the transfer of species from these clades to six novel genera of Bacillaceae family, namely *Alkalihalobacillus* gen. nov., *Cytobacillus* gen. nov., *Mesobacillus* gen. nov., *Neobacillus* gen. nov., *Metabacillus* gen. nov., and *Peribacillus* gen. nov. [23]. Moreover, as a result of the creation of these new genera, 103 erroneously assigned species, that were insufficiently related to the genus *Bacillus*, were assigned to the new genera. The results above constitute an important step in elucidating the taxonomy of the *Bacillus* spp. and related genera. However, as indicated above, comprehensive studies are still needed for the correct classification of *Bacillus* spp. and related species.

3. Occurrence of Spore-Forming Bacteria in Arable Soils

Bacteria of the genus *Bacillus* and related genera are widely distributed in the environment, e.g., in soil, air, water, animals, plants, or sediments [50–53]. This group of bacteria plays a particularly important role in the soil, including the decay of matter [54], promotion of plant growth, and protection against phytopathogens.

A very good and widely used tool for assessing the abundance of bacteria is the next-generation sequencing (NGS), including 16S rRNA genes sequencing. However, due to the still existing limitations of sequencing technologies, most studies present the abundance of bacteria at high taxonomic levels, i.e., phyla or orders, and rarely present the abundance of bacteria at the genus level, which is a subject to much greater error [55,56].

The phylum Firmicutes is one of the dominant phyla in cultivation soils. Its relative abundance in the soil ranges from 2% to about 20% depending on agrotechnical practices used, including crop rotation systems and fertilization type [57–61]. In general, the Firmicutes type is more abundant in soils from crop rotation than in soils from continuous cropping [62–64]. The reason for these patterns is probably a greater influence of crop residues and decomposing roots in the soil from crop rotation compared to monoculture soil. For instance, in a greenhouse experiment, Li et al. [59] detected a higher number of sequences assigned to the phylum Firmicutes in soil (Mollisol with sandy loam texture) derived from rotation (tomato/potato-onion) compared to monoculture (tomato). The same patterns were noted for the genus *Bacillus*. The abundance values obtained by the authors at the level of the phylum Firmicutes and the genus *Bacillus* did not exceed 10%. However, there are also cases where more Firmicutes are detected in monocultures than in rotations, or in longer monocultures than shorter ones. For example, in the soil from the Morrow Plots experiment (USA), the relative abundance of the phylum Firmicutes ranged from a few to a maximum of 14%. Its abundance was dependent on soil management; in this case, the highest value was recorded in soil from a maize monoculture, while the lowest abundance of sequences assigned to the phylum Firmicutes was noted in soil from a maize-soybean rotation [6]. Similarly, Zhao et al. [60] observed a significantly increased number of sequences belonging to phylum Firmicutes in soil from 15- and 22-year continuous cropping of cucumber in comparison with cucumber grown for only one year. Earlier, Zhao et al. [65] noted similar patterns in continuous cropping of coffee. However, these authors did not find specific reasons for this phenomenon [60,65]. Hence, further studies are needed to find parameters that have a considerable role in shaping the abundance of the phylum Firmicutes including *Bacillus* spp. and related genera, e.g., identifying detailed correlations between physical and chemical properties of the soil and the abundance of bacteria belonging to the phylum Firmicutes in differently managed soils. For example, Alami et al. [66] observed robust correlations between the phylum Firmicutes and the physicochemical properties of arable soil including continuous cropping of maize and cabbage continuous cropping of cabbage (Hubei province, China); total phosphorus, available potassium, and available boron contents were positively correlated with the phylum Firmicutes. Furthermore, a study on the effect of continuous cotton cultivation (20 years) on the bacterial communities of the soil showed a positive correlation between the number of the OTUs of the phylum Firmicutes and the EC of the soil [66].

In addition, fertilization also affects the abundance of the phylum Firmicutes in the soil. Particularly because most members of the phylum Firmicutes are considered copiotrophs which are fast-growing microorganisms that prefer environments rich in C and N [67]. For instance, Li et al. [68] also found a several percent abundance of the phylum Firmicutes in fertilized soil (rice-rape rotation), and the highest number of OTUs belonging to the phylum was found in soil fertilized with NPKS (NPK + straw). Similar values were also found by Zeng et al. [58] who observed an abundance of the phylum Firmicutes at an average of 7% (the highest value was 10%) in soil fertilized with nitrogen fertilizer. Dang et al. [69] observed a significant increase in the abundance of Firmicutes in soil fertilized with manure (compared to the controls) across the globe, and detected a positive correlation between the SOC content and the abundance of the phylum. Furthermore, Francioli et al. [70] noted more OTUs assigned to the phylum Firmicutes in farmyard manure (FYM) fertilized soils compared to mineral fertilization (in a long-term fertilization trial). Hartmann et al. [71] also observed higher abundance of the phylum Firmicutes in long-term FYM fertilization in comparison with mineral fertilization. Similar findings were noted in a study on the effects of various treatments on the microbial community of bulk and rhizosphere soil [72]. Importantly, it was also found that manure fertilization is a factor influencing the bacterial community (including the abundance of Firmicutes) more strongly than the method of cultivation, including monoculture and crop rotation [31,73].

In conclusion, it should be noted that the abundance of bacteria of the phylum Firmicutes in soil may also be influenced by other agronomic treatments such as the use of plant protection agents. Thus, the study results may also have been caused by the heterogeneity of agricultural practices, as previously recorded by Soman et al. [6]. Moreover, the discrepancies in studies in this aspect may be an effect of the diversity of soils around the world, e.g., in terms of physical properties.

4. Cellulolytic Properties of *Bacillus* Spp. and Related Genera

4.1. Cellulases

Cellulose is the most common (bio)polymer on earth, made of glucose linked by β-1,4-glycosidic bonds. It contains two types of regions—crystalline and amorphous regions [74]. Hence, an important group of microbes that are participating in the element's circulation in the soil are microorganisms that decompose cellulose [75]. Soil properties such as pH, organic carbon content, nitrogen content, and moisture impact microbial cellulose degradation. The process of cellulose degradation depends on the presence of a complex of enzymes belonging to the class of O-glycoside hydrolases, including the three main cellulases [74]. Cellulolytic enzymes include: (i) endo-β-1,4-glucanases (EC 3.2.1.4) whose mechanism of action is based on random degradation of β-1,4-glycosidic bonds in amorphous regions of cellulose–endoglucanase activity is measured using cellulose derivatives, for instance, semi-soluble carboxymethylcellulose (CMC); the enzyme that degrades CMC is carboxymethylcellulase (CMCase); (ii) exo-1,4-β-glucanases (EC 3.2.1.91) that separate single molecules of glucose and cellobiose from reducing or non-reducing ends of the cellulose. Exoglucanases include e.g., avicelase–microcrystalline–cellulose (Avicel) degrading enzyme; and (iii) β-glucosidase whose mechanism of action is the conversion of cellobiose into glucose (EC 3.2.1.21) [76,77]. The synergistic cooperation of the above-mentioned enzymes and, in particular, the presence of a processive exoglucanase is required for cellulose degradation [Figure 2].

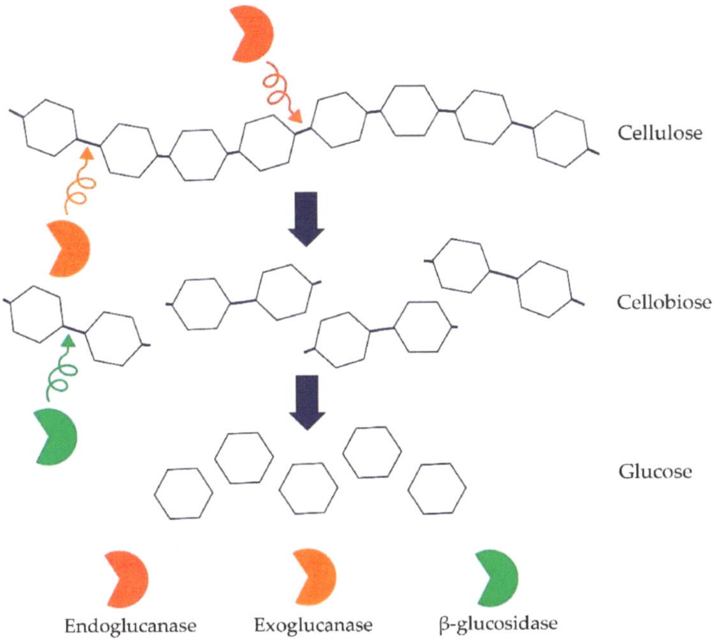

Figure 2. Mechanism of action of cellulases (own figure).

Cellulases have a modular structure that contains catalytic modules (CM) that act synergistically and/or with non-catalytic modules, i.e., substrate-binding modules [78]. In the case of cellulases, substrate-binding modules are called cellulose–binding modules (also called carbohydrate-binding module) (CBMs) and they can have affinity for amorphous or crystalline cellulose as well as binding to other similar polymers composed of carbon chains [79]. Due to the enzyme's ability to bind to cellulase, the local concentration of the enzyme increases, resulting in better substrate degradation efficiency. Some CBMs also have a structural function of stabilizing the catalytic module or altering its activity, for instance by inserting a substrate molecule into a substrate pocket [79]. Although the binding of cellulase by CBM is very stable, the enzyme can still diffuse across the substrate surface and, in some cases, CBM can also catalyze the breaking of non-covalent bonds located between the cellulose chains of crystalline cellulose [80].

4.2. Structural Diversity of Cellulases

Cellulases belong to the glycoside hydrolases (GH). The classification of GH is based on similarities in amino acid sequence and is included in the carbohydrate-active enzymes (CAZy) database. The CAZy database contains the CAZy families and subfamilies and is very dynamically updated. Due to the large differences in amino acid sequences, the GH group is remarkably heterogeneous and is divided into as many as 165 families [81]. The enzymes involved in cellulose degradation are classified in the following families of glycoside hydrolases: endoglucanases in families 5, 6, 7, 8, 12, 44, 45, 48, 51, 64, 71, 74, 81, 87, 124, 128, exoglucanases 5, 6, 7 and 4, and β-glucosidases in families 1, 3, 4, 17, 30, and 116 [82,83].

Sequence diversity can be related to a distinct modular architecture. It has been shown that the domain architecture in fungi is not very complex. However, in bacterial cellulases, there are many combinations of domain architectures, even though most sequences consist of a single catalytic domain [84]. In terms of carbohydrate-binding modules, CBM2 is related to the cellulolytic GH families and is found in the following families–GH5, GH12, GH44, GH45, GH48, GH51, and GH74. So far, it has been shown that the common CBM2 domain (in bacterial cellulases) in most cases binds cellulose, and less often chitin and xylan. Importantly, CBM2 is often found together with other accessory domains including CBM3 and CBM4, as well as catalytic domains [85]. In the terms of bacteria of the genus *Bacillus*, *B. licheniformis* possesses the H1AD14 gene encoding an endoglucanase belonging to the GH9 family and the cellulase has a CBM3 domain that is attached to the C-terminal end and plays a significant role in substrate degradation [86]. CBM3 has also been detected in a cellulase belonging to the GH9 family in *B. pumilus* [87]. Interestingly, Honda et al. [88] found that a unique chitinase domain in *B. thuringiensis* enabled binding to both crystalline chitin and cellulose, indicating that CBMs with affinity to multiple substrates could contribute to the increased occurrence of multifunctional hydrolytic enzymes [88]. On the other hand, previously, in the *B. subtilis* IFO 3034, an endoglucanase was detected that possessed a microcrystalline cellulose-binding domain but was unable to degrade microcrystalline cellulose [89]. Also in species related to *Bacillus* spp. CBMs were detected; in the *Paenibacillus lautus* BHU3 as many as four domains were detected, including CBM6, CBM46, CBM56, and CBM9, showing affinity for amorphous cellulose [90]. The CBM9 was also found in the genome of *P. dendritiformis* CRN1 [91].

Furthermore, cellulases belonging to different families have various protein fold structures, including the $(\beta/\alpha)8$ barrel fold, which is found in the GH5, GH44 and GH51 families, modified α/β barrel in family GH6, β-jelly roll—GH7 and GH12, the 7-fold β-propeller (GH74), $(\alpha/\alpha)6$ barrel—GH8, GH9 and GH48, the superhelical fold—GH124, and modified β barrel (GH45) [81]. Importantly, within a single GH family, structures are globally conserved, but sequences can be remarkably different. For example, GH5, one of the largest GH families, is currently divided into 166 subfamilies on the basis of sequence similarity, with only eight residues conserved across the family, including two catalytic glutamic acid residues [92]. In conclusion, GH families exhibiting different classes of protein folds have evolved to bind and degrade the same substrate, indicating that cellulolytic enzymes may have evolved independently and may be derived from many evolutionary origins, but have converged functionally [81]. Similar patterns regarding the evolution of cellulases can be inferred from the large number of cellulose-binding domains.

4.3. Cellulases Genes

Referring to the number of cellulases, it can be concluded that cellulolytic enzymes are highly diverse, which is further manifested in the large number of genes that are responsible for encoding these hydrolases. The number of genes encoding cellulases exceeds 100 [83,93]. As mentioned earlier, the reason for such a large number of cellulase-encoding genes may be due to independent evolution [81]. In fungi, the genes encoding cellulases in bacteria are located on a chromosome [94]. The spatial organization of these genes may differ between microorganism species, for example, in the bacterial species *Clostridium thermocellum* there is a random distribution, whereas in *C. cellulovorans* "clustered" distribution in a cluster occurs [95,96]. The cellulosome gene cluster in *C. cellulovorans* is about 22 kpz in size and contains nine genes encoding cellulosome domains with a putative transposon gene in the flanking region. A similar organization was also detected in the chromosome of the bacterial species *C. acetobutylicum* and *C. cellulolyticum*, suggesting the presence of a common bacterial ancestor of the clostridia [97]. In contrast, in fungi the genes encoding cellulases are usually distributed randomly, in which case each gene has its own transcriptional regulation. Only in exceptional cases, e.g., in *Phanerochaete chrysosporium* (fungus), the cellulase genes form a three-gene cluster [94].

In terms of *Bacillus* spp., in the genome of *B. licheniformis* [98] detected two clusters of genes involved in the cellulose decomposition. For instance, in the genome of the strain *B. subtilis* 168 no equivalents of the cluster were found. The enzymes encoded by the first gene cluster are likely endoglucanases belonging to the GH9 and GH5 families, and the probable cellulase–1,4-β-cellobiosidase belonging to GH48 and the potential β-mannanase belonging to GH5. Importantly, β-mannanase (GH5) and endoglucanase (GH9) contain carbohydrate-binding modules. In addition, with the exception of 1,4-β-cellobiosidase belonging to GH48, all gene proteins encoded have secretory signal peptides and all have homologs with *Bacillus* spp. but other than *B. subtilis* [98]. Researchers also detected a second cluster—encoding a probable β-glucosidase (from the family GH1). In addition, a second β-glucosidase gene (from the family GH3) was found at an unrelated locus in the genome. Importantly, the presence of these genes in the *B. licheniformis* genome indicate the possibility of complete degradation of cellulose [98]. Furthermore, 4 genes responsible for encoding β-glucosidase and 1 gene encoding endoglucanase were noted in strain *B. amyloliquefaciens* TL106. The β-glucosidases encoded by these genes belong to the GH1 and GH73 families, and the endoglucanase belongs to the GH5 family [99]. Moreover, Carbonaro et al. [100] analyzed the genome of *Alicyclobacillus mal*i FL18 to find new cellulose–degrading enzymes. The analysis revealed four genes belonging to the GH1, GH9, GH51, and GH94 families, of which GH1 and GH94 legitimately hydrolyse short oligosaccharides, and a gene from GH1 encodes a β-glucosidase. In addition, the *A. mali* FL-18 genome also contained genes encoding two probable arabinofuranosidases, which belong to GH51. Interestingly, *A. acidocaldarius*, which is a close relative of the aforementioned species, also possesses two endoglucanases–CelA belonging to the GH9 family and CelB from the GH51 family [101,102]. At the same time, other authors have detected a large number of genes encoding various GH enzymes in the *P. polymyxa* genome, including cellulases belonging to GH 1, 3, and 5 [103].

Finally, it is worth adding that most of the cellulases described in metagenomic studies (different environments) have less than 70% homology with known cellulolytic enzymes, and some of them have no significant similarity to other glycosyl hydrolases, indicating that large numbers of new cellulolytic enzymes are still being found [104]. Moreover, approximately 40% of sequenced bacterial genomes contain at least one cellulase gene, but only 4% of these bacteria are known as true cellulase bacteria due to low cellulase diversity or a lack of gene expression [83].

4.4. Cellulosomes

Some bacteria exhibiting cellulolytic activity are capable of synthesizing and secreting enzyme multicomplexes called cellulosomes; the secreted proteins outside the bacterial cell take the form of spherical structures. Sometimes, individual cellulosomes are joined together to form so-called polycellulosomes. A single complex may contain up to a dozen proteins with different activities, including endoglucanase, cellobiase or hemicellulase, and lichenase [105,106]. Interestingly, most of the research on cellulosomes concerns the cellulosomes of the phylum Firmicutes [83]. First studies on cellulosomes were carried out on the anaerobic *Clostridium thermocellum*. As shown, *C. thermocellum* is capable of synthesizing an enzyme complex of more than 2000 kDa, which consists of fourteen different proteins with molecular weights ranging from 45 kDa to 210 kDa [107]. For instance, 15 genes encoding the presence of endoglucanases, two genes responsible for the expression of xylanases, two genes encoding cellobiase, and one gene encoding lichenase were detected in *C. thermocellum* strain NC1B 10682 [108]. Cellulosomes have also been detected in *Bacillus* spp. and related genera. However, to date, little research has been conducted on these genera. For instance, *B. megaterium* was found to be capable of producing a cellulosome (celluloxylanosomes) exhibiting avicelase, CMCase, and xylanase activity. In addition, van Dyk et al. [105] noted that the *B. licheniformis* SVD1 strain was capable of synthesizing multi-enzyme complex (MEC) with hemi-cellulolytic activity. The total molecular mass of the complex was about 2000 kDa. The enzymes included in the MEC hydrolyzed such compounds as xylan, mannose, pectin, and carboxymethylcellulose. However, the MEC was not able to bind Avicel cellulose and, despite several similarities to the cellulosome, was ultimately not identified as such [105,109]. Waeonukul et al. [110] studied an enzymatic complex from *P. curdolanolyticus* B-6 (in culture on Avicel microcrystalline cellulose). A single cellulosome had the ability to hydrolyze the Avicel cellulose and insoluble xylan. The researchers noted that the complex included such enzymes as avicelase, CMCase, cellobiohydrolase, β-glucosidase, xylanase, α-L-arabinofuranosidase, and β-xylosidase.

The total mass of the multicomplex was about 1600 kDa. Importantly, the isolated cellulosome degraded lignocellulose efficiently. In terms of the genus *Paenibacillus*, using transmission microscopy, cellulosome production was detected in *P. polymyxa* strains EG2 and EG14 [111,112]. Besides, using scanning electron microscopy, protuberances were observed indicating cellulosome production on the cell surface in the thermophilic strain *Brevibacillus* sp. JXL [113].

4.5. Cellulase Activity

Cellulose decomposition starts when cellulase adsorbs to cellulose. Referring to previous subsections, it should be stated that bacteria of the genus *Bacillus* and related genera are capable of producing several types of cellulases including CMCase, FPase, or Avicelase. Different bacterial species have distinct activities of cellulolytic enzymes, and significant differences within the same species or strains may also occur due to discrepancies in culture conditions of the studies conducted on the topic. For instance, Acharya and Chaudhary [114] observed a CMCase activity of 0.300 U mL^{-1} in *Bacillus licheniformis* MVS1 (medium with beef extract). While Shajahan et al. [115], using response surface methodology in *Bacillus licheniformis* NCIM 5556, recorded a CMCase activity of 42.99 U mL^{-1} (medium contained CMC—19.21 g L^{-1}, $CaCl_2$—25.06 mg L^{-1}, Tween 20—2.96 mL L^{-1}, and temperature 43.35 °C).

The type of cellulose, medium composition, temperature and pH are most important for cellulase activity [2]. So far, depending on the strains, it has been found that the type of cellulose used as substrate induces cellulolytic activity to a different extent. For instance, Sadhu et al. [116] observed that carboxymethylcellulose better induced Avicelase and CMCase production by *Bacillus* sp. MTCC10046 compared to other substrates including sucrose, starch, glucose, or maltose. Also, Akaracharanya et al. [117] recorded higher cellulase activity of *Bacillus* sp. P3–1 and P4–6 in culture based on CMC medium, compared to culture with cellulose powder-containing medium. Similar patterns were also reported by Thomas et al. [118] who observed that CMCase activity by *Bacillus* sp. SV1 was higher in the CMC medium, compared to Avicel cellulose-containing medium and other carbon sources, including mannitol, glycerol, lactose, or chitin. In addition, CMCase and Avicelase activities were also obtained by Dobrzynski et al. [2] who noted the highest activity of the two enzymes in the cultures of *Bacillus* sp. 8E1A with CMC. However, in the case of FPase (cellulose saccharifying enzyme), the highest activity value was recorded for the culture of the studied strain with Avicel cellulose. Mihajlovski et al. [119] also reported slightly higher FPase activity in *P. chitinolyticus* CKS1 in a medium supplemented with Avicel compared to cultures with CMC. Similarly, in the case of thermophilic *Bacillus* sp. K-12, Kim and Kim [120] noted that FPase activity was higher when the strain studied by the authors was cultured in Avicel microcrystalline cellulose medium compared to other carbon sources. Interestingly, in contrast to previously cited reports, the strain *Bacillus* sp. K-12 also had high CMCase and Avicelase activity in cultures with Avicel cellulose. It is worth mentioning that the differences between studies may result from a number of factors including culture conditions.

Another important factor that affects the activity of cellulases produced by *Bacillus* spp. and related genera is temperature. According to the studies cited below, the optimum temperature range for cellulase activity ranges from 20 °C to 80 °C, depending on the strain and type of enzyme. For instance, Kazeem et al. [121] observed that a temperature of 20 °C is optimal for the production of FPases in the strain *B. licheniformis* 2D55. Cellulases produced by *B. pseudomycoides* (grown on sugarcane bagasse medium) have a slightly higher optimal temperature −40 °C (within 72 h of incubation) [122]. Interestingly, Li et al. [123] detected optimal cellulase activity in the thermophilic strain at 50 °C, and below this value the activity of enzymes significantly decreased. On the other hand, optimum temperature values for cellulase activity exceeding 70 °C have been recorded for activity of CMCase and Avicelase produced by *Geobacillus thermoleovorans* T4 (70 °C) and CMCase produced by *Bacillus* sp. DUSELR13 (75 °C) [124,125]. Similar patterns for *Bacillus* sp. 8E1A were observed by Dobrzyński et al. [2]. Importantly, thermophilic cellulases can potentially be used in various industries including textile, biofuel, and agriculture [2].

In terms of the optimal pH for cellulase activity, the range of values is as wide as for temperature; according to current reports, the highest activity of cellulases produced by *Bacillus* spp. and related genera is recorded in the pH range from 3 to 10. For example, Mihajlovski et al. [119] observed that the avicelase produced by the strain was most active at about pH 5. Similar results were reported by Seo et al. [126] whose *B. licheniformis* strain produced cellulases with high activity in the pH range of 4.0–6.0. While, in a study by Dobrzynski et al. [2], the highest CMCase and Avicelase activities were noted at pH 7.0 and FPase at 6.0. Interestingly, the highest cellulase activities produced by the bacteria of the genus *Bacillus* were also detected at pH 9.0 [127]. Previously, similar patterns were also obtained, as shown in Table 2.

Table 2. Optimum temperature and pH for celullolytic activity.

Strains	Egzoenzymes	Temperature Optimum	pH Optimum	References
Anoxybacillus sp. 527	Avicelase	70 °C	6.0	[113]
Anoxybacillus flavithermus EHP2	CMCase	75 °C	7.5	[26]
Bacillus sp. *K1*	CMCase	50 °C	6.0	[128]
Bacillus sp. KSM 330	CMCase Avicelase	45 °C	5.2	[129]
Bacillus sp. No.1139	CMCase	50 °C	9.0	[130]
B. licheniformis	CMCase	65 °C	6.0	[131]
B. subtilis YJ1	CMCase Avicelase	50–60 °C	6.0	[132]
Paenibacillus sp. *B39*	CMCase	60 °C	6.5	[133]
Paenibacillus terrae ME27–1	CMCase	50 °C	5.5	[134]

Importantly, the differences between the optimal conditions for the activity of cellulolytic enzymes result from the large variety of cellulases produced by the spore-forming bacteria of the genus *Bacillus* and related genera.

Moreover, the activity of cellulases is also affected by other parameters of the media or solutions. Gaur and Tiwari [135] found that the cellulase activity of *B. vallismortis* RG-07 was stimulated by Tween-60, Ca^{2+}, mercaptoethanol, and NaClO. While the cellulase activity of *Lysinibacillus xylanilyticus* was stimulated by the presence of $CaCl_2$ nanoparticles in medium [136].

Importantly, some of the spore-forming strains of cellulolytic bacteria are already being used to convert lignocellulosic waste. For instance, the activity of *P. polymyxa* ND24 was studied in a 5-L laboratory bioreactor where the cellulosic substrate in the medium was sugarcane bagasse; the strain showed the highest endoglucanase activity after 72 h of incubation. The sugarcane hydrolysate was then used for biogas production; the authors suggest that the obtained results support the use of *P. polymyxa* ND24 for cost-effective bioprocessing of lignocellulosic biomass [137]. In turn, other authors have used strains from the genus *Bacillus* to treat rice straw in order to increase the biomethane fermentation efficiency. The study, using multiple strains, demonstrated that the use of mixtures of different bacterial strains was more effective than the use of single bacterial strains, due to an increase in the pool of cellulases present in the process. Finally, the authors concluded that the choice of a mixture of strains from the genus *Bacillus*, which decompose lignocelluloses, can be robust catalysts for the processing of biomass from these wastes [138].

However, despite such a large number of bacterial strains of the genus *Bacillus* and related ones that produce cellulases, there is still little research on the practical aspect of their use, including the utilization and conversion of lignocellulosic biomass. Nevertheless, potentially, cellulolytic bacteria of the genus *Bacillus* spp. and their cellulases can be used: (i) in the textile industry (for instance for biostoning of jeans); (ii) in biorefining; (iii) in biogas and biofuel production; (iv) in agriculture including biodegradation of lignocellulosic waste and biocontrol of fungal phytopathogens; (v) in the paper industry (coadditive in pulp bleaching); (vi) in detergents (cellulose-based detergents); (vii) in the food industry including release of the antioxidants from fruit and vegetables, and improved texture and quality of bakery products; (viii) and for improving carotenoids extraction or improving olive oil extraction [139,140].

5. Promoting Plant Growth by the Bacteria of the Genus *Bacillus* and Related Genera

Bacteria of the genus *Bacillus* and related genera are also classified as plant growth-stimulating bacteria [53,141–143]. Bacteria from this group are capable of promoting plant growth either directly or indirectly. Mechanisms of direct promotion of plant growth include i.a. production of phytohormones including indole-3-acetic acid (IAA), cytokinins, and gibberellins, production of nitrogenase thanks to which bacteria fix atmospheric nitrogen (N) and make it available to plants, and the possibility of solubilizing phosphorus. Indirect mechanisms, on the other hand, include for instance production of antibiotics including cyclic lipopeptides, and enzymes degrading fungal cell walls [144–147].

So far, plant growth-promoting abilities have been detected in a very large number of bacteria belonging to the genus *Bacillus* or related genera. Bacteria from this group have promoted plant growth both under controlled and field conditions. Because of the greater value of studies under field conditions, several examples of such studies are presented in the review. For instance, inoculation of rice seedlings with *B. pumilus* TUAT-1 supplemented with N fertilizer led to an increase in height, biomass, and chlorophyll content of rice plants [148]. Besides, Ali et al. [149] showed that *B. cereus* (potassium solubilizing strain) increased the plant's height and shoots' dry weight. Importantly, compared to plants that were not inoculated, the application of the strain resulted in an increase of about 20% in potato yield. Moreover, the application of *Paenibacillus triticisoli* BJ-18 led to an increase in N, P, and organic matter contents in soil and enhanced nitrogenase activity and wheat yield [150]. Interestingly, in comparison to the control, the application with the strain also increased the biodiversity of rhizosphere bacterial communities and led to an increase in the abundance of the genus *Paenibacillus* in the inoculated soil, which also resulted in a high abundance of genes encoding nitrogenases. Furthermore, the inoculation with *P. triticisoli* BJ-18 also increased the abundance of native plant growth-stimulating bacteria of the genera *Bacillus* and *Podospora* [150].

Besides Okoroafor et al. [151], after applying *B. velezensis* FZB42 (formerly B. *amyloliquefaciens* FZB4) in maize and common sunflower cultivations, detected over 20% increase in biomass production in each of the crops. Moreover, inoculation with the tested preparation increased the bioavailability of soil elements. Interestingly, the study on winter wheat cultivation by Stepien et al. [152] is an example of a field experiment with *Bacillus* and related bacteria. The researchers demonstrated that the combination of mineral fertilization and three bacteria-*Paenibacillus azotofixans*, *B. megaterium*, and *B. subtilis*-significantly increased wheat grain yield compared to the application of mineral fertilization alone. In addition, the bacteria significantly increased the leaf greenness index SPAD at two time points, and together with NPK fertilization, significantly increased the content of two forms of nitrogen ($N\text{-}NO_3$ and $N\text{-}NH_4$) and phosphorus in the soil.

Another example of research using a bacterial consortium with *Bacillus* spp. is an experiment using *B. cereus* AR156, *B. subtilis* SM21, and *Serratia* sp. XY21 (BBS) strains applied to phytophthora-infested sweet pepper [153]. Compared to the control, the application of BBS reduced the occurrence of phytophthora blight and enhanced the fruit quality and soil properties. BBS also significantly increased the abundance of the bacterial genera *Burkholderia*, *Comamonas*, and *Ramlibacter*, which were negatively correlated with disease severity; moreover, the abundance of these genera were associated with organic carbon, ammonia nitrogen, potassium, and available phosphorus. These patterns suggest that changing the bacterial community improved the soil properties and reduced the phytopathogen development.

Importantly, there are still not enough studies in field conditions, especially those showing the effect of the inoculants used on the native microbiota whose biodiversity and taxonomic composition have the greatest influence on the biochemical processes of the soil. Finally, field studies with a wide range of parameters will bring inoculants closer to commercialization. However, there are already quite a number of commercial preparations containing *Bacillus* and related bacteria, for example biofertilizers, biofungicides, or biopesticides (listed in Table 3).

Table 3. Commercial preparations containing *Bacillus* and related bacteria.

Bacteria	Application	Mechanism	Commercial Biopreparation	Reference
B. subtilis C-3102	biofertilizer	for example: IAA production	Thervelics®	[154]
B. subtilis	biofertilizer	phosphate solubilization	BCMF	[155]
B. megaterium (combination with *Azotobacter chroococcum*, *Azospirillum brasilense*)	biofertilizer	phosphate solubilization	Azoter®	[156]
P. azotofixans, *B. megaterium* and *B. subtilis*	biofertilizer	nitrogen fixation	no information available	[152]
B. velezensis D747	biofungicide	cyclic lipopeptides	Double Nickel 55™	[157]
B. velezensis FZB42	biofungicide	antibiotic substances (polyketides and lipopeptides)	Taegro®	[157]
B. velezensis QST 713	antifungal and antibacterial product	antibiotic substances	Serenade®ASO	[158]
B. thuringiensis var. kurstakivar	biopesticide	crystal proteins (Cry) production	BT-Biox WP®	[159]
B. firmus I-1582	biopesticide	protection against nematode infection	VOTiVO®	[160,161]

6. Conclusions

In summary, bacteria of the genus *Bacillus* and related genera constitute an important group of bacteria that populate soil and other environments in large numbers, but their taxonomy is still inadequately defined, due to, among other things, their great diversity and the selection of insufficiently suitable molecular and biochemical techniques to determine their relationship. Among this group of bacteria, cellulolytic bacteria are one of the most important, but knowledge about their occurrence in the soil environment is still limited, which is caused by methodological difficulties faced by scientists studying it. Most studies on the presence of cellulolytic bacteria in the soil are limited to determining the abundance of genes encoding cellulase, which, due to the diversity of these genes, makes it impossible to determine the abundance of individual groups of cellulolytic bacteria.

Moreover, despite dozens of isolates of *Bacillus* and related bacteria showing cellulolytic activity, still few of these bacterial strains are used, for example, to degrade lignocellulosic waste. Importantly, the amount of lignocellulosic waste generated by agriculture and other industries is steadily increasing, which, in an era of progressive agriculture and other industries generating large amounts of such waste, poses a huge environmental problem. Therefore, researchers should focus on studying the cellulolytic bacteria, e.g., in biogasification processes or other conversions, which could contribute to the commercialization of these bacteria.

Author Contributions: Conceptualization, J.D.; methodology, J.D.; software, B.W.; writing—original draft preparation, J.D. and E.B.G.; writing—review and editing, J.D., B.W. and E.B.G.; visualization, B.W.; supervision, J.D. and E.B.G.; project administration, J.D.; funding acquisition, B.W. All authors have read and agreed to the published version of the manuscript.

Funding: This research received no external funding.

Institutional Review Board Statement: Not applicable.

Acknowledgments: Many thanks to Katarzyna Rafalska for help in revising the English language and Aleksandra Wróbel for help in data visualization.

Conflicts of Interest: The authors declare no conflict of interest.

References

1. Nicholson, W.L.; Munakata, N.; Horneck, G.; Melosh, H.J.; Setlow, P. Resistance of bacillus endospores to extreme terrestrial and extraterrestrial environments. *Mol. Biol. Rev.* **2000**, *64*, 548–557. [CrossRef] [PubMed]
2. Dobrzyński, J.; Wróbel, B.; Górska, E.B. Cellulolytic Properties of a Potentially Lignocellulose-Degrading *Bacillus* sp. 8E1A Strain Isolated from Bulk Soil. *Agronomy* **2022**, *12*, 665. [CrossRef]
3. Yakovleva, G.; Kurdy, W.; Gorbunova, A.; Khilyas, I.; Lochnit, G.; Ilinskaya, O. Bacillus pumilus proteome changes in response to 2, 4, 6-trinitrotoluene-induced stress. *Biodegradation* **2022**, *33*, 593–607. [CrossRef] [PubMed]
4. Zhang, Q.; Lin, Y.; Shen, G.Z.; Zhang, H.H.; Lyu, S.X. Siderophores of Bacillus pumilus promote 2-keto-L-gulonic acid production in a vitamin C microbial fermentation system. *J. Basic Microbiol.* **2022**, *62*, 833–842. [CrossRef]
5. Gamrat, R.; Gałczyńska, M.; Brysiewicz, A. Preliminary assessment of the impact of soil microorganisms on greenhouse gas emissions expressed in CO_2 equivalent and grass biomass. *J. Water Land Dev.* **2022**, *53*, 108–117. [CrossRef]
6. Soman, C.; Li, D.; Wander, M.M.; Kent, A.D. Long-Term Fertilizer and Crop-Rotation Treatments Differentially Affect Soil Bacterial Community Structure. *Plant Soil* **2017**, *413*, 145–159. [CrossRef]
7. Zielewicz, W.; Swędrzyński, A.; Dobrzyński, J.; Swędrzyńska, D.; Kulkova, I.; Wierzchowski, P.S.; Wróbel, B. Effect of Forage Plant Mixture and Biostimulants Application on the Yield, Changes of Botanical Composition, and Microbiological Soil Activity. *Agronomy* **2021**, *11*, 1786. [CrossRef]
8. Zhang, X.-Z.; Zhang, Y.-H.P. Cellulases: Characteristics, sources, production, and applications. In *Bioprocessing Technologies in Biorefinery for Sustainable Production of Fuels, Chemicals, and Polymers*; Yang, S.-T., El-Enshasy, H.A., Thongchul, N., Eds.; John Wiley & Sons, Inc.: Hoboken, NJ, USA, 2013; pp. 131–146.
9. Poulsen, H.V.; Willink, F.W.; Ingvorsen, K. Aerobic and anaerobic cellulase production by *Cellulomonas uda*. *Arch. Microbiol.* **2016**, *198*, 725–735. [CrossRef] [PubMed]
10. Deka, D.; Bhargavi, P.; Sharma, A.; Goyal, D.; Jawed, M.; Goyal, A. Enhancement of cellulase activity from a new strain of *Bacillus subtilis* by medium optimization and analysis with various cellulosic substrates. *Enzym. Res.* **2011**, *2011*, 151656. [CrossRef]
11. Irfan, M.; Mushtaq, Q.; Tabssum, F.; Shakir, H.A.; Qazi, J.I. Carboxymethyl cellulase production optimization from newly isolated thermophilic *Bacillus subtilis* K-18 for saccharification using response surface methodology. *AMB Express* **2017**, *7*, 29. [CrossRef] [PubMed]
12. Matsubara, H.; Goto, K.; Matsumura, T.; Mochida, K.; Iwaki, M.; Niwa, M.; Yamasato, K. *Alicyclobacillus acidophilus* sp. nov., a novel thermo-acidophilic, omega-alicyclic fatty acid-containing bacterium isolated from acidic beverages. *Int. J. Syst. Evol. Microbiol.* **2005**, *55*, 1681–1685, Erratum in *Int. J. Syst. Evol. Microbiol.* **2005**, *55 Pt 5*, 2233. [CrossRef]
13. Morana, A.; Esposito, A.; Maurelli, L.; Ruggiero, G.; Ionata, E.; Rossi, M.; Cara, F. A novel thermoacidophilic cellulase from *Alicyclobacillus acidocaldarius*. *Protein Pept. Lett.* **2008**, *15*, 1017–1021. [CrossRef] [PubMed]
14. Potprommanee, L.; Wang, X.-Q.; Han, Y.-J.; Nyobe, D.; Peng, Y.-P.; Huang, Q.; Liu, J.; Liao, Y.-L.; Chang, K.-L. Characterization of a thermophilic cellulase from *Geobacillus* sp. HTA426, an efficient cellulase-producer on alkali pre-treated of lignocellulosic biomass. *PLoS ONE* **2017**, *12*, e0175004. [CrossRef] [PubMed]
15. Khianngam, S.; Pootaeng-on, Y.; Techakriengkrai, T.; Tanasupawat, S. Screening and identification of cellulase producing bacteria isolated from oil palm meal. *J. Appl. Pharm. Sci.* **2014**, *4*, 90. [CrossRef]
16. Fritze, D. Taxonomy of the genus Bacillus and related genera: The aerobic endospore-forming bacteria. *Phytopathology* **2004**, *94*, 1245–1248. [CrossRef] [PubMed]
17. Harirchi, S.; Sar, T.; Ramezani, M.; Aliyu, H.; Etemadifar, Z.; Nojoumi, S.A.; Yazdian, F.; Awasthi, M.K.; Taherzadeh, M.J. *Bacillales*: From Taxonomy to Biotechnological and Industrial Perspectives. *Microorganisms* **2022**, *10*, 2355. [CrossRef] [PubMed]
18. Harwood, C.R. *Bacillus subtilis* and its relatives: Molecular biological and industrial workhorses. *Trends Biotechnol.* **1992**, *10*, 247–256. [CrossRef]
19. Zeigler, D.R. The genome sequence of *Bacillus subtilis* subsp. *spizizenii* W23: Insights into speciation within the *B. subtilis* complex and into the history of *B. subtilis* genetics. *Microbiology* **2011**, *157*, 2033–2041. [CrossRef] [PubMed]
20. Franco-Duarte, R.; Černáková, L.; Kadam, S.; Kaushik, K.S.; Salehi, B.; Bevilacqua, A.; Corbo, M.R.; Antolak, H.; Dybka-Stępień, K.; Leszczewicz, M.; et al. Advances in Chemical and Biological Methods to Identify Microorganisms—From Past to Present. *Microorganisms* **2019**, *7*, 130. [CrossRef] [PubMed]
21. Boyd, M.A.; Antonio, M.A.; Hillier, S.L. Comparison of API 50 CH strips to whole-chromosomal DNA probes for identification of *Lactobacillus* species. *J. Clin. Microbiol.* **2005**, *43*, 5309–5311. [CrossRef]
22. Klingler, J.M.; Stowe, R.P.; Obenhuber, D.C.; Groves, T.O.; Mishra, S.K.; Pierson, D.L. Evaluation of the Biolog automated microbial identification system. *Appl. Environ. Microbiol.* **1992**, *58*, 2089. [CrossRef]
23. Patel, S.; Gupta, R.S. A phylogenomic and comparative genomic framework for resolving the polyphyly of the genus *Bacillus*: Proposal for six new genera of *Bacillus* species, *Peribacillus* gen. nov., *Cytobacillus* gen. nov., *Mesobacillus* gen. nov., *Neobacillus* gen. nov., *Metabacillus* gen. nov. and *Alkalihalobacillus* gen. nov. *Int. J. Syst. Evol. Microbiol.* **2020**, *70*, 406–438. [CrossRef] [PubMed]
24. Ash, C.; Farrow, J.A.E.; Wallbanks, S.; Collins, M.D. Phylogenetic heterogeneity of the genus *Bacillus* revealed by comparative analysis of small-subunit-ribosomal RNA sequences. *Lett. Appl. Microbiol.* **1991**, *13*, 202–206. [CrossRef]
25. Shida, O.; Takagi, H.; Kadowaki, K.; Komagata, K. Proposal for two new genera, *Brevibacillus* gen. nov. and *Aneurinibacillus* gen. nov. *Int. J. Syst. Evol. Microbiol.* **1996**, *46*, 939–946. [CrossRef] [PubMed]

26. Ahmed, I.; Yokota, A.; Yamazoe, A.; Fujiwara, T. Proposal of *Lysinibacillus boronitolerans* gen. nov. sp. nov., and transfer of *Bacillus fusiformis* to *Lysini bacillus fusiformis* comb. nov. and *Bacillus sphaericus* to *Lysini bacillus sphaericus* comb. nov. *Int. J. Syst. Evol. Microbiol.* **2007**, *57*, 1117–1125. [CrossRef] [PubMed]
27. Wainø, M.; Tindall, B.J.; Schumann, P.; Ingvorsen, K. *Gracilibacillus* gen. nov., with description of *Gracilibacillus* halotolerans gen. nov., sp. nov.; transfer of *Bacillus dipsosauri* to *Gracilibacillus* dipsosauri comb. nov., and *Bacillus salexigens* to the genus *Salibacillus* gen. nov., as *Sali bacillus salexigens* comb. nov. *Int. J. Syst. Bacteriol.* **1999**, *49*, 821–831. [CrossRef] [PubMed]
28. Yoon, J.H.; Lee, K.C.; Weiss, N.; Kho, Y.H.; Kang, K.H.; Park, Y.H. *Sporosarcina aquimarina* sp. nov., a bacterium isolated from seawater in Korea, and transfer of *Bacillus globisporus* (Larkin and Stokes 1967), *Bacillus psychrophilus* (Nakamura 1984) and *Bacillus pasteurii* (Chester 1898) to the genus *Sporosarcina* as *Sporosarcina* globispora comb. nov., *Sporosarcina* psychrophila comb. nov. and *Sporosarcina* pasteurii comb. nov., and emended description of th. *Int. J. Syst. Evol. Microbiol.* **2001**, *51*, 1079–1086. [CrossRef]
29. Vos, P.; Garrity, G.M.; Jones, D.; Krieg, N.R.; Ludwig, W.; Rainey, F.A.; Schleifer, K.; Whitman, W.B. (Eds.) *Bergey's Manual of Systematic Bacteriology: Volume 3: The Firmicutes*; Springer: New York, NY, USA, 2010.
30. Yokota, A.; Fujii, T.; Goto, K. *Alicyclobacillus: Thermophilic Acidophilic Bacilli*; Springer: Tokyo, Japan, 2007.
31. Dobrzyński, J.; Wierzchowski, P.S.; Stępień, W.; Górska, E.B. The Reaction of Cellulolytic and Potentially Cellulolytic Spore-Forming Bacteria to Various Types of Crop Management and Farmyard Manure Fertilization in Bulk Soil. *Agronomy* **2021**, *11*, 772. [CrossRef]
32. De Maayer, P.; Aliyu, H.; Cowan, D.A. Reorganising the order Bacillales through phylogenomics. *Syst. Appl. Microbiol.* **2018**, *42*, 178–189. [CrossRef] [PubMed]
33. Gupta, R.S.; Patel, S. Robust Demarcation of the Family Caryophanaceae (Planococcaceae) and Its Different Genera Including Three Novel Genera Based on Phylogenomics and Highly Specific Molecular Signatures. *Front. Microbiol.* **2020**, *10*, 2821. [CrossRef]
34. Gupta, R.S.; Patel, S.; Saini, N.; Chen, S. Robust Demarcation of 17 Distinct *Bacillus* Species Clades, Proposed as Novel *Bacillaceae* Genera, by Phylogenomics and Comparative Genomic Analyses: Description of *Robertmurraya kyonggiensis* sp. Nov. and Proposal for an Emended Genus *Bacillus* Limiting It Only to the Members of the Subtilis and Cereus Clades of Species. *Int. J. Syst. Evol. Microbiol.* **2020**, *70*, 5753–5798. [PubMed]
35. Liu, Y.; Lai, Q.; Du, J.; Shao, Z. Genetic diversity and population structure of the *Bacillus cereus* group bacteria from diverse marine environments. *Sci. Rep.* **2017**, *7*, 689. [CrossRef] [PubMed]
36. NamesforLife Bacterial and Archaeal Nomenclature. Available online: https://www.namesforlife.com/search (accessed on 20 August 2022).
37. Parte, A.C.; Carbasse, J.S.; Meier-Kolthoff, j.; Reimer, L.C.; Göker, M. List of Prokaryotic Names with Standing in Nomenclature (Lpsn) Moves to the Dsmz. *Int. J. Syst. Evol. Microbiol.* **2020**, *70*, 5607–5612. [CrossRef] [PubMed]
38. Tindall, B.J. Names at the rank of class, subclass and order, their typification and current status: Supplementary information to Opinion 79. Judicial Commission of the International Committee on Systematics of Prokaryotes. *Int. J. Syst. Evol. Microbiol.* **2014**, *64*, 3599–3602. [CrossRef] [PubMed]
39. Huang, W.C.; Hu, Y.; Zhang, G.; Li, M. Comparative genomic analysis reveals metabolic diversity of different *Paenibacillus* groups. *Appl. Microbiol. Biotechnol.* **2020**, *104*, 10133–10143. [CrossRef] [PubMed]
40. Logan, N.A.; De Vos, P. Genus Bacillus Cohn 1872. In *Bergey's Manual of Systematic Bacteriology*; De Vos, P., Garrity, M., Jones, D., Krieg, R., Ludwig, W., Eds.; Springer: New York, NY, USA, 2009; pp. 21–128.
41. Logan, N.A.; Berge, O.; Bishop, A.H.; Busse, H.J.; De Vos, P.; Fritze, D.; Heyndrickx, M.; Kämpfer, P.; Rabinovitch, L.; Salkinoja-Salonen, M.S.; et al. Proposed minimal standards for describing new taxa of aerobic, endospore-forming bacteria. *Int. J. Syst. Evol. Microbiol.* **2009**, *59*, 2114–2121. [CrossRef]
42. Patel, S.; Gupta, R.S. Robust demarcation of fourteen different species groups within the genus *Streptococcus* based on genome-based phylogenies and molecular signatures. *Infect. Genet. Evol.* **2018**, *66*, 130–151. [CrossRef]
43. Rooney, A.P.; Price, N.P.J.; Ehrhardt, C.; Swezey, J.L.; Bannan, J.D. Phylogeny and molecular taxonomy of the *Bacillus subtilis* species complex and description of *Bacillus subtilis* subsp. *inaquosorum* subsp. nov. *Int. J. Syst. Evol. Microbiol.* **2009**, *59*, 2429–2436. [CrossRef]
44. La Duc, M.T.; Satomi, M.; Agata, N.; Venkateswaran, K. *gyrB* as a phylogenetic discriminator for members of the Bacillus anthracis-cereus-thuringiensis group. *J. Microbiol. Methods* **2004**, *56*, 83–394. [CrossRef]
45. Wang, L.T.; Lee, F.L.; Tai, C.J.; Kasai, H. Comparison of *gyrB* gene sequences, 16S rRNA gene sequences and DNA-DNA hybridization in the *Bacillus subtilis* group. *Int. J. Syst. Evol. Microbiol.* **2007**, *57*, 1846–1850. [CrossRef]
46. Verma, A.; Ojha, A.K.; Pal, Y.; Kumari, P.; Schumann, P.; Gruber-Vodicka, H.; Dastager, D.G.; Natarajan, R.K.; Mayilraj, S.; Krishnamurthi, S. An investigation into the taxonomy of "*Bacillus aminovorans*" and its reclassification to the genus *Domibacillus* as *Domibacillus aminovorans* sp. nov. *Syst. Appl. Microbiol.* **2017**, *40*, 458–467. [CrossRef] [PubMed]
47. Rössler, D.; Ludwig, W.; Schleifer, K.H.; Lin, C.; McGill, T.J.; Wisotzkey, J.D.; Jurtshuk, P., Jr.; Fox, G.E. Phylogenetic diversity in the genus *Bacillus* as seen by 16S rRNA sequencing studies. *Syst. Appl. Microbiol.* **1991**, *14*, 266–269. [CrossRef]
48. Klenk, H.P.; Lapidus, A.; Chertkov, O.; Copeland, A.; Del Rio, T.G.; Nolan, M.; Lucas, S.; Chen, F.; Tice, H.; Cheng, J.-F.; et al. Complete genome sequence of the thermophilic, hydrogen-oxidizing *Bacillus tusciae* type strain (T2) and reclassification in the new genus, *Kyrpidia* gen. nov. as *Kyrpidia tusciae* comb. nov. and emendation of the family *Alicyclobacillaceae* dA costa and rainey, 2010. *Stand. Genomic Sci.* **2011**, *5*, 121–134. [CrossRef]

49. Harwood, C.R.; Mouillon, J.-M.; Pohl, S.; Arnau, J. Secondary metabolite production and the safety of industrially important members of the *Bacillus subtilis* group. *FEMS Microbiol. Rev.* **2018**, *42*, 721–738. [CrossRef] [PubMed]
50. Kovács, Á.T. *Bacillus subtilis*. *Trends Microbiol.* **2019**, *27*, 724–725. [CrossRef]
51. Sansinenea, E.; Ortiz, A. Secondary metabolites of soil *Bacillus* spp. *Biotechnol. Lett.* **2011**, *33*, 1523–1538. [CrossRef]
52. Zhang, L.; Jin, M.; Shi, X.; Jin, L.; Hou, X.; Yu, Y.; Liu, B.; Cao, J.; Quan, C. Macrolactin metabolite production by *Bacillus* sp. ZJ318 isolated from marine sediment. *Appl. Biochem. Biotechnol.* **2022**, *194*, 2581–2593. [CrossRef]
53. Dobrzyński, J.; Jakubowska, Z.; Dybek, B. Potential of *Bacillus pumilus* to Directly Promote Plant Growth. *Front. Microbiol.* **2022**, *13*, 1069053. [CrossRef] [PubMed]
54. Smagacz, J.; Martyniuk, S. Soil properties and crop yields as influenced by the frequency of straw incorporation in a rape-wheat-triticale rotation. *J. Water Land Dev.* **2023**, *56*, 1–6. [CrossRef]
55. Mignard, S.; Flandrois, J.P. 16S rRNA sequencing in routine bacterial identification: A 30-month experiment. *J. Microbiol. Methods* **2006**, *67*, 574–581. [CrossRef] [PubMed]
56. Winand, R.; Bogaerts, B.; Hoffman, S.; Lefevre, L.; Delvoye, M.; Van Braekel, J.; Fu, Q.; Roosens, N.H.; De Keersmaecker, S.C.; Vanneste, K. Targeting the 16S rRNA Gene for Bacterial Identification in Complex Mixed Samples: Comparative Evaluation of Second (Illumina) and Third (Oxford Nanopore Technologies) Generation Sequencing Technologies. *Int. J. Mol. Sci.* **2020**, *21*, 298. [CrossRef]
57. Hartmann, M.; Widmer, F. Community Structure Analyses Are More Sensitive to Differences in Soil Bacterial Communities than Anonymous Diversity Indices. *Appl. Environ. Microbiol.* **2006**, *72*, 7804–7812. [CrossRef] [PubMed]
58. Zeng, J.; Liu, X.; Song, L.; Lin, X.; Zhang, H.; Shen, C.; Chu, H. Nitrogen fertilization directly affects soil bacterial diversity and indirectly affects bacterial community composition. *Soil Biol. Biochem.* **2016**, *92*, 41–49. [CrossRef]
59. Li, N.; Gao, D.; Zhou, X.; Chen, S.; Li, C.; Wu, F. Intercropping with Potato-Onion Enhanced the Soil Microbial Diversity of Tomato. *Microorganisms* **2020**, *8*, 834. [CrossRef] [PubMed]
60. Zhao, Y.; Mao, X.; Zhang, M.; Yang, W.; Di, H.J.; Ma, L.; Liu, W.; Li, B. Response of Soil Microbial Communities to Continuously Mono-Cropped Cucumber under Greenhouse Conditions in a Calcareous Soil of North China. *J. Soils Sediments* **2020**, *20*, 2446–2459. [CrossRef]
61. Wierzchowski, P.S.; Dobrzyński, J.; Mazur, K.; Kierończyk, M.; Wardal, W.J.; Sakowski, T.; Barszczewski, J. Chemical Properties and Bacterial Community Reaction to Acidified Cattle Slurry Fertilization in Soil from Maize Cultivation. *Agronomy* **2021**, *11*, 601. [CrossRef]
62. Li, H.; Wang, J.; Liu, Q.; Zhou, Z.; Chen, F.; Xiang, D. Effects of Consecutive Monoculture of Sweet Potato on Soil Bacterial Community as Determined by Pyrosequencing. *J. Basic Microbiol.* **2019**, *59*, 181–191. [CrossRef] [PubMed]
63. Zhu, S.; Wang, Y.; Xu, X.; Liu, T.; Wu, D.; Zheng, X.; Tang, S.; Dai, Q. Potential Use of High-Throughput Sequencing of Soil Microbial Communities for Estimating the Adverse Effects of Continuous Cropping on Ramie (*Boehmeria nivea* L. Gaud). *PLoS ONE* **2018**, *13*, e0197095. [CrossRef]
64. Yang, L.; Tan, L.; Zhang, F.; Gale, W.J.; Cheng, Z.; Sang, W. Duration of Continuous Cropping with Straw Return Affects the Composition and Structure of Soil Bacterial Communities in Cotton Fields. *Can. J. Microbiol.* **2018**, *64*, 167–181. [CrossRef] [PubMed]
65. Zhao, Q.; Xiong, W.; Xing, Y.; Sun, Y.; Lin, X.; Dong, Y. Long-Term Coffee Monoculture Alters Soil Chemical Properties and Microbial Communities. *Sci. Rep.* **2018**, *8*, 6116. [CrossRef] [PubMed]
66. Alami, M.M.; Pang, Q.; Gong, Z.; Yang, T.; Tu, D.; Zhen, O.; Yu, W.; Alami, M.J.; Wang, X. Continuous Cropping Changes the Composition and Diversity of Bacterial Communities: A Meta-Analysis in Nine Different Fields with Different Plant Cultivation. *Agriculture* **2021**, *11*, 1224. [CrossRef]
67. Lienhard, P.; Terrat, S.; Prévost-Bouré, N.C.; Nowak, V.; Régnier, T.; Sayphoummie, S.; Panyasiri, K.; Tivet, F.; Mathieu, O.; Levêque, J.; et al. Pyrosequencing Evidences the Impact of Cropping on Soil Bacterial and Fungal Diversity in Laos Tropical Grassland. *Agron. Sustain. Dev.* **2014**, *34*, 525–533. [CrossRef]
68. Li, J.; Gan, G.; Chen, X.; Zou, J. Effects of Long-Term Straw Management and Potassium Fertilization on Crop Yield, Soil Properties, and Microbial Community in a Rice–Oilseed Rape Rotation. *Agriculture* **2021**, *11*, 1233. [CrossRef]
69. Dang, P.; Li, C.; Lu, C.; Zhang, M.; Huang, T.; Wan, C.; Wang, H.; Chen, Y.; Qin, X.; Liao, Y.; et al. Effect of fertilizer management on the soil bacterial community in agroecosystems across the globe. *Agric. Ecosyst. Environ.* **2021**, *326*, 107795. [CrossRef]
70. Francioli, D.; Schulz, E.; Lentendu, G.; Wubet, T.; Buscot, F.; Reitz, T. Mineral vs. Organic Amendments: Microbial Community Structure, Activity and Abundance of Agriculturally Relevant Microbes Are Driven by Long-Term Fertilization Strategies. *Front. Microbiol.* **2016**, *7*, 1446. [CrossRef] [PubMed]
71. Hartmann, M.; Frey, B.; Mayer, J.; Mäder, P.; Widmer, F. Distinct Soil Microbial Diversity under Long-Term Organic and Conventional Farming. *ISME J.* **2015**, *9*, 1177–1194. [CrossRef] [PubMed]
72. Hartman, K.; van der Heijden, M.G.A.; Wittwer, R.A.; Banerjee, S.; Walser, J.-C.; Schlaeppi, K. Cropping Practices Manipulate Abundance Patterns of Root and Soil Microbiome Members Paving the Way to Smart Farming. *Microbiome* **2018**, *6*, 14.
73. Guo, Z.; Wan, S.; Hua, K.; Yin, Y.; Chu, H.; Wang, D.; Guo, X. Fertilization regime has a greater effect on soil microbial community structure than crop rotation and growth stage in an agroecosystem. *Appl. Soil Ecol.* **2020**, *149*, 103510. [CrossRef]
74. Bhardwaj, N.; Kumar, B.; Agrawal, K.; Verma, P. Current perspective on production and applications of microbial cellulases: A review. *Bioresour. Bioprocess.* **2021**, *8*, 95. [CrossRef]

75. Thapa, S.; Mishra, J.; Arora, N.; Mishra, P.; Li, H.; O'Hair, J.; Bhatti, S.; Zhou, S. Microbial cellulolytic enzymes: Diversity and biotechnology with reference to lignocellulosic biomass degradation. *Rev. Environ. Sci. Biol. Technol.* **2020**, *19*, 621–648.
76. Horn, S.; Vaaje-Kolstad, G.; Westereng, B.; Eijsink, V.G. Novel enzymes for the degradation of cellulose. *Biotechnol. Biofuels* **2012**, *5*, 45. [CrossRef] [PubMed]
77. Singhania, R.R.; Ruiz, H.A.; Awasthi, M.K.; Dong, C.D.; Chen, C.W.; Patel, A.K. Challenges in cellulase bioprocess for biofuel applications. *Renew. Sustain. Energy Rev.* **2021**, *151*, 111622. [CrossRef]
78. Zverlov, V.V.; Schwarz, W.H. Bacterial cellulose hydrolysis in anaerobic environmental subsystems—*Clostridium thermocellum* and *Clostridium stercorarium*, thermophilic plant-fiber degraders. *Ann. N. Y. Acad. Sci.* **2008**, *1125*, 298–307. [CrossRef]
79. dos Santos, J.R.; de Souza Moreira, L.R.; Filho, E.X.F. Cellulose-degrading enzymes: Key players in biorefinery development. *Biologia* **2023**, *78*, 1759–1772. [CrossRef]
80. Maki, M.; Leung, K.T.; Qin, W. The prospects of cellulase-producing bacteria for the bioconversion of lignocellulosic biomass. *Int. J. Biol. Sci.* **2009**, *5*, 500. [CrossRef] [PubMed]
81. Liu, L.; Huang, W.C.; Liu, Y.; Li, M. Diversity of cellulolytic microorganisms and microbial cellulases. *Int. Biodeterior. Biodegrad.* **2021**, *163*, 105277. [CrossRef]
82. Himmel, M.E.; Xu, Q.; Luo, Y.; Ding, S.Y.; Lamed, R.; Bayer, E.A. Microbial enzyme systems for biomass conversion: Emerging paradigms. *Biofuels* **2010**, *1*, 323–341. [CrossRef]
83. Koeck, D.E.; Pechtl, A.; Zverlov, V.V.; Schwarz, W.H. Genomics of cellulolytic bacteria. *Curr. Opin. Biotechnol.* **2014**, *29*, 171–183. [CrossRef]
84. Nguyen, S.N.; Flores, A.; Talamantes, D.; Dar, F.; Valdez, A.; Schwans, J.; Berlemont, R. GeneHunt for rapid domain-specific annotation of glycoside hydrolases. *Sci. Rep.* **2019**, *9*, 10137. [CrossRef] [PubMed]
85. Sidar, A.; Albuquerque, E.D.; Voshol, G.P.; Ram, A.F.; Vijgenboom, E.; Punt, P.J. Carbohydrate binding modules: Diversity of domain architecture in amylases and cellulases from filamentous microorganisms. *Front. Bioeng. Biotechnol.* **2020**, *8*, 871. [CrossRef] [PubMed]
86. Konar, A.; Aich, S.; Katakojwala, R.; Datta, S.; Mohan, S.V. A processive GH9 family endoglucanase of *Bacillus licheniformis* and the role of its carbohydrate-binding domain. *Appl. Microbiol. Biotechnol.* **2022**, *106*, 6059–6075. [CrossRef] [PubMed]
87. Ogonda, L.A.; Saumonneau, A.; Dion, M.; Muge, E.K.; Wamalwa, B.M.; Mulaa, F.J.; Tellier, C. Characterization and engineering of two new GH9 and GH48 cellulases from a Bacillus pumilus isolated from Lake Bogoria. *Biotechnol. Lett.* **2021**, *43*, 691–700. [CrossRef] [PubMed]
88. Honda, S.; Kunii, T.; Nohara, K.; Wakita, S.; Sugahara, Y.; Kawakita, M.; Oyama, F.; Sakaguchi, M. Characterization of a Bacillus thuringiensis chitinase that binds to cellulose and chitin. *AMB Express* **2017**, *7*, 51. [CrossRef] [PubMed]
89. Park, J.S.; Nakamura, A.; Horinouchi, S.; Beppu, T. Identification of the cellulose-binding domain of a *Bacillus subtilis* endoglucanase distinct from its catalytic domain. *Biosci. Biotechnol. Biochem.* **1993**, *57*, 260–264. [CrossRef] [PubMed]
90. Yadav, S.; Dubey, S.K. Cellulose degradation potential of *Paenibacillus lautus* strain BHU3 and its whole genome sequence. *Bioresour. Technol.* **2018**, *262*, 124–131. [CrossRef] [PubMed]
91. Srivastava, S.; Dafale, N.A.; Purohit, H.J. Functional genomics assessment of lytic polysaccharide mono-oxygenase with glycoside hydrolases in *Paenibacillus dendritiformis* CRN18. *Int. J. Biol. Macromol.* **2020**, *164*, 3729–3738. [CrossRef] [PubMed]
92. Bianchetti, C.M.; Brumm, P.; Smith, R.W.; Dyer, K.; Hura, G.L.; Rutkoski, T.J.; Phillips, G.N., Jr. Structure, dynamics, and specificity of endoglucanase D from *Clostridium cellulovorans*. *J. Mol. Biol.* **2013**, *425*, 4267–4285. [CrossRef]
93. Duan, C.J.; Feng, J.X. Mining metagenomes for novel cellulase genes. *Biotechnol. Lett.* **2010**, *32*, 1765–1775. [CrossRef] [PubMed]
94. Tomme, P.R.; Warren, A.J.; Gilkes, N.R. Cellulose hydrolysis by bacteria and fungi. *Adv. Microb. Physiol.* **1995**, *37*, 1–81.
95. Guglielmi, G.; Beguin, P. Cellulase and hemicellulase genes of *Clostridium thermocellum* from five independent collections contain few overlaps and are widely scattered across the chromosome. *FEMS Microbiol. Lett.* **1998**, *161*, 209–215. [CrossRef] [PubMed]
96. Belaich, J.P.; Belaich, A.; Fierobe, H.P.; Gal, L.; Gaudin, C.; Pagès, S.; Reverbel-Leroy, C.; Tardif, C. The cellulolytic system of *Clostridium cellulolyticum*. In *Genetics, Biochemistry and Ecology of Cellulose Degradation*; Ohmiya, K., Hayashi, K., Sakka, Y., Kobayashi, S., Karita, T., Eds.; Kimura Uni Publishers: Tokyo, Japan, 1999; pp. 479–487.
97. Tamaru, Y.; Karita, S.; Ibrahim, A.; Chan, H.; Doi, R.H. A large gene cluster for the Clostridium cellulovorans cellulosome. *J. Bacteriol.* **2000**, *182*, 5906–5910. [CrossRef] [PubMed]
98. Rey, M.W.; Ramaiya, P.; Nelson, B.A.; Brody-Karpin, S.D.; Zaretsky, E.J.; Tang, M.; Lopez de Leon, A.; Xiang, H.; Gusti, V.; Clausen, I.G.; et al. Complete genome sequence of the industrial bacterium *Bacillus licheniformis* and comparisons with closely related *Bacillus* species. *Genome Biol.* **2004**, *5*, R77. [CrossRef] [PubMed]
99. Shang, Z.; Liu, S.; Duan, Y.; Bao, C.; Wang, J.; Dong, B.; Cao, Y. Complete genome sequencing and investigation on the fiber-degrading potential of *Bacillus amyloliquefaciens* strain TL106 from the tibetan pig. *BMC Microbiol.* **2022**, *22*, 186. [CrossRef] [PubMed]
100. Carbonaro, M.; Aulitto, M.; Gallo, G.; Contursi, P.; Limauro, D.; Fiorentino, G. Insight into CAZymes of *Alicyclobacillus mali* FL18: Characterization of a New Multifunctional GH9 Enzyme. *Int. J. Mol. Sci.* **2023**, *24*, 243. [CrossRef]
101. Eckert, K.; Schneider, E. A Thermoacidophilic Endoglucanase (CelB) from *Alicyclobacillus acidocaldarius* Displays High Sequence Similarity to Arabinofuranosidases Belonging to Family 51 of Glycoside Hydrolases. *Eur. J. Biochem.* **2003**, *270*, 3593–3602. [CrossRef]

102. Eckert, K.; Zielinski, F.; lo Leggio, L.; Schneider, E. Gene Cloning, Sequencing, and Characterization of a Family 9 Endoglucanase (CelA) with an Unusual Pattern of Activity from the Thermoacidophile *Alicyclobacillus acidocaldarius* ATCC27009. *Appl. Microbiol. Biotechnol.* **2002**, *60*, 428–436.
103. Eastman, A.W.; Heinrichs, D.E.; Yuan, Z.C. Comparative and genetic analysis of the four sequenced *Paenibacillus polymyxa* genomes reveals a diverse metabolism and conservation of genes relevant to plant-growth promotion and competitiveness. *BMC Genomics* **2014**, *15*, 581. [CrossRef]
104. Kuhad, R.C.; Deswal, D.; Sharma, S.; Bhattacharya, A.; Jain, K.K.; Kaur, A.; Pletschke, B.I.; Singh, A.K.; Karp, M. Revisiting cellulase production and redefining current strategies based on major challenges. *Renew. Sustain. Energy Rev.* **2016**, *55*, 249–272. [CrossRef]
105. Van Dyk, J.S.; Sakka, M.; Sakka, K.; Pletschke, B.I. The cellulolytic and hemi-cellulolytic system of *Bacillus licheniformis* SVD1 and the evidence for production of a large multi-enzyme complex. *Enzyme Microb. Technol.* **2009**, *45*, 372–378. [CrossRef]
106. Orencio-Trejo, M.; De la Torre-Zavala, S.; Rodriguez-Garcia, A.; Avilés-Arnaut, H.; Gastelum-Arellanez, A. Assessing the performance of bacterial cellulases: The use of *Bacillus* and *Paenibacillus* strains as enzyme sources for lignocellulose saccharification. *BioEnergy Res.* **2016**, *9*, 1023–1033. [CrossRef]
107. Lamed, R.; Setter, E.; Kenig, R.; Bayer, E.A. Cellulosome: A discrete cell surface organelle of *Clostridium thermocellum* which exhibits separate antigenic, cellulose-binding and various cellulolytic activities. *Biotechnol. Bioeng. Symp.* **1983**, *13*, 163–181.
108. Hazlewood, G.P.; Romaniec, M.P.M.; Davidson, K.; Grepinet, O.; Beguin, P.; Millet, J.; Raynaud, O.; Aubert, J.P. A catalogue of Clostridium thermocellum endoglucanase, B-glucosidase, and xylanase genes cloned in Escherichia coli. *FEMS Microbiol. Lett.* **1988**, *51*, 231–236. [CrossRef]
109. Muras, A.; Romero, M.; Mayer, C.; Otero, A. Biotechnological applications of Bacillus licheniformis. *Crit. Rev. Biotechnol.* **2021**, *41*, 609–627. [CrossRef] [PubMed]
110. Waeonukul, R.; Kyu, K.J.; Sakka, K.; Ratanakhanokchai, K. Isolation and characterization of a multienzyme complex (cellulosome) of the *Paenibacillus curdlanolyticus* B-6 grown on Avicel under aerobic conditions. *J. Biosci. Bioeng.* **2009**, *107*, 610–614. [CrossRef] [PubMed]
111. Górska, E.; Tudek, B.; Russel, S. Degradation of cellulose by nitrogen-fixing strain of Bacillus polymyxa. *Acta Microbiol. Pol.* **2001**, *50*, 129–137. [PubMed]
112. Górska, E.B.; Jankiewicz, U.; Dobrzynski, J.; Russel, S.; Pietkiewicz, S.; Kalaji, H.; Gozdowski, D. Degradation and colonization of cellulose by diazotrophic strains of Paenibacillus polymyxa isolated from soil. *J. Bioremediat. Biodegrad.* **2015**, *6*, 271. [CrossRef]
113. Liang, Y.; Yesuf, J.; Schmitt, S.; Bender, K.; Bozzola, J. Study of cellulases from a newly isolated thermophilic and cellulolytic Brevibacillus sp. strain JXL. *J. Ind. Microbiol. Biotechnol.* **2009**, *36*, 961–970. [CrossRef] [PubMed]
114. Acharya, S.; Chaudhary, A. Optimization of fermentation conditions for cellulases production by *Bacillus licheniformis* MVS1 and *Bacillus* sp. MVS3 isolated from Indian hot spring. *Braz. Arch. Biol. Technol.* **2012**, *55*, 497–503. [CrossRef]
115. Shajahan, S.; Moorthy, I.G.; Sivakumar, N.; Selvakumar, G. Statistical modeling and optimization of cellulase production by *Bacillus licheniformis* NCIM 5556 isolated from the hot spring, Maharashtra, India. *King Saud Univ. Sci.* **2017**, *29*, 302–310. [CrossRef]
116. Sadhu, S.; Ghosh, P.K.; Aditya, G.; Maiti, T.K. Optimization and strain improvement by mutation for enhanced cellulase production by *Bacillus* sp. (MTCC10046) isolated from cow dung. *J. King Saud Univ. Sci.* **2014**, *26*, 323–332. [CrossRef]
117. Akaracharanya, A.; Taprig, T.; Sitdhipol, J.; Tanasupawat, S. Characterization of cellulase producing *Bacillus* and *Paenibacillus* strains from Thai soils. *J. Appl. Pharm. Sci.* **2014**, *4*, 6. [CrossRef]
118. Thomas, L.; Ram, H.; Singh, V.P. Inducible cellulase production from an organic solvent tolerant *Bacillus* sp. SV1 and evolutionary divergence of endoglucanase in different species of the genus *Bacillus*. *Braz. J. Microbiol.* **2018**, *49*, 429–442. [CrossRef]
119. Mihajlovski, K.R.; Carević, M.B.; Dević, M.L.; Šiler-Marinković, S.; Rajilić-Stojanović, M.D.; Dimitrijević-Branković, S. Lignocellulosic waste material as substrate for Avicelase production by a new strain of *Paenibacillus chitinolyticus* CKS1. *Int. Biodeterior. Biodegrad.* **2015**, *104*, 426–434. [CrossRef]
120. Kim, D.S.; Kim, C.H. Production and characterization of crystalline cellulose-degrading cellulase components from a thermophilic and moderately alkalophilic bacterium. *J. Microbiol. Biotechnol.* **1992**, *2*, 7–13.
121. Kazeem, M.O.; Shah, U.K.M.; Baharuddin, A.S.; Abdul Rahman, N.A. Enhanced cellulase production by a novel thermophilic *Bacillus licheniformis* 2D55: Characterization and application in lignocellulosic saccharification. *Bioresources* **2016**, *11*, 5404–5423. [CrossRef]
122. Pramanik, S.K.; Mahmud, S.; Paul, G.K.; Jabin, T.; Naher, K.; Uddin, M.S.; Zaman, S.; Saleh, M.A. Fermentation optimization of cellulase production from sugarcane bagasse by Bacillus pseudomycoides and molecular modeling study of cellulase. *Curr. Res. Microb. Sci.* **2021**, *2*, 100013. [CrossRef] [PubMed]
123. Li, W.; Zhang, W.-W.; Yang, M.-M.; Chen, Y.-L. Cloning of the thermostable cellulase gene from newly isolated *Bacillus subtilis* and its expression in *Escherichia coli*. *Mol. Biotechnol.* **2008**, *40*, 195–201. [CrossRef] [PubMed]
124. Rastogi, G.; Bhalla, A.; Adhikari, A.; Bischoff, K.M.; Hughes, S.R.; Christopher, L.P.; Sani, R.K. Characterization of thermostable cellulases produced by *Bacillus* and *Geobacillus* strains. *Bioresour. Technol.* **2010**, *101*, 8798–8806. [CrossRef] [PubMed]
125. Tai, S.-K.; Lin, H.-P.P.; Kuo, J.; Liu, J.-K. Isolation and characterization of a cellulolytic *Geobacillus thermoleovorans* T4 strain from sugar refinery wastewater. *Extremophiles* **2004**, *8*, 345–349. [CrossRef] [PubMed]

126. Seo, J.K.; Park, T.S.; Kwon, I.H.; Piao, M.Y.; Lee, C.H.; Ha, J.K. Characterization of cellulolytic and xylanolytic enzymes of *Bacillus licheniformis* JK7 isolated from the rumen of a native Korean goat. *Asian-Australas. J. Anim. Sci.* **2013**, *26*, 50–58. [CrossRef] [PubMed]
127. Tabssum, F.; Irfan, M.; Shakir, H.A.; Qazi, J.I. RSM based optimization of nutritional conditions for cellulase mediated Saccharification by *Bacillus cereus*. *J. Biol. Eng.* **2018**, *12*, 7. [CrossRef]
128. Paudel, Y.P.; Qin, W. Characterization of novel cellulase-producing bacteria isolated from rotting wood samples. *Appl. Biochem. Biotechnol.* **2015**, *177*, 1186–1198. [CrossRef]
129. Ozaki, K.; Ito, S. Purification and properties of an acid endo-1, 4-β-glucanase from *Bacillus* sp. KSM-330. *Microbiology* **1991**, *137*, 41–48. [CrossRef] [PubMed]
130. Fukumori, F.; Kudo, T.; Horikoshi, K. Purification and properties of a cellulase from alkalophilic *Bacillus* sp. no. 1139. *Microbiology* **1985**, *131*, 3339–3345. [CrossRef]
131. Bischoff, K.M.; Rooney, A.P.; Li, X.L.; Liu, S.; Hughes, S.R. Purification and characterization of a family 5 endoglucanase from a moderately thermophilic strain of *Bacillus licheniformis*. *Biotech. Lett.* **2006**, *28*, 1761–1765. [CrossRef]
132. Yin, L.J.; Lin, H.H.; Xiao, Z.R. Purification and characterization of a cellulase from *Bacillus subtilis* YJ1. *J. Mar. Sci. Technol.* **2010**, *18*, 19. [CrossRef]
133. Wang, C.M.; Shyu, C.L.; Ho, S.P.; Chiou, S.H. Characterization of a novel thermophilic, cellulose-degrading bacterium *Paenibacillus* sp. strain B39. *Lett. Appl. Microb.* **2008**, *47*, 46–53. [CrossRef]
134. Liang, Y.L.; Zhang, Z.; Wu, M.; Wu, Y.; Feng, J.X. Isolation, screening, and identification of cellulolytic bacteria from natural reserves in the subtropical region of China and optimization of cellulase production by *Paenibacillus terrae* ME27-1. *BioMed Res. Int.* **2014**, *2014*, 512497. [CrossRef] [PubMed]
135. Gaur, R.; Tiwari, S. Isolation, production, purification and characterization of an organic-solvent-thermostable alkalophilic cellulase from *Bacillus vallismortis* RG-07. *BMC Biotechnol.* **2015**, *15*, 19. [CrossRef] [PubMed]
136. Yousef, N.; Mawad, A.; Abeed, A. Enhancement the Cellulase Activity Induced by Endophytic Bacteria Using Calcium Nanoparticles. *Curr. Microbiol.* **2019**, *76*, 346–354. [CrossRef] [PubMed]
137. Bohra, V.; Tikariha, H.; Dafale, N.A. Genomically Defined *Paenibacillus polymyxa* ND24 for Efficient Cellulase Production Utilizing Sugarcane Bagasse as a Substrate. *Appl. Biochem. Biotechnol.* **2019**, *187*, 266–281. [CrossRef] [PubMed]
138. Shah, T.A.; Charles, C.; Orts, W.J.; Tabassum, R. Biological pretreatment of rice straw by ligninolytic Bacillus s strains for enhancing biogas production. *Environ. Prog. Sustain. Energy* **2018**, *38*, e13036. [CrossRef]
139. Kuhad, R.C.; Gupta, R.; Singh, A. Microbial cellulases and their industrial applications. *Enzyme Res.* **2011**, *2011*, 280696. [CrossRef] [PubMed]
140. Ejaz, U.; Sohail, M.; Ghanemi, A. Cellulases: From Bioactivity to a Variety of Industrial Applications. *Biomimetics* **2021**, *6*, 44. [CrossRef]
141. Kaushal, M.; Kumar, A.; Kaushal, R. *Bacillus pumilus* strain YSPMK11 as plant growth promoter and biocontrol agent against *Sclerotinia sclerotiorum*. *3 Biotech* **2017**, *7*, 90–99. [CrossRef] [PubMed]
142. Wróbel, M.; Śliwakowski, W.; Kowalczyk, P.; Kramkowski, K.; Dobrzyński, J. Bioremediation of Heavy Metals by the *Genus Bacillus*. *Int. J. Environ. Res. Public Health* **2023**, *20*, 4964. [CrossRef]
143. Kulkova, I.; Dobrzyński, J.; Kowalczyk, P.; Bełżecki, G.; Kramkowski, K. Plant Growth Promotion Using *Bacillus cereus*. *Int. J. Mol. Sci.* **2023**, *24*, 9759. [CrossRef] [PubMed]
144. Lipková, N.; Cinkocki, R.; Maková, J.; Medo, J.; Javoreková, S. Characterization of endophytic bacteria of the genus bacillus and their influence on the growth of maize (*Zea mays*) in vivo. *J. Microbiol. Biotechnol. Food Sci.* **2021**, *10*, e3602. [CrossRef]
145. Shahid, I.; Han, J.; Hanooq, S.; Malik, K.A.; Borchers, C.H.; Mehnaz, S. Profiling of metabolites of *bacillus* spp. and their application in sustainable plant growth promotion and biocontrol. *Front. Sustain. Food Syst.* **2021**, *5*, 605195. [CrossRef]
146. Mirskaya, G.V.; Khomyakov, Y.V.; Rushina, N.A.; Vertebny, V.E.; Chizhevskaya, E.P.; Chebotar, V.K.; Chesnokov, Y.V.; Pishchik, V.N. Plant Development of Early-Maturing Spring Wheat (*Triticum aestivum* L.) under Inoculation with *Bacillus* sp. V2026. *Plants* **2022**, *11*, 1817. [CrossRef] [PubMed]
147. Dobrzyński, J.; Jakubowska, Z.; Kulkova, I.; Kowalczyk, P.; Kramkowski, K. Biocontrol of fungal phytopathogens by *Bacillus pumilus*. *Front. Microb.* **2023**, *14*, 1194606. [CrossRef]
148. Win, K.T.; Oo, A.Z.; Ohkama-Ohtsu, N.; Yokoyama, T. *Bacillus pumilus* Strain TUAT-1 and Nitrogen Application in Nursery Phase Promote Growth of Rice Plants under Field Conditions. *Agronomy* **2018**, *8*, 216. [CrossRef]
149. Ali, A.M.; Awad, M.Y.M.; Hegab, S.A.; Gawad, A.M.A.E.; Eissa, M.A. Effect of Potassium Solubilizing Bacteria (*Bacillus cereus*) on Growth and Yield of Potato. *J. Plant Nutr.* **2021**, *44*, 411–420. [CrossRef]
150. Li, Y.; Wang, M.; Chen, S. Application of N2-fixing *Paenibacillus triticisoli* BJ-18 changes the compositions and functions of the bacterial, diazotrophic, and fungal microbiomes in the rhizosphere and root/shoot endosphere of wheat under field conditions. *Biol. Fertil. Soils* **2021**, *57*, 347–362. [CrossRef]
151. Okoroafor, P.U.; Ikwuka, G.; Zaffar, N.; Ngalle Epede, M.; Mensah, M.K.; Haupt, J.; Golde, A.; Heilmeier, H.; Wiche, O. Field Studies on the Effect of Bioaugmentation with *Bacillus amyloliquefaciens* FZB42 on Plant Accumulation of Rare Earth Elements and Selected Trace Elements. *Minerals* **2022**, *12*, 409. [CrossRef]

152. Stępień, A.; Wojtkowiak, K.; Kolankowska, E. Effect of Commercial Microbial Preparations Containing *Paenibacillus azotofixans*, *Bacillus megaterium* and *Bacillus subtilis* on the Yield and Photosynthesis of Winter Wheat and the Nitrogen and Phosphorus Content in the Soil. *Appl. Sci.* **2022**, *12*, 12541. [CrossRef]
153. Zhang, L.N.; Wang, D.C.; Hu, Q.; Dai, X.Q.; Xie, Y.S.; Li, Q.; Liu, H.M.; Guo, J.H. Consortium of plant growth-promoting rhizobacteria strains suppresses sweet pepper disease by altering the rhizosphere microbiota. *Front. Microbiol.* **2019**, *10*, 1668. [CrossRef]
154. Jamily, A.S.; Koyama, Y.; Win, T.A.; Toyota, K.; Chikamatsu, S.; Shirai, T.; Yasuhara, T. Effects of inoculation with a commercial microbial inoculant Bacillus subtilis C-3102 mixture on rice and barley growth and its possible mechanism in the plant growth stimulatory effect. *J. Plant Protect. Res.* **2019**, *59*, 193–205. [CrossRef]
155. Qiu, F.; Liu, W.; Chen, L.; Wang, Y.; Ma, Y.; Lyu, Q.; Zheng, Y. Bacillus subtilis biofertilizer application reduces chemical fertilization and improves fruit quality in fertigated Tarocco blood orange groves. *Sci. Hortic.* **2021**, *281*, 110004. [CrossRef]
156. Artyszak, A.; Gozdowski, D. The Effect of Growth Activators and Plant Growth-Promoting Rhizobacteria (PGPR) on the Soil Properties, Root Yield, and Technological Quality of Sugar Beet. *Agronomy* **2020**, *10*, 1262. [CrossRef]
157. Ngalimat, M.S.; Yahaya, R.S.R.; Baharudin, M.M.A.-a.; Yaminudin, S.M.; Karim, M.; Ahmad, S.A.; Sabri, S. A Review on the Biotechnological Applications of the Operational Group Bacillus amyloliquefaciens. *Microorganisms* **2021**, *9*, 614. [CrossRef] [PubMed]
158. Matzen, N.; Heick, T.M.; Jørgensen, L.N. Control of powdery mildew (*Blumeria graminis* spp.) in cereals by Serenade® ASO (Bacillus amyloliquefaciens (former subtilis) strain QST 713). *Biol. Control* **2019**, *139*, 104067. [CrossRef]
159. Montes-Bazurto, L.G.; Bustillo-Pardey, A.E.; Morales, A. New Alternative to Control *Stenoma impressella* (Lepidoptera: Elachistidae) Using *Bacillus thuringiensis* Commercial Formulations in Oil Palm Crops. *Agronomy* **2022**, *12*, 883. [CrossRef]
160. Ghahremani, Z.; Escudero, N.; Beltrán-Anadón, D.; Saus, E.; Cunquero, M.; Andilla, J.; Sorribas, F.J. Bacillus firmus strain I-1582, a nematode antagonist by itself and through the plant. *Front. Plant Sci.* **2020**, *11*, 796. [CrossRef]
161. Mendis, H.C.; Thomas, V.P.; Schwientek, P.; Salamzade, R.; Chien, J.-T.; Waidyarathne, P.; Kloepper, J.; De La Fuente, L. Strain-specific quantification of root colonization by plant growth promoting rhizobacteria Bacillus firmus I-1582 and Bacillus amyloliquefaciens QST713 in non-sterile soil and field conditions. *PLoS ONE* **2018**, *13*, e0193119. [CrossRef] [PubMed]

MDPI AG
Grosspeteranlage 5
4052 Basel
Switzerland
Tel.: +41 61 683 77 34

Agriculture Editorial Office
E-mail: agriculture@mdpi.com
www.mdpi.com/journal/agriculture

www.ingramcontent.com/pod-product-compliance
Lightning Source LLC
Chambersburg PA
CBHW041408210926
51850CB00013B/3208
* 9 7 8 3 7 2 5 8 3 3 2 1 4 *